Humean Laws for Human Agents

Humean Laws for Human Agents

Edited by

MICHAEL TOWNSEN HICKS, SIEGFRIED JAAG,
AND CHRISTIAN LOEW

OXFORD
UNIVERSITY PRESS

OXFORD
UNIVERSITY PRESS

Great Clarendon Street, Oxford, OX2 6DP,
United Kingdom

Oxford University Press is a department of the University of Oxford.
It furthers the University's objective of excellence in research, scholarship,
and education by publishing worldwide. Oxford is a registered trade mark of
Oxford University Press in the UK and in certain other countries

Published in the United States of America by Oxford University Press
198 Madison Avenue, New York, NY 10016, United States of America

British Library Cataloguing in Publication Data
Data available

Library of Congress Control Number: 2022950538

ISBN 978-0-19-289381-9

DOI: 10.1093/oso/9780192893819.001.0001

Printed and bound in the UK by
Clays Ltd, Elcograf S.p.A.

Contents

Preface vii

Notes on Contributors ix

Introduction: Humeanism and the Pragmatic Turn 1
Michael Townsen Hicks, Siegfried Jaag, and Christian Loew

1. Humean Laws of Nature: The End of the Good Old Days 16
Craig Callender

2. Humean Disillusion 42
Jenann Ismael

3. Knowing the Powers 66
Wolfgang Schwarz

4. Naturalism, Functionalism, and Chance: Not a Best Fit for the Humean 86
Alison Fernandes

5. Generalizing the Problem of Humean Undermining 108
Heather Demarest and Elizabeth Miller

6. Are Humean Laws Flukes? 128
Barry Loewer

7. The Package Deal Account of Naturalness 145
Harjit Bhogal

8. Properties for and of Better Best Systems 168
Markus Schrenk

9. Predictive Infelicities and the Instability of Predictive Optimality 193
Chris Dorst

10. Best-System Laws, Explanation, and Unification 215
Thomas Blanchard

11. A Discourse on Methods; or, Humean Metaphysics of Science without Best Systems 237
John T. Roberts

12. Humean Reductionism about Essence 258
Ned Hall

Index 287

Preface

The idea for this volume originated from our collaboration at the University of Cologne and a series of workshops on laws of nature organized by Barry Loewer (in Budapest) and Chris Dorst and Ned Hall (in Cambridge, Massachusetts). We would like to thank Ned Hall and Barry Loewer for their help in bringing about the volume. Thanks also to Andreas Hüttemann, Alexander Reutlinger, and Alastair Wilson for helpful advice, to Shawn Demarest for allowing us to use one of her paintings as cover art, and to Peter Momtchiloff for his editorial assistance. We are also grateful to the Deutsche Forschungsgemeinschaft (DFG) for funding Siegfried Jaag's work in the research group 'Inductive Metaphysics' (FOR 2495), and to the European Research Council (ERC) for funding Michael Townsen Hicks's contribution as part of the project 'A Framework for Metaphysical Explanation in Physics' (FraMEPhys; #757295). We would especially like to thank all the contributors for their collaboration and excitement about the volume.

Notes on Contributors

Harjit Bhogal is an Assistant Professor of Philosophy at the University of Maryland, College Park.

Thomas Blanchard is an Assistant Professor of Philosophy at the University of Cologne.

Craig Callender is a Professor of Philosophy at the University of California, San Diego.

Heather Demarest is an Assistant Professor of Philosophy at the University of Colorado Boulder.

Chris Dorst is an Assistant Professor of Philosophy at the University of Florida.

Alison Fernandes is an Assistant Professor of Philosophy at Trinity College Dublin.

Ned Hall is a Professor of Philosophy at Harvard University.

Michael Townsen Hicks is a Lecturer in Philosophy at the University of Glasgow.

Jenann Ismael is a Professor of Philosophy at Johns Hopkins University.

Siegfried Jaag is a Research Fellow in Philosophy at Düsseldorf University and Visiting Professor of Philosophy at the University of Tübingen.

Christian Loew is a Senior Lecturer (Associate Professor) in Philosophy at Umeå University.

Barry Loewer is a Professor of Philosophy at Rutgers University.

Elizabeth Miller is an Assistant Professor of Philosophy at Brown University.

John T. Roberts is a Professor of Philosophy at the University of North Carolina at Chapel Hill.

Markus Schrenk is a Professor of Philosophy at Düsseldorf University.

Wolfgang Schwarz is a Lecturer in Philosophy at the University of Edinburgh.

Introduction

Humeanism and the Pragmatic Turn

Michael Townsen Hicks, Siegfried Jaag, and Christian Loew

A central question in the philosophy of science is: *What is a law of nature?* Different answers to this question define an important schism: Humeans, in the wake of David Hume, hold that the laws of nature are nothing over and above what actually happens and reject irreducible facts about natural modality (Lewis 1983, 1994; cf. Miller 2015). According to non-Humeans, by contrast, the laws are metaphysically fundamental (Maudlin 2007) or grounded in primitive modal structures, such as dispositional essences of powerful properties (Bird 2007), necessitation relations (Armstrong 1983), or primitive subjunctive facts (Lange 2009).

This volume focuses on recent developments in the discussion of Humeanism, specifically on pragmatic versions of the view that put the needs of limited agents like us front and center. These views are perhaps best understood in contrast to their immediate ancestor, the Humean view defended in the work of David Lewis. Lewis provided a set of instructions for obtaining laws of nature from entirely non-modal ingredients. The ingredients are specified by Lewis's thesis of *Humean Supervenience* (HS). According to HS, the world fundamentally is nothing over and above the total pattern of instantiations of perfectly natural intrinsic properties at space–time points (or their point-sized occupants) and all other facts supervene on this global pattern (see Lewis 1986, p. ix). This pattern is usually called the 'Humean Mosaic'. HS is Humean since natural properties are freely recombinable: no property instantiation has any modal implications for the instantiation of any other property.

Lewis's Best Systems Account of Laws (BSA), inspired by earlier work of Mill (1843/1967) and Ramsey (1928/1990), specifies the instructions for obtaining laws from these non-modal ingredients. The BSA is a variant of the regularity theory of laws. But unlike earlier versions, the lawhood of a generalization is not determined in isolation. Instead, a generalization is a law only if it is a theorem or axiom in an axiomatic systematization of the Humean mosaic that strikes the best balance between simplicity and informativeness. Some systems will be very informative about the Humean mosaic but contain many or long axioms; others

Michael Townsen Hicks, Siegfried Jaag, and Christian Loew, *Introduction: Humeanism and the Pragmatic Turn* In: *Humean Laws for Human Agents*. Edited by: Michael Townsen Hicks, Siegfried Jaag, and Christian Loew, Oxford University Press.
© Oxford University Press 2023. DOI: 10.1093/oso/9780192893819.003.0001

will be simple but less informative. We are instructed to choose the system which best balances the two, and the laws of nature are generalizations in this system.

Note that for the Lewisian BSA to work as advertised, it is crucial to place restrictions on eligible predicates. If any predicate is allowed into a best system, we can easily make any systematization extremely simple: "Given system S, let F be a predicate that applies to all and only things at worlds where S holds. Take F as primitive, and axiomatise S (or an equivalent thereof) by the single axiom $\forall x F x$" (Lewis 1983, p. 367). This problem, among others, prompted Lewis to include perfectly natural properties in his formulations of HS and the BSA. Predicates in a best system must then refer to perfectly natural properties, an elite class of properties that he thought physics was in the business of discovering. This restriction rules out constructs such as the predicate F in the above example (see Lewis 1983; and the contributions of Bhogal (ch. 7), Callender (ch. 1), Loewer (ch. 6), and Schrenk (ch. 8) to this volume).

Lewis's BSA deservedly had a major influence. It promises to furnish Humeans with a reductive account of all natural modalities. Chances come in a single package with the laws because their inclusion can make axiomatic systems simpler by condensing information about frequencies (Lewis 1980, 1994; Schwarz 2014, 2015). Facts about chances, then, reduce to facts about how we can best systematize the world's events, including the frequency of different sorts of events. And Lewis then argued that all other natural modalities, including counterfactuals, causation, and dispositions, can be accounted for in a reductive hierarchy with laws of nature at the bottom (see Lewis 1986).

Moreover, the BSA draws a principled distinction between laws and non-laws that is designed to mesh well with scientific practice. Lewis maintained that strength and simplicity are the very standards that physics itself uses in discovering laws (see Lewis 1983, p. 367, and 1986, p. 123; cf. Earman 1986, p. 88; and Loewer 2007, p. 320). Of course, he might be wrong about this, but the BSA can be flexibly adapted to whatever standards physics actually uses (see Loewer 2007, forthcoming, and this volume, ch. 6). And while Lewis thought that only generalizations can be laws, even this restriction can be lifted (see Albert 2000, 2015; Loewer 2012). With these amendments, the BSA outputs as laws those claims which science in fact chooses (cf. Roberts 2008, p. 331). We then seem to get a realist account of laws, where laws are objective and mind-independent, from only minimal metaphysical ingredients.

But there are also problems with Lewis's account. Many recent developments of Humeanism about laws have originated from two worries about Lewis's BSA. First, some philosophers have objected to positing perfectly natural properties. By requiring that predicates in successful scientific theories refer to natural properties, Lewis's BSA puts *a priori* restrictions on science, tying the BSA to a metaphysical posit that empiricist-minded philosophers worry is undetectable.

There, the thought goes, is no reason to expect that actual scientific laws will conform to this restriction (see van Fraassen 1989; Loewer 2007; Cohen and Callender 2009; Demarest 2017), and some reason to think that a satisfying Humean account of quantum mechanics might relax it (see Miller 2014; Bhogal and Perry 2017).

Second, a theory of laws of nature should explain why discovering the laws is such an important goal of scientific inquiry. Non-Humeans appear to have an answer to this question by making nomic facts metaphysically distinguished from other facts. It is less clear that Lewis's BSA offers a good answer: Why should it be such an important goal of science to discover generalizations that maximize virtues like strength and simplicity? It is not enough to say that these are the virtues that science in fact values. We would like to know *why* it makes sense to care about them in the first place (see Hall 2015, p. 268).

The two worries seem to pull in opposite directions: The first worry, about natural properties, is that the Lewisian BSA posits too much metaphysical structure, thus creating a chasm between science and metaphysics. The second worry is that it posits too little metaphysical structure and so cannot explain why science makes certain crucial distinctions. Despite this tension, many Humeans have recently argued that these and other difficulties can be addressed in a unified way by introducing pragmatic elements into our recipe for laws (see Callender 2017; Hall 2015; Dorst 2018; Hicks 2018; Jaag and Loew 2020; and Loewer 2007, forthcoming).

Pragmatic Humeans dispense with some metaphysical posits of the BSA. As mentioned above, Lewis posits natural properties to provide a distinguished vocabulary for best systems. Pragmatic Humeans, instead, argue that the language a best system is framed in is determined by its practical usefulness (Loewer 2007, forthcoming, and this volume, ch. 6; Cohen and Callender 2009; Ismael 2015; Jaag and Loew 2020). One way of making this change has the further benefit that it naturally extends the BSA to provide the laws of the special sciences (see the Better Best System Account of Schrenk 2007, this volume, ch. 8; and Cohen and Callender 2009, 2010). Similarly, Loewer's 'Package Deal Account' (2007, forthcoming, and this volume, ch. 6; see also Bhogal, this volume, ch. 7) defers to scientific practice for delineating the properties going into the best system and lifts Lewis's requirement that these properties must be intrinsic.

Moreover, pragmatic considerations can be used to motivate why laws are crucial to scientific practice. The question 'why do scientists aim to discover laws' for best systems theorists becomes the question 'why do scientists care about discovering facts that jointly maximize certain virtues'. Pragmatic Humeans answer this question in two parts: First, they argue that the features that best systems maximize are not strength and simplicity but more fine-grained features that make for useful laws (Dorst 2018; Earman 1986; Hall 2015, Hicks 2018; Jaag and

Loew 2020; Loewer 2007, forthcoming). Second, pragmatic Humeans motivate these features by showing how they make the resulting laws useful for 'getting around in the world' (see Albert 2015, p. 23; Beebee 2000, p. 547). In fact, Humeans may be able to explain why the laws have the characteristic features they do have (Callender 2017, and this volume, ch.1; Dorst 2018, this volume, ch. 9; Hicks 2018; Jaag and Loew 2020).

Lewis (1994) himself resisted incorporating pragmatic considerations into his BSA. Positing perfectly natural properties and maintaining that simplicity and strength are completely objective were meant to assure that the resulting laws are also objective. Many pragmatic Humeans, however, see this insistence on objectivity as a missed opportunity. To their eyes it seems entirely natural, given a Humean metaphysics, that what distinguishes laws from non-laws has something to do with us. In this vein, Hall (2015, p. 268) asks: if the distinction between laws and non-laws is not part of the world's fundamental structure, then '[h]ow could the details of our peculiar human situation *not* be relevant to this matter [i.e. how we ought to draw the distinction]?'

The current volume provides a venue for: (i) developing and critically examining pragmatic Humean accounts of laws; (ii) addressing fundamental and longstanding problems for Humeanism from the pragmatic perspective; and (iii) exploring alternatives to Humeanism that, while deviating from its metaphysics, maintain some of its spirit and incorporate pragmatic elements.

Craig Callender's opening chapter views Humeanism's pragmatic turn in a larger context. According to Callender, pragmatic best systems theories signal the 'end of the good old days' of meta-metaphysical naivety for Humeans. Callender argues that pragmatic best systems theories are best understood as 'ideal advisor' rather than 'ideal observer' accounts. The goal of best systems is no longer to systematize the Humean mosaic in accordance with certain detached, objective virtues (as Lewis may have thought) but to condense information in a way that makes it useful for limited, embedded creatures like us.

This pragmatic reorientation makes best systems theories rather similar to anti-realist, projectivist Humean theories according to which law statements express, for example, intentions to make certain inferences. The only remaining difference between pragmatic, ideal advisor best systems theories and projectivist theories then seems to be that the former maintain that law statements are truth-apt by expressing generalizations. But Callender points out that an analogous discussion in meta-ethics has taught us that this distinction is fickle and may even vanish if one adopts a minimalist theory of truth (this is the 'Problem of Creeping Minimalism' in meta-ethics). What reasons then do we have for choosing one version of Humeanism over the other?

At this point, Callender has good news and bad news: The good news is that rather than having to work out the different Humean positions from scratch, we can 'leapfrog' much discussion by drawing from the rich analogy with meta-ethics. The bad news is that the result is somewhat unsatisfying as the meta-ethical discussion is complex and controversial. Callender's final advice for Humeans is to largely set aside questions about realism. Humeans should focus on the motivations for being Humean rather than on working out the appropriate semantics of law-talk. And here, he thinks, systems theorists and projectivists have been driven by different guiding ideas, which, however, may nicely complement each other. It is here that Callender sees potential for fruitful future work.

In the following chapter, Jenann Ismael also foresees an end of the good old days for Humeanism. But while Callender views this development with optimism, Ismael attests disillusion. She argues that a Humean metaphysics, as standardly understood, is fatally unable to explain how nomic facts, such as laws and chances, can guide credences about the future for limited agents. This failure would, obviously, be especially problematic for pragmatic Humeans, who put the practical relevance of knowledge of natural modalities at the center of their project.

Humeans maintain free recombination: no property instantiation has any implications for what properties are instantiated elsewhere. Moreover, it is plausible (both on Humean and general grounds) that our universe is indefinitely extendible: it has no intrinsic maxima or borders. Ismael argues that these posits entail that an agent who observes only a finite region of the universe cannot confirm hypotheses about laws and chances. After all, laws and chances supervene on the global pattern of property instantiation; but any merely finite region could be embedded in *any* global, infinite pattern. Humeanism's core commitment—no necessary connections between distinct existences—then puts facts about natural modality beyond the epistemic grasp of finite agents. Ismael argues that any kind of epistemic principle that might make nomic facts discoverable by limited agents—such as that unobserved regions resemble observed regions—would be at odds with a Humean metaphysics.

Ismael's response is to give up the free recombination of properties. This move, since it admits necessary connections between distinct existences, puts her into the anti-Humean camp. Nonetheless, Ismael's view maintains the central motivation of pragmatic Humean positions, viz., that a theory of laws and chances ought to explain their role in helping limited agents get around in the world. In addition, Ismael also has reservations about most anti-Humean positions that tend to reify possibilities by positing primitive entities. She, instead, advocates a position (for which she considers the labels 'neo-anti-Humean' and 'anti-neo-Humean') that takes scientific practice at face value by taking restrictions on what possibilities there are as built into the very structure of reality.

Wolfgang Schwarz's contribution turns the tables and argues that it is non-Humean accounts that confront a deep problem about our knowledge of natural modalities. Schwarz agrees with Lewis that it is not enough to posit 'unHumean whatnots' (Lewis 1994, p. 239)—such as irreducible laws, powers, potentialities, or chances—to explain modal facts; in addition, a plausible story is needed about how such unHumean posits could play the familiar roles of laws, powers, potentialities, or propensities. And an integral part of that role concerns the methods by which these natural modalities can be discovered. Focusing on dispositions and chance, Schwarz argues that non-Humeans lack a plausible epistemology of natural modalities. He calls the epistemological worry he raises the *access problem*.

In a nutshell, the access problem is that observation and experiment only tell us what does happen, but they do not directly reveal what might, must, or would happen in non-actual circumstances. If modal phenomena are reducible to facts about occurrent non-modal events, as Humeans claim, then it is no surprise that observing occurrent events provides information about modality. By contrast, if modal facts are primitives that do not supervene on occurrent facts, as non-Humeans have it, then knowledge of modal facts would seem to require an inexplicable leap from observations of one kind of fact (occurrent facts) to conclusions about an entirely different kind of fact (modal facts). Schwarz shows that if the world has primitive modal elements this gap creates skeptical scenarios: There are different a priori conceivable ways in which these modal elements might be arranged, many of which are perceptually indistinguishable.

Schwarz examines three attempts of solving the access problem: (i) appealing to a 'thin' conception of knowledge; (ii) claiming that epistemic norms, such as the Principal Principle (which links chance to credence), are primitive; and (iii) reframing the anti-Humean position as a doctrine about ideology rather than ontology. He argues that none of these proposals is ultimately satisfactory, and therefore Humean pragmatic accounts of natural modality, and in particular about chance, are to be preferred.

The next chapter, by Alison Fernandes, further builds on the themes introduced in the contributions by Ismael (ch. 2) and Schwarz (ch. 3). Like Ismael, Fernandes is concerned with a naturalistic, science-friendly account of natural modalities. And like Ismael, Fernandes argues that—despite what many Humeans believe—such an account will not be Humean. But while Ismael is concerned with the epistemology of chance, Fernandes argues that Humeanism falls short when it comes to the explanatory role of chance.

Humean accounts of chance are often presented as being in a unique position to explain why agents should align their credence that an outcome will occur with what they believe its chance to be. According to Fernandes, this is false advertisement. First, she argues that Humean accounts of chance rely on *a priori* reasoning in the form of indifference principles. This reliance on a priori principles, according to Fernandes, is in tension with the naturalistic ambitions of Humean accounts. And, second, by reducing chances to relative frequencies, Humean accounts provide a metaphysical guarantee that chances align with facts about relative frequencies. But this alignment creates a mismatch with scientific practice: when scientists reason about chances, they take seriously the possibility that chances and relative frequencies can come apart radically.

Fernandes concludes that philosophers who want a science-friendly account of chance need to look beyond Humean accounts. She thinks that it is a mistake to think that a scientifically respectable account of chance-reasoning needs to provide a non-circular analysis of chance. Instead, she argues that agents should align their credences with what they believe the chances to be because doing so guarantees a high chance of success; Fernandes argues that, though circular, this sort of reasoning is virtuous.

Heather Demarest and Elizabeth Miller continue to build on this theme. They extend some of the worries addressed in this volume by Ismael (ch. 2), Schwarz (ch. 3), and Fernandes (ch. 4) to causal and dispositional modalities we find in the special sciences. Like Ismael, their concerns arise from the fact that, for Humeans, modal facts depend on the global distribution of occurrent properties. This feature generates various 'undermining problems'. One well-known sort of undermining problem, discussed by Ismael and Fernandes, concerns chances: since the chances, for a Humean, are determined by the world's frequencies, there are no worlds in which the total sequence of events is unlikely. But such worlds are apparently compatible with the chances. This creates an unpalatable mismatch between what the chances allow and what the Humean takes to be possible; though apparently compatible with the chances, unlikely events 'undermine' their status as the chances. Demarest and Miller take this issue beyond its usual setting by showing that analogous worries arise for Humean accounts of counterfactuals. They start by showing that an influential proposal due to David Albert (2000) and Barry Loewer (2012) (called the 'Mentaculus') can dodge some undermining worries regarding chance: Since the Mentaculus takes chances to be a measure of sets of possibilities, rather than statements about the frequencies within a world, they have more tools to make sense of nested claims like 'even if there were a long string of heads, the chance of the next flip would still be 0.5'.

But Demarest and Miller argue that there are serious complications when it comes to extending this account to other modalities, including causal and dispositional modality. These complications generate puzzles for Humean accounts in the special sciences. For Humeans hold that ascriptions of dispositional properties are partially made true by the laws, and that some physically possible worlds have different physical laws. So, at such worlds with different laws, these properties don't exist, or at least aren't had by the same things. This means that, at many physically possible worlds, there are no predators, giraffes, or soluble molecules, even if there are things that are qualitatively identical to our predators, giraffes, and soluble molecules (cf. Schrenk, this volume, ch. 8). If we ask what prey animals would do if there were no predators, we may have no worlds to look at.

Demarest and Miller argue that the Humean can make use of King's (2007) distinction between truth in a world and truth at a world, where the former tells us what is true according to the world, and the latter uses our laws and higher-level kinds to determine which of our sentences the world makes true. This provides at least one avenue out of these undermining problems for Humeans.

After this series of Humean critical papers, Barry Loewer, in his contribution, defends Humeanism by arguing that it can overcome some of the most entrenched criticisms. Loewer's preferred kind of Humeanism also falls into the class of pragmatic views. He builds on his earlier 'Package Deal Account' (PDA), in which the laws and the fundamental properties are determined jointly: for Loewer, whatever package of laws and properties best meets scientific criteria of theory choice delivers both the laws and fundamental properties of the world.

Loewer argues that this joint package can provide novel responses to two extant criticisms of Humeanism. First, Humeans are accused of explanatory circularity. The laws, for Humeans, are determined by, and so metaphysically explained by, the Humean mosaic. But the laws in turn explain parts of the mosaic: their instances. To many, this looks like a tight explanatory circle. Loewer argues that the PDA gives us a neat way to sever the circle: since the laws and properties are determined together, we need not accept that the laws are explained by their instances. Both where the borders of the mosaic's tiles are and which patterns of tiles count as laws are a result of the systematizing procedure.

Second, many authors have argued that, if Humeanism is true, it is astoundingly unlikely for the world to have any order at all. Recently, a number of authors have further argued that Humeanism is self-undermining: since there are many more disordered worlds than ordered ones, they claim, Humeans should expect our world to soon descend into chaos. This sort of reasoning occupies a few of our other authors, including Ismael (this volume, ch. 2) and Schwarz (this volume, ch. 3). Loewer points out that the PDA gives us an avenue to reject a key premise of this argument: since, on his view, worlds do not come equipped with a preferred property structure, there is no reason to accept that every combination of properties is possible. The combinatorial reasoning which underlies the objection rests on the idea that the properties are prior to the laws, a claim that Loewer rejects. Like Ismael and Fernandes, Loewer's view flirts with anti-Humeanism: if the systematizing procedure determines the world's property structure, it might well turn out that some properties are necessarily connected to others.

The next two entries present challenges for developing the BSA in a pragmatic way that arise from doing away with Lewis's ingredients, i.e. an antecedently given structure of perfectly natural non-modal properties.

Harjit Bhogal elaborates on Loewer's PDA. He interprets the PDA as not just a reductive theory of laws but also a reductive theory of natural properties. His central aim is to explore whether this account provides a notion of naturalness that can play the various roles Lewis and others have intended it to play. He focuses on three roles of natural properties: their metaphysical role in characterizing the Humean mosaic, their role in identifying what the relevant data is that theories need to be informative about, and their role in fixing the language that the axioms need to be formulated in.

In the course of examining to what extent the PDA can account for these roles, he maps out the connections between naturalness and laws on the traditional BSA and examines how those connections have to be adjusted in order to develop various versions of the PDA. He argues that the PDA can only be developed by letting pragmatic considerations take center stage. The PDA is classified as an instance of a broader and recently popular approach to Humeanism that focuses on the role of 'ideal observers' or 'ideal scientists' (cf. Callender, this volume, ch. 1). But viewing the PDA as an ideal-scientist view makes it hard to answer the question of why we should care about what the ideal scientist says (any more than we should care about, say, an ideal astrologer), and consequently, why we should care about the laws and natural properties on the PDA approach.

Furthermore, Bhogal argues that while some other versions of the PDA might be feasible, they don't provide as much as its proponents might hope for. In particular, according to Bhogal, such a version of the PDA needs to be agnostic about the underlying metaphysical structure of the world and thus fails with respect to their central metaphysical role, i.e. specifying reality's fundamental nature in an entirely non-modal Humean way.

Like Loewer (ch. 6) and Bhogal (ch. 7), Markus Schrenk also takes up the theme of employing pragmatic considerations in determining the predicates going into the BSA. But his focus is on the Better Best System Account (BBSA) of laws and in particular on applying it to the laws in the special sciences (see Cohen and Callender 2009, 2010; and Schrenk 2008, 2014). Like proponents of the PDA, defenders of the BBSA deny that a best system is confined to the Lewisian ingredients, i.e. a privileged set of natural properties; instead, they hold that systematizations can be executed for any pragmatically chosen set of properties. Unlike Loewer, however, Schrenk and other proponents of the BBSA think that each science can appeal to its own division of properties, rather than reducing these properties to a privileged set determined by physics. So, instead of only providing fundamental physical laws, the BBSA is supposed to be able to deliver also the laws for the various special sciences such as chemistry and biology.

Schrenk provides a new way of articulating best systems accounts as a function of a chosen set of predicates, a corresponding property distribution and set of theoretical virtues. The main aim of his chapter is addressing several challenges and worries of the BBSA thus construed: A first concerns the extent of the anthro-pocentricity of laws resulting from the pragmatic choice of predicates. A second results from the observation that the predicates figuring in the theories of the special sciences might be dispositional, i.e. already equipped with a nomological profile. However, this seems to be in conflict with the idea that nomological facts are determined by the systematization procedure. A third class of challenges arises with respect to the boundaries between the different sets of properties that demarcate the sciences. A fourth worry is that the BBSA thus construed might depict the whole of science as a patchwork of unrelated, maybe even contradict-ory, theories instead of a hierarchical unified system. Fifth and finally, there is a related issue concerning scientific progress: as a scientific discipline develops, it might host different sets of properties. Systems analyses for different property sets, however, might well be incommensurable.

Like Schrenk (ch. 8), Chris Dorst responds to a series of objections to prag-matic regularity theories of laws. First, Dorst illustrates the neo-Humean view as a sort of ideal-advisor view (see Callender, this volume, ch. 1): imagine there were a hotline for dealing with novel physical systems. What sort of information would the hotline request before dispensing advice? And what sort of advice would we want the hotline to dispense? Dorst's framing builds on Callender's discussion of the importance of idealized agents to Humean views. But Dorst presents a novel challenge for them: if the utility of laws is evidence for an ideal-agent view, are the ways in which laws fail to be useful to us evidence against them? Dorst considers four apparent problems: first, quantum indeterminacy yields laws that are less informative than we might like. Second, quantum nonlocality means that, to gen-erate ideally strong predictions, we need distant information we do not have access to. Third, as Ismael (2019) has argued, special relativity implies that the total causal history of an event is not accessible at any point before the event hap-pens. So we cannot gain the information necessary to predict an event before it happens. And finally, the exceptionlessness of the laws might seem surprising from the perspective of a limited agent that would do just as well with laws that work most of the time and around here.

Dorst provides four strategies in responding to these challenges. Two involve accommodating them within the neo-Humean framework: Dorst points out that features that make laws useful trade off, and argues that some of these constraints involve trade-offs in the face of an unkind world. He further argues that the apparent non-optimalities can have surprising benefits. But Dorst's other two strategies are interestingly related. Dorst claims that some of the desiderata for laws might be historical accidents, driven by the philosophical or theological biases of scientists who look for laws. This process of conceptual development may continue: as we use the laws to gain information about the world, we gain new epistemic handles, and shed the constraints that we—and consequently the laws—were previously subject to. Dorst calls this 'the instability of predictive optimality', and addresses it with a mixed strategy, aiming to both mitigate some of its counterintuitive consequences and exploit those implications which make the laws more useful for us, or agents like us.

Thomas Blanchard considers a similar worry to Dorst's, but comes to a different conclusion. Like Dorst (this volume, ch. 9), Blanchard worries that some features of physical theorizing cannot be accounted for merely in terms of their predictive utility. Physicists look for exceptionless, comprehensive laws: laws which subsume or explain everything. Like Dorst, Blanchard sees this as a problem for the pragmatic Humean: why should we look for totally comprehensive laws if restricted ones will serve us just as well? Rather than attempting to accommodate this practice within the pragmatic framework as it stands, Blanchard argues that we should alter the framework. He proposes an account of laws—or fundamental laws anyway—which takes them to be maximally explanatory principles.

Blanchard then develops a Humean-friendly account of explanation along the lines of Friedman's (1974) unificationist proposal. He shows how unificationism can explain the physicists' yearning for a maximally comprehensive theory of everything. Blanchard's development of the proposal will be a boon to Humeans, who frequently appeal to unification as a goal of explanation, but less frequently show how a specific unificationist proposal ties in to the Humean account of laws.

Pragmatic considerations are not entirely absent: first, Blanchard argues that striving for maximal unification is a goal of physics, but not every science. By appealing to distinct and often predictively directed goals, Blanchard argues, we can explain the less programmatic structure of the special sciences. By suggesting that the special sciences and physics might have different goals, and so different criteria for lawhood, Blanchard's view dovetails with the BBSA, which argues that the special sciences build their laws out of different ingredients, but use the same procedure (see Schrenk, this volume, ch. 8). If Blanchard is right, the recipe of physics might not be the same as that of chemistry. The chapter concludes by arguing that a unificationist goal in physics yields more than just useful laws at the end of inquiry: it motivates an approach to theory building which tends to uncover novel truths about the world in addition to useful laws.

John Roberts's contribution lays a comprehensive groundwork for a novel Humean theory of laws, counterfactuals, and causation that, like Blanchard's (ch. 10), is an alternative to best systems theories. Roberts's theory is unabashedly pragmatist, but the pragmatism is different from the one we have encountered so far in this volume. Dorst (ch. 9), Callender (ch. 1), and Loewer (ch. 6) consider versions of Humeanism that are pragmatic in that they are partly appealing to limited agents. Robert's pragmatism traces directly back to Peirce's pragmatic theory of truth. Peirce thought that truth pertains to those statements that we arrive at at the limit of scientific inquiry.

Roberts's key primitive is that of an effective method. A method consists of a means, enabling conditions, and an end. In Peircean fashion, Roberts defines a method as effective 'just in case it is one of the methods we would classify as effective, in the limit of scientific inquiry, assuming that conditions are optimal' (p. 250) More precisely, they are the methods we get when we take our initial opinions about effective methods and update them, taking the entire Humean mosaic as our evidence. This account is Humean because effective methods are solely determined by the Humean mosaic plus the standards ideal science uses for discovering effective methods.

With a Humean account of effective methods in hand, Roberts sets out to define all other natural modalities in terms of effective methods. In his chapter, he shows how this can be done for counterfactuals and laws. Very roughly, he shows that the truth conditions for an important subclass of counterfactuals—viz. semifactuals, which have a true consequent—can be defined in terms of effective methods. And he then shows that all other counterfactuals are equivalent to semifactuals. Robert's account of laws is inspired by Lange's (2009) counterfactual stability account of laws. But rather than appealing to counterfactuals, he defines laws as those propositions that form an 'unviolatable set', that is, a set that is such that no effective method can render one of them false.

Ned Hall develops a Humean reductionist account of essence that in important aspects resembles Humean reductionism about laws of nature. He invites us to contemplate the differences between particular existential claims like 'There are exactly seventeen black holes within 20 light years of our solar system' and generalizations like 'Every black hole at equilibrium is completely physically characterized by just three parameters: its mass, its angular momentum, and its charge'. Assuming that both claims are true (although the former is not), the latter is significantly more important than the former. The question now is what explains this difference in importance.

Hall presents two fundamentally different approaches to explaining this: a *metaphysical account* and an *epistemic-utility account*. According to the metaphysical approach, the generalization is more important because it reveals something about the *essence* of the kind *black hole* whereas the particular fact does not. According to the epistemic approach, the reason why the generalization is more significant is that knowing it *facilitates inquiry* to a vastly greater extent than does knowing the particular claim. In the case of black holes, the generalization helps us to make various novel predictions and give various explanations about black holes in a way the particular claim does not.

Hall then works towards an epistemic account of essence by drawing inspiration from the debate between Humeans and non-Humeans about laws of nature.

He thinks the core question in this debate is whether one should give a *metaphysical* or an *epistemic* account of what laws are. In a parallel way, he argues that a key question we face about essences is whether to give a metaphysical or, rather, an epistemic account of them. A benefit of connecting laws and essences according to Hall is that enough progress has been made on the former to provide valuable guidance for investigating the latter.

Finally, he outlines the contours of a Humean reductionist account of essence where the epistemic notion of 'inquiry about *Xs*' (where *X* is some kind) takes center stage. Hall then presents important consequences that naturally fall out of his approach. Here is a little selection: First, the *importance* of the concept of 'essence' can be reductively explained without leading into a Humean eliminativism about essence. Second, essences turn out to be interest-dependent and capable of vagueness. Third, the view helps clarify the distinction between essences of *kinds* and essences of *individuals*. Fourth, it naturally motivates a strikingly deflationary treatment of essentiality about origins. And finally, fifth, the view neatly solves a puzzle about the essences of arbitrary fusions.

Taken together, these chapters provide a road map for developing pragmatic Humeanism further. As the contributions to this volume show, there is not just one path but a series of divergences. Humeans can seek to strengthen their account of laws by adding pragmatic considerations, as Callender (ch. 1), Schwarz (ch. 3), and Dorst (ch. 9) suggest; but doing so may have far-reaching ramifications. First, their pragmatic motivations might drive them away from the austere Humean metaphysics, as they have Loewer (ch. 6), Ismael (ch. 2), and Fernandes (ch. 4). This may create problems of the sort Bhogal (ch. 7), and Demarest and Miller (ch. 5) point to, or of the variety that Schrenk (ch. 8) and Dorst (ch. 9) raise in order to allay. These considerations may require Humeans to move even further from the Lewisian orthodoxy and to replace the BSA entirely with something more focused on the needs of agents or scientists, as in the new positive accounts of laws and chances developed by Blanchard (ch. 10), Fernandes (ch. 4), and Roberts (ch. 11). And finally, it may provide the resources for Humeans to give new positive accounts of those things non-Humeans take to be basic, as in Hall (ch. 12) argues. Opportunities abound, and the way before us is full of potential for interesting and novel work.

References

Albert, D. Z. (2000). *Time and Chance*. Cambridge, MA: Harvard University Press.

Albert, D. Z. (2015). *After Physics*. Cambridge, MA: Harvard University Press.

Armstrong, D. (1983). *What is a Law of Nature?* Cambridge: Cambridge University Press.

Beebee, H. (2000). 'The Non-Governing Conception of Laws of Nature', *Philosophy and Phenomenological Research*, 61(3), pp. 571–94.

Bhogal, H., and Perry, Z. (2017). 'What the Humean Should Say About Entanglement', *Noûs*, 51(1), pp. 74–94.

Bird, A. (2007). *Nature's Metaphysics: Laws and Properties*. Oxford: Oxford University Press.

Callender, C. (2017). *What Makes Time Special?* Oxford: Oxford University Press.

Cohen, J., and Callender, C. (2009). 'A Better Best System Account of Lawhood', *Philosophical Studies*, 145(1), pp. 1–34.

Cohen, J., and Callender, C. (2010). 'Special Sciences, Conspiracy and the Better Best System Account of Lawhood', *Erkenntnis*, 73, pp. 427–47.

Demarest, H. (2017). 'Powerful Properties, Powerless Laws', in Jonathan D. Jacobs (ed.), *Causal Powers*. Oxford: Oxford University Press, pp. 38–53.

Dorst, C. (2018). 'Toward a Best Predictive System Account of Laws of Nature', *British Journal for the Philosophy of Science*, 70(3), pp. 877–900.

Earman, J. (1986). *A Primer on Determinism*. Dordrecht: Reidel.

Friedman, M. (1974). 'Explanation and Scientific Understanding', *Journal of Philosophy*, 71, pp. 5–19.

Hall, N. (2015). 'Humean Reductionism about Laws of Nature', in Loewer, B., and Schaffer, J. (eds.), *The Blackwell Companion to David Lewis*. Oxford: Blackwell, pp. 262–77.

Hicks, M. T. (2018). 'Dynamic Humeanism', *British Journal for the Philosophy of Science*, 69(4), pp. 983–1007.

Ismael, J. (2015). 'How to be Humean', in Loewer, B., and Schaffer, J. (eds.) *The Blackwell Companion to David Lewis*. Oxford: Blackwell, pp. 188–205.

Ismael, J. (2019). 'Determinism, Counterpredictive Devices, and the Impossibility of Laplacean Intelligences', *The Monist* 102(4), pp. 478–98.

Jaag, S., and Loew, C. (2020). 'Making Best Systems Best for Us', *Synthese*, 197, pp. 2525–50.

King, Jeffrey C. (2007). *The Nature and Structure of Content*. Oxford: Oxford University Press.

Lange, M. (2009). *Laws and Lawmakers: Science, Metaphysics, and the Laws of Nature*. Oxford: Oxford University Press.

Lewis, D. K. (1980). 'A Subjectivist's Guide to Objective Chance', in Harper, W. L., Stalnaker, R., and Pearce, G. (eds.), *IFS. The University of Western Ontario Series in Philosophy of Science*, vol. 15. Dordrecht: Springer.

Lewis, D. K. (1983). 'New Work for a Theory of Universals', *Australasian Journal of Philosophy*, 61, pp. 343–77.

Lewis, D. K. (1986). *Philosophical Papers Vol. II*. Oxford: Oxford University Press.

Lewis, D. K. (1994). 'Humean Supervenience Debugged', *Mind*, 103(412), pp. 473–90.

Loewer, B. (2007). 'Laws and Natural Properties', *Philosophical Topics*, 35(1/2), pp. 313–28.

Loewer, B. (2012). 'Two Accounts of Laws and Time', *Philosophical Studies*, 160(1), pp. 115–37.

Loewer, B. (forthcoming). 'The Package Deal Account of Laws and Properties (PDA)', *Synthese*, https://doi.org/10.1007/s11229-020-02765-2.

Maudlin, T. (2007). *The Metaphysics Within Physics*. Oxford: Clarendon Press.

Mill, J. S. (1843/1967). *A System of Logic. Ratiocinative and Inductive. Being a Connected View of the Principles of Evidence and the Methods of Scientific Investigation. Collected Works VII & VIII*. Toronto: Toronto University Press.

Miller, E. (2014). 'Quantum Entanglement, Bohmian Mechanics, and Humean Supervenience', *Australasian Journal of Philosophy*, 92(3), pp. 567–83.

Miller, E. (2015). 'Humean Scientific Explanation', *Philosophical Studies*, 172(5), pp. 1311–32.

Ramsey, F. P. (1928/1990). 'Universals of Law and of Fact', in Mellor, D. (ed.) *Philosophical Papers*. Cambridge: Cambridge University Press, pp. 140–4.

Roberts, J. T. (2008). *The Law-Governed Universe*. Oxford: Oxford University Press.

Schrenk, M. (2007). *The Metaphysics of Ceteris Paribus Laws*. Frankfurt: Ontos.

Schrenk, M. (2008). 'A Lewisian Theory for Special Science Laws', in Bohse, H., and Walter, S. (eds.), *Selected Contributions to GAP. 6, Sixth International Conference of the Society for Analytical Philosophy, Berlin, 11.–14. September 2006*. Paderborn: Mentis.

Schrenk, M. (2014). 'Better Best Systems and the Issue of CP-Law', *Erkenntnis*, 79, pp. 1787–99.

Schwarz, W. (2014). 'Proving the Principal Principle', in Wilson, A. (ed.), *Chance and Temporal Asymmetry*. Oxford: Oxford University Press, pp. 81–99.

Schwarz, W. (2015). 'Best System Approaches to Chance', in Alan Hájek and Christopher Hitchcock (eds.), *The Oxford Handbook of Probability and Philosophy*, Oxford: Oxford University Press, 423–39.

Van Fraassen, B. C. (1989). *Laws and Symmetry*. Oxford: Oxford University Press.

1

Humean Laws of Nature

The End of the Good Old Days

Craig Callender

In the Good Old Days philosophers knew where they stood regarding moral real-ism. Irrealists and realists each came in two varieties. Irrealists could pick between emotivism and error theory. Realists, meanwhile, had a choice between non-naturalism and a naturalistic Ideal Observer theory. If someone claimed that cheating is bad, you would ask yourself whether that statement is truth-evaluable, and, if so, determine what makes it true or false. Your answers located you in conceptual space. Life was simple. Turning to the topic of laws of nature, we are still—naively—living in the Good Old Days. The sepia-toned geography is the same. Irrealists can choose between projectivism and a "no-laws" counterpart to error theory, and realists divide between non-Humean governing views and Humean systems theories that are counterparts to Ideal Observer theory. For philosophers of science, life is still simple, the days warm and easy.[1]

Unfortunately, those days are over in meta-ethics and I will show that they are also finished for theories of laws of nature. Focusing on theories that find inspir-ation in David Hume's thought (systems theory and projectivism), I'll begin with a problem for systems theory. I'll demonstrate that the natural resolution of that problem explicitly parallels moves made in meta-ethics, moves that led to the end of the Good Old Days. Just as meta-ethics now faces an uncertain future—one where the difference between realism and irrealism is unclear—Humean theories of laws face this same predicament. Life is now complicated. How do we progress? I tentatively suggest that we move forward by looking back, back to the Really Old Days, namely, Hume's original theory.

---

[1] The phrase, metaphor, and conceptual geography in ethics are all due to the excellent Dreier (2004). Ayer (1956) is an example of projectivist/expressivist style accounts of both morality and laws of nature. Mackie is an example of a moral error theorist, and van Fraassen (1989) might be considered a nomic counterpart. Firth (1952) defends Ideal Observer theory in ethics, and Mill (1843), Ramsey (1990), and Lewis (1973) do in laws. Moore (1903) is a non-reductionist realist about moral proper-ties, and Armstrong (1983) is one about nomic properties.

Craig Callender, *Humean Laws of Nature: The End of the Good Old Days* In: *Humean Laws for Human Agents.*
Edited by: Michael Townsen Hicks, Siegfried Jaag, and Christian Loew, Oxford University Press. © Oxford University Press 2023.
DOI: 10.1093/oso/9780192893819.003.0002

1.1 The Best Systems Theory of Laws of Nature

The best systems theory of laws (Mill 1843, Ramsey 1929, Lewis 1973) holds that the laws of nature are a kind of elegant summary of the non-nomic facts of the world. It is Humean in the sense that, like Hume, it denies that there are necessary connections in the world. On this theory, some true generalizations qualify as laws not due to metaphysical facts that these generalizations represent but, rather, because they express especially powerful compact summaries of the world. David Lewis imagined the set of non-nomic facts as a great mosaic of perfectly natural fundamental properties such as mass and charge distributed across space–time. The laws, he said, are the axioms that systematize this mosaic while optimally balancing the two virtues of theoretical simplicity and predictive strength.

I suggest that we regard Lewis's formula—a trade-off between simplicity and strength—as merely a first pass at characterizing how science discovers projectable generalizations, not the final word. Science cares about scores of theoretical virtues. Simplicity alone can be understood in dozens of distinct ways that often compete against one another. General relativity contains many more equations than Newtonian gravitation, yet it posits one fewer force—which is simpler?[2] Strength is just as complicated. We should understand the best system as that theory that optimizes whatever metric science actually employs when judging theoretical goodness and not get too bogged down in Lewis's gesture at this metric. Like Hall (2015), I agree that the "central, nonnegotiable idea" behind systems theory is that science's "implicit standards for judging lawhood are in fact constitutive of lawhood."

Is there such a metric? Assuming so is a substantial assumption, one that Feyerabend (1975), for instance, may have denied. But the assumption is defensible. When we look at physics, we see this metric hard at work. Quark models were proposed in the early 1970s. Many were empirically adequate. Some were ruled out for not constraining the data enough and others for constraining it too tightly. A delicate balance was sought. We see this balancing act, which is especially clear in curve-fitting, operate throughout science. It is behind the criticism that the Ptolemaic model was too complex and today in the complaint that superstring theory is too unconstrained. Science uses a kind of rough implicit standard in picking theories and the generalizations central to these theories.

Once we accept that there is such a metric function, Humeans have essentially solved the original problem of lawhood, the one bequeathed to us from Hempel (1965). That problem asked us to distinguish intuitively "accidental" true universal generalizations from intuitively "necessary" true universal generations.

---

[2] See, e.g., Woodward (2014).

Syntactically, nothing distinguishes 'all gold spheres are less than 1 mile in diameter' from 'all uranium spheres are less than 1 mile in diameter', even though the first is intuitively accidental and the second intuitively necessary (because uranium is unstable). For positivists with few tools besides syntactic structure, this is a puzzle. But for the Humean who says that the metric actually used is constitutive of lawhood, there is a solution. Although the example is somewhat contrived and we must solve it indirectly, we can appreciate that the statement about gold spheres—albeit true and simple—is not a simple corollary of a statement that plays a central role in the theoretical system describing our world, whereas the statement about uranium is a corollary of such a statement (ultimately, quantum theory).

Two toy theories may help us present this account as an Ideal Observer theory.

First, consider the silly theory that declares that the laws of nature are whatever Steven Weinberg, the Nobel Prize-winning physicist, says are the laws. Weinberg uses some implicit metric in his law judgements. He would probably not declare either the gold or the uranium statements as laws, but he will distinguish some true statements as laws and others as not. Hempel's problem is therefore answered. However, as smart as Weinberg is, this theory isn't remotely satisfactory. Weinberg simply doesn't have all the data, nor will he.

Second, consider actual computer programs that "discover" the laws of nature (e.g., Schmidt and Libson 2009). These are programs trained on data from simple systems, e.g., double pendulums. Accurate predictions and compact summaries are rewarded. A genetic algorithm is employed, choosing the best of the failures, modifying, and trying again. Using this method, the (overhyped[3]) Eureka Machine in 2009 produced a conservation law and Newton's Second Law. In the Eureka Machine we have our metric explicitly coded into the program and one can understand the genetic algorithm as an actual best system competition played out in real time. Here we have a timeless program that can handle an indefinite amount of data. Of course, the Eureka program and subsequent ones don't come close to encoding the creativity and insight of Copernicus, Darwin, and Einstein. It's doubtful that we'll ever have an actual program that does because we have no idea how to program the metric science uses.

To make systems theory more vivid, Hall anthropomorphizes the systematizer. Imagine someone who shares the criteria our scientists prize. How would she describe the mosaic if she had perfect understanding of these standards, unlike the Eureka Machine, but also full information about the mosaic, unlike Weinberg? Hall insists that this anthropomorphizing heuristic is entirely dispensable, that it is only a narrative device. That's right. It's just another way of expressing systems

---

[3] *The Guardian* reported. "'Eureka machine' puts scientists in the shade by working out laws of nature": see https://www.theguardian.com/science/2009/apr/02/eureka-laws-nature-artificial-intelligence-ai; but the program is essentially one that performs regression analysis.

theory. The point I want to emphasize is that the "person" behind this heuristic must be hypothetical, an Ideal Observer. The laws are the axioms the Ideal Observer uses when she systematizes the mosaic. Seen this way, we can appreciate how the laws are objective—they are not relative to Weinberg or Eureka or anyone else—and how they do not share the limitations of an actual person or program. What decides the laws is the best system. That system is the hypothetical system we employ at the ideal limit of science. The Ideal Observer anthropomorphizes this hypothetical element.

1.2 The Problem of Alien Laws

Framing systems theory via an Ideal Observer helps make obvious specific connections to ethics. Before we get there, however, we must confront a set of problems that have grown over time for systems theory. Putting the theory in terms of an Ideal Observer helps here too, as it better allows us to see that all these individual problems are particular instances of a more general problem, one I'll dub the *problem of alien laws*.

Here is the general problem: wouldn't an Ideal Observer declare as laws propositions that no human scientist would ever find acceptable? Indeed, shouldn't we *expect* an Ideal Observer to declare as laws propositions that are alien to us? By "alien" I mean not merely that these winning propositions might be surprising or counter-intuitive, but worse, that they would not play the roles laws play in actual science, e.g., supporting counterfactuals, prediction, and explanation.

To have something to work with, let's stick with the original best system. The Ideal Observer cares only about balancing Lewis's strength and simplicity in a theoretical system. Her goal is to recover as much of the Humean mosaic as possible while optimizing these two virtues. Scrutinizing the properties of our best current candidates for laws, we are led to wonder whether she would produce propositions with these properties. Or would she instead produce something alien to us, something lacking some of the crucial properties we favor in our best current candidates? Summarizing and condensing a vast literature, we can ask, why would the Ideal Observer find laws that...[4]

1. apply to both systems and sub-systems?
2. yield results even when supplied with only initial data?
3. are Markovian?
4. contain a division between initial conditions and dynamics?

---

[4] See Jaag and Loew (2020), Eddon and Meachem (2015), Hicks (2018), Dorst (2019), Ismael (2015), Hall (2015), and more. In these articles one can find some of the questions below and others in the same spirit. Jaag and Loew, Dorst, and Hicks all identify the same kind of pattern I do here.

5. permit various types of error tolerance?
6. can be approximately solved by tractable mathematics?
7. are different from $(x)Fx$, where F is the predicate true of all and only those non-nomic events that exist?

The Ideal Observer's goal is to recover as much as it can about the mosaic while balancing the two virtues of simplicity and strength. Why would it "care" about these other features? None obviously increase simplicity or strength individually or the optimization of these virtues. Let me expand on this point.

Regarding 1, the Ideal Observer works with only one system, the grand mosaic. Most propositions we've ever contemplated as fundamental laws—e.g., Hamilton's equation, Schrödinger's equation, Maxwell's equations, Einstein's field equations—have the notable feature of working for systems *and* subsystems. Classical mechanics describes the projectile motion of cannonballs, but also small bits of cannonballs. Quantum mechanics works for water but also hydrogen. Relativity works for the solar system but also the earth–moon system. There is simply no reason for this to occur from a systems perspective. So it's surprising, even suspicious, that what we take to be candidate laws so often work for subsystems.

Question 2 reminds us that the Ideal Observer is not limited in space or time and yet we are. One way to appreciate the huge difference this makes is by noticing that most candidate fundamental laws are described by hyperbolic partial differential equations (Callender 2017). These equations have the property that we can get something useful out of them (i.e., a non-trivial domain of dependence) by plugging information at a moment into them. We don't need to input temporally non-local data into them. Most types of equations do not have this feature. Instead of what's called a Cauchy Problem, where one puts data across a spacelike hypersurface and marches it forward or backward in time, why shouldn't the Ideal Observer produce a Dirichlet Problem, where one puts the system in a "box" and solves for the inside? That could be very simple and powerful, yet a four-dimensional version of such a problem—as opposed to an actual box— would be utterly unusable for human beings.

Relatedly, question 3 asks why are most of the laws Markovian? A time-dependent system is Markovian if and only if the distribution of future outcomes depends only on the present state of the system. In such systems the present screens off the past, so we do not now need input from the year 1837 to correctly predict the future path of (say) a satellite. This property of the laws is very convenient for creatures like us who lack detailed knowledge of the past (and must pay to store it when we have it). And it's a puzzle why the Ideal Observer should produce such laws given that there are indefinitely many intuitively "simple" equations that are non-Markovian. Thanks to the thermodynamic arrow of time human beings are "stuck in time." Conveniently, the laws tend to accommodate this limitation by being hyperbolic and Markovian.

Question 4 is raised by Hall (2015). Our laws tend to allow free or nearly free initial conditions and only constrain the dynamics (probabilistically or deterministically). We can use the same laws for cannonball trajectories no matter the mass of the ball, angle or direction of the cannon, and so on. Not only does there seem no reason for the Ideal Observer to allow free initial conditions; that seems downright *against the spirit of her job*, which is to increase strength.

Question 5 points to a family of questions. The Ideal Observer needs to recover the mosaic, nothing more. Yet our laws have the curious and again convenient feature of not being too finicky. When sending the Rover to Mars, do we need to input into the laws the full decimal expansion of its weight? Suppose we only go down to the nearest femtogram after 533 kilograms. Does the Rover go to the Moon rather than Mars? No. Plenty of possible theories, however, are such that small errors in initial data lead to disastrously large differences in solution. Not most of our theories. To put the point sharply, many of our fundamental equations have well-posed initial value formulations. A well-posed problem is one where the mathematical model for the system has three features: the existence of a solution, uniqueness of solution, and the solution varies continuously with the initial data. The first two features make perfect sense for an Ideal Observer because she prizes strength. The third feature is a mystery. For an Ideal Observer, existence and uniqueness should be enough.

Question 6 reminds us that the Ideal Observer is very good at math. Our laws are often hard to work with. Analytic solutions rarely exist. One can barely do any quantum mechanics with the naked Schrödinger equation. Fortunately, they all seem to have the nice property of admitting good approximations by more tractable math. One can plug in, say, the WKB ansatz into the Schrödinger equation and get sensible empirically verified predictions across a large domain, e.g., in quantum tunneling. To have that feature, and others like it, the laws have to have very specific forms, forms that again seem hard to connect to simplicity and strength.

Question 7 is familiar from the philosophical literature and never raised with respect to the other problems, but we can usefully see it as expressing this same problem. Lewis (1983) notices that if we allow *any* type of predicate into a system, then $(x)Fx$ will be as strong as possible and as simple as can be. Lewis uses this worry as part of an argument for restricting the lawful predicates to natural kinds. The worry is the same as ours, however, for the concern is that $(x)Fx$ is an alien law. Unlike our laws, it lacks any modal latitude. As a result, if $(x)Fx$ is a law, it's hard to see it playing any of the law roles (supporting counterfactuals, playing a role in explanation, connecting to causation, being used in science by human beings with finite resources, and so on). As this and many of the examples show, we dearly prize modal latitude, but it's hard to understand why an Ideal Observer would.[5]

In sum, it's suspiciously convenient for us human beings that the candidate law statements are so nice to us and have the above features.

In reply one may try to link simplicity or strength to one or more of the above properties. Descartes said that the way God preserves straight line motion is simple because "He always preserves the motion in the precise form in which it is occurring at the very moment when he preserves it, without taking into account of how it was moving [a moment before]" (1644, II, 39). In this passage it sounds like Descartes is linking simplicity to the Markov property. Even if this were plausible, as one goes through the list this type of move becomes more and more a reach. There is no reason why the Ideal Observer, as presently characterized, should be expected to produce such marvelously practicable features when it devises laws.

The supercomputer Deep Thought is tasked with computing the Ultimate Question of Life, the Universe and Everything in Douglas Adams's *Hitchhiker's Guide to the Galaxy*. Famously, after 7.5 million years of computation, the computer reveals the answer, 42. What makes this funny is that the answer is so alien to us. We have no idea how 42 could answer the question of the meaning of life. The problem of alien laws is similar. We worry—indeed, we expect—that the Ideal Observer will spit out the counterpart for laws of 42.

1.3 Meta-ethical Interlude 1

While these worries about the best system come up here and there, putting them all together in terms of the Ideal Observer helps us see the general pattern. And once one sees the pattern, one quickly realizes that we've previously encountered this worry. It arises in Hume scholarship (see Beebee 2016, Radcliffe 1994, Sayre-McCord 1994) and prominently in meta-ethics. According to Ideal Observer theory in ethics (Sidwick 1907, Firth 1952), ethical expressions are claims about the attitudes of a hypothetical observer who is fully informed and rational. Like Ideal Observer theory for laws of nature, it is a cognitivist (law claims are truth-evaluable) and realist theory.

With the problem of alien laws fresh in mind, the reader can quickly appreciate the comparable problem for Ideal Observer theory in ethics. In Firth's theory, the Ideal Observer is omniscient, disinterested, dispassionate, immune to subconscious effects, and perfectly consistent—but otherwise normal!

---

[5] Real-life examples similar to this can happen when machine learning tools are used to discover laws. Sometimes the laws produced are ones no human being would ever use or want. Presumably these laws are often the result of overfitting the data. That is perhaps another way to put some of the above worries: shouldn't the Ideal Observer overfit the data?

These features of Ideal Observers raise epistemological and motivational problems. Epistemologically, the idealization is so drastic that we can have little faith that we'll approximately know what the Ideal Observer thinks. Real human beings are riddled with inconsistencies, subconscious biases, and passions right to the core. Removing these and making me omniscient, I have little idea of what preferences the Ideal Observer will want satisfied. We also expect that the good for us will be motivating. Here too we have trouble. Even if we solved the epistemological hurdle and learned what the Ideal Observer wants, why should what moves this godlike creature also move us? Often I'm inclined to find more information, but clearly the Ideal Observer isn't. The internalist link to motivation seems threatened if the advice is no different than what I might learn from reading a book. As with laws of nature, the idealization process leaves us expecting that the pronouncements of ethical Ideal Observers will be alien to us.

As a result of these well-known and widely accepted criticisms, Railton (1986) proposed that we replace the Ideal Observer with what is now sometimes called an Ideal Advisor. On this theory, the hypothetical entity's reactions don't determine the good for you, but rather she recommends what is best for non-ideal you. If you imagine the hypothetical being as a kind of guardian angel sitting on your shoulder, the idea is that your good is not determined by what the guardian angel wants but rather by what she wants for you. Now the hypothetical entity considers not merely what information you lack and acts rationally but also takes account of your psychological traits, motivational system, and even the way you have lived your life. Suppose it is best for you that you believe in climate science. Studies of climate change deniers show that "more information"—more climate science— isn't always what is needed for someone to change their mind. What climate change deniers sometimes require is having the facts presented in a certain way. Appreciating that need may change the advice to you, which is a feature that the Ideal Advisor but not Ideal Observer may exploit. For Railton (1986, p. 23), we hold the non-moral features of a person "as nearly constant as possible when asking what someone like *him* would come to desire." That means that if the Ideal Advisor is to be a normative authority for you, she had better take account of *you*, warts and all.[6]

Before leaving, let me mention another relevant worry raised about Ideal Observer theory in ethics. Brandt famously pressed another question:

The facts of ethnology and psychological theory suggest that there could (causally) be two persons, both "Ideal Observers" in Firth's sense, who would have different or even opposed reactions...with respect to the same act, say on account of past conditioning, as different system of desires, etc.

(Brandt 1955, p. 26)

---

[6] Whether this move is successful is debatable; see Rosati (1995) and Sobel (1994) for criticism.

Need Ideal Observers agree? Brandt claims no, arguing that any assumed convergence will beg the question. Theorists appealing to a hypothetical thinker need to decide whether to embrace this relativism to evaluative perspective or not. This question will come up with laws of nature too.

1.4 Ideal Advisor Theory

The challenges faced by the best system theory, we now see, are essentially counterparts to those faced by Firth's Ideal Observer theory. The idealization process left Firth's Ideal Observer godlike and not human enough. We expect its recommendations to seem alien to us, and therefore, not motivating. The same is true of the Ideal Observer for laws. An Ideal Observer or even just an algorithm seeking to balance simplicity and strength would declare as laws propositions that no human scientist would find acceptable. Worse, these laws wouldn't play the law-role that motivated the project in the first place. Just as "42" is not an action-guiding answer to the question of the meaning of life, neither are the propositions likely delivered by the Ideal Observer action-guiding in science.

How to answer this worry seems clear. In an amusing parable, David Albert imagines an audience with God where God agrees to tell you about the world. God starts listing all the facts, whereupon

> you explain to God that you're actually a bit pressed for time, that this is not all you have to do today, that you are not going to be in a position to hear out the whole story. And you ask if maybe there's something meaty and pithy and helpful and informative and short that He might be able to tell you about the world which (you understand) would not amount to everything, or nearly everything, but would nonetheless still somehow amount to a lot. Something that will serve you well, or reasonably well, or as well as possible, in making your way about in the world.   (Albert 2015, p. 23)

God then replies (presumably) with something like the Schrödinger equation, an algorithm with which we finite fallible creatures can make successful predictions. The Ideal Observer has now turned into Ideal Advisor for someone in our predicament. The laws aren't what God Themself would use, but they are what They would recommend to creatures such as us.

To a Humean, the answer to all our questions 1–7 above is staring at us from the mirror: the laws are partly about us. Laws are useful, and they're laws partly *because* they're useful.

Once we view the laws as designed for us—given by an Ideal Advisor who knows us, warts and all—then the reason why the laws have all of the features mentioned is crystal clear. We always deal with subsystems in science. These are what matter to us. This chip, that chemical solution, those electrons.... In cosmology we sometimes aspire to describe the entire system, but even then we never know whether we're dealing with the system as opposed to a subsystem in a larger cosmos. We gather information in spatiotemporally limited regions. If our predictions varied significantly by requiring data from distant spatial locations or from distant times past, we wouldn't be able to use our theories. If they didn't permit various types of error tolerance and allow for mathematically tractable approximations, then again laws would be impractical. We need approximate predictions from equations that are solvable in polynomial time—the shorter, the better. Due to the arrow of time, we don't know the future but want to predict it from whatever state we find ourselves in. That's why we need great latitude in initial conditions and want little latitude in dynamics. We don't know what situation we'll be in, but once we do we want to narrow down what will happen. (The Ideal Observer, by contrast, doesn't suffer from a temporal knowledge asymmetry.) And we care about predicates that we can measure, intervene on, and so forth, not useless gruesome predicates like F.

The key to all of our puzzles is that we see laws as "partially prepared solutions to frequently encountered problems" (Ismael 2015, p. 197) or Albert's "meaty and pithy and helpful and informative and short" statements that "serve you well, or reasonably well, or as well as possible, in making your way about in the world." We move from an Ideal Observer theory to an Ideal Advisor theory. And that is in fact the way the literature has reacted. Long ago, Earman (1984) advocated thinking of strength as strength *for us*. Elsewhere Cohen and Callender (2009) eliminate predicates like F in Question 7 not for reasons of metaphysics but because such predicates aren't useful to us. Hicks (2018) modifies the best system competition so that it rewards the ability to be confirmed by experiment, thereby helping explain why we have laws that work for subsystems and that divide initial conditions from dynamics. Dorst (2019) changes the desiderata so as to emphasize prediction, producing "principles that are predictively useful to creatures like us." Along the way he shows how this alteration explains a host of features besides 1–7 above (e.g., why we prize symmetries) that together explain why scientists would care about system laws. Jaag and Loew (2020), focusing on these issues but also modal latitude, argue that the criteria for laws must maximize their "cognitive usefulness for creatures like us." In each case, the best system competition is modified by adding to strength and simplicity pragmatic criteria, resulting in laws that we might care about.

These moves to make the best system best for us are the counterparts (for essentially the same reasons) of the changes transforming an Ideal Observer into an Ideal Advisor in ethics. Not all the resulting system theories are the same, of course; for instance, Hicks's Ideal Observer is perfectly rational but not omniscient. Zooming out, however, they are each moves in a pragmatic direction, moves that bring the Ideal Observer down to earth. As in ethics, I think they're all moves in the right direction.

I want to highlight one further development that also has a counterpart in ethics. As mentioned, Brandt asked, why are we confident that Ideal Observers would converge on the same propositions? In the case of laws we can similarly ask why we think they would converge on the same set of laws? This question is especially pressing once Ideal Observers become advisors to the limited creatures that we are. The best advice depends on the audience. If the audience changes, so then should the advice. For laws of nature, the question is sometimes broached by asking whether all scientific communities need discover the same laws we do.

This question gets buried in the details of the systems approach. For example, the best system in its Lewisian formulation demands a preferred language. Lewis insists on perfectly natural properties. Since what is perfectly natural is in principle impossible to know, Cohen and Callender (2009) and Loewer (2007) loosen things up and allow an indefinite number of languages. What chooses the language is us and our theorizing, not the world. Since simplicity, strength, and their balance are all "immanent" notions—that is, they depend on the predicates being used—it's plausible that different systems formulated in different languages will yield different laws. A system using green and blue may have different laws than one using grue and bleen. If there are no right or wrong languages but only more or less useful ones, then the laws are hostage to what language is pragmatically best. If what is pragmatically best doesn't converge on one language, then one admits that the laws are relative to language and system. Different communities of scientists, using the language that is best for them, might arrive at different laws.

Laws can also become relative to system if one believes that the standards of simplicity, strength, balance, or any other additional theoretical virtue vary with need. The history of science arguably displays change in our standards for a good theory (Doppelt 1978). Maybe this change can happen at a time too? Maybe it depends on the science involved—biology, chemistry, or economics?

Advocates of the best system such as Taylor (1993), Halpin (2003), and Cohen and Callender (2009) all advocate making the laws relative to either language or metric or both (for discussion, see Eddon and Meacham 2015). One possible benefit of this move is that it arguably makes understanding the special sciences easier from a systems perspective (Callender and Cohen 2010). In any case, the more pragmatic the theory goes, i.e., the more the features of the audience being advised matter, the more pressure there will be to allow relativity.

In sum, for good reasons, the best systems theory has progressed from an Ideal Observer theory to an Ideal Advisor theory. In so doing, it also moves closer to its historical rival, projectivism, as we'll now see.

1.5 Meta-ethical Interlude 2

Shifting to an Ideal Advisor theory of laws of nature is a natural transition for the Humean. It more or less solves the problem of alien laws. Once we've made that transition, however, a new parallel with meta-ethics appears and this time the best response is less clear.

To appreciate this new parallel, I need to sketch a quick potted history of meta-ethics. Back in the Good Old Days, Hume's thought was developed in two ways. There were non-cognitivist accounts of moral discourse which are the heirs of Hume's internalist tendencies. Think here of Ayer's (1956) emotivism, Stevenson's (1944) expressivism, and R. M. Hare's (1952) prescriptivism. And there were cognitivist accounts that accommodate some of Hume's more externalist claims. One might have naturalistic theories like Firth's Ideal Observer theory in mind as examples.

These theories were each subjected to many well-known criticisms. Ideal Observer theory came under attack by challenges coming from the more "subjective" aspects of our moral language. Meanwhile the emotivist/prescriptivist/expressivist/projectivist strand faced challenges from the more "objective" aspects of our moral language. For instance, we speak as if there is a fact of the matter when two people disagree about a moral claim, e.g., eugenics is bad. We do not shrug such disagreement away as we do when someone likes pickles and someone else doesn't. Also, there are problems in developing the semantics for a non-cognitivist position, e.g., the Frege–Geach problem.

Due to these challenges, each strand of thought developed in sophistication. We saw that Ideal Observers became Ideal Advisors. And on the (let's abbreviate this strand to) expressivist side, Gibbard (1990)'s norm expressivism and Blackburn (1993)'s quasi-realism were developed. Both expressivists propose semantics that better handle the objectivist functions of moral language than previous versions of the theory. In fact, expressivists felt that their non-cognitivism freed them to do better than cognitivists like Firth. They were able to mimic what a realist about moral properties like Moore would say.

That brought the two historical rivals—Moorean non-naturalism and Humeanism—closer to each other. But if we focus on the Humean views, to the degree that contemporary Ideal Advisor theories can reproduce what non-naturalists can say, then this result also brings the two Humean views closer together. Railton's Ideal Advisor takes you and all your features into account when determining the good for you. But Gibbard points out that the rules for expressing yourself morally are much more constrained than had been appreciated. These twin moves left the two Humean theories, expressivism and Ideal Advisor theory, with little distance between them.

However, it was always still possible to distinguish the positions thanks to the cognitivism versus non-cognitivism divide. For the non-cognitivist, moral statements were not truth-evaluable, did not represent moral properties, and so on. For cognitivists they did, for moral statements expressed truths about what the hypothetical observer would want for you. Then one day expressivists adopted minimalism about truth, representation, propositions, and properties, and this convenient distinction potentially vanished. *Real* trouble in distinguishing all these historical rivals began, including the two Humean views considered here. This position, expressivism + minimalism, allows expressivists to mimic the language of moral realism. Minimalism entails that anything with content found in a 'that' clause can be said to be truth-apt, represent, and so on. The sentence 'keeping a promise is good' represents that act as good and is true iff that act is good. Minimalism even allows us to say that the property goodness exists. Coupled to minimalism, contemporary expressivists can accept that ethical claims are beliefs that represent mind-independent facts. Blackburn (2015) famously accepts all three defining tenets of Richard Boyd's (1988) moral realism.

Dreier (2004) labels this the Problem of Creeping Minimalism. Coming from the other side, Price (2015) dubs it the Problem of Creeping Cognitivism. Who it's a problem for depends on one's default perspective. From outside the debate, it seems a problem for everyone because one now wonders what all the fuss was about in meta-ethics regarding moral realism. Gibbard, the arch expressivist, announces at the beginning of his book (Gibbard 2003, p. x [preface]) that he is ambivalent about whether there is an issue at stake or not.

Of course there are replies to the Problem of Creeping, both by Dreier and others. But it's fair to say that there is no widely accepted answer to the Problem of Creeping. Whether the moral realism debate is best described as in a state of *ennui* or *détente* is not clear; what's clear is only that one must resort to French to describe it.

The two great traditions emanating from Hume have been developed to the point where few if anyone can tell them apart. Hybrid views abound: cognitive expressivism, cognitive and non-cognitive sentimentalism, and more. All these views agree that there are no Moorean non-natural properties in the world but that nonetheless contract cheating on university essay assignments is truly very evil. After that they fragment into dozens of views differing mostly over questions about the meaning and function of moral language.

1.6 Nomic Projectivism

For the final piece of our story, we begin with Ramsey's about-face on laws. Ramsey developed Mill's system theory in 1928, explicitly connecting it to an ideal future scientific theory where we know everything non-nomic. Interestingly, under a year later, Ramsey switched to a form of projectivism about laws of nature. Hints of projectivism can be found in both Hume and Pierce. In Ramsey (1929) this idea is developed. A law is no longer viewed as a summary of events but instead a recommendation about one's confidence in a way of inferring future events. Laws are "not judgments but rules for judging 'If I meet a φ, I shall regard it as a Ψ.' This cannot be negated but it can be disagreed with by one who does not adopt it" (Ramsey 1990, p. 149). The crucial insight we incorporated into Ideal Observer theories is front and center: laws are guides to the future. Like the warning "prepare yourself, winter is coming," they are not truth evaluable, even if rules exist for their use.

Despite Ramsey's switch, few followed. Projectivism about the nomic is discussed by Blackburn (1986) but is still mostly associated with Ayer's theory. That theory, if it makes an appearance anywhere, is typically found only in undergraduate courses where it is "counter-exampled" to death. Sophisticated developments of projectivist theories of laws are rarely discussed.

But they exist. Inspired by Ayer, Ramsey, and Blackburn, Barry Ward develops projectivism in detail through an impressive series of articles (e.g., 2002, 2003). What is interesting about Ward's theory is that he explicitly models it on the most detailed form of ethical expressivism available at the time he was writing, namely, Gibbard's norm expressivism. In Gibbard's theory, when one makes an evaluative judgement, one is accepting a norm that permits or forbids the relevant action. It is an endorsement. Because these norms play social roles, there are rules and logic behind how they function. One goal of such norms is social cooperation. Gibbard modifies possible world semantics to provide a semantics for normative judgements. Using this semantics, he is able to recover the logic underlying most normative thought and language. Ward takes over Gibbard's apparatus wholesale. For him, as for Ramsey, a law of nature is an endorsement, but for Ward the goal of law discourse is not only prediction but also explanation. Saying a generalization is a law is a recommendation that using it will be fruitful to both. The theory is non-cognitivist about language involving laws of nature; but like Gibbard's theory, by focusing on the function of such language and using a modified form of Gibbard's semantics, one can again recover the logic underlying modal discourse.

Note the parallels. For the new "pragmatic" systems theorists, one demands that the Ideal Advisor produce laws that are useful for actual scientists to use in explanation, prediction, and experiment. Dorst, Hicks, and Jaag and Loew explicitly build this kind of criterion into the systems view. For the new projectivist, the laws are what you would advise someone to use if they care about explanation, prediction, and experiment. There are differences between the two views, but they are not great.[7]

---

[7] Apart from the difference I'm about to discuss (cognitivism), another is that the projectivist doesn't expect Humean supervenience to hold for the content of law assertions. Humean laws are

1.7 Leapfrogging to Creepiness

Sometimes in political science writers speak of countries "leapfrogging" over the Industrial Age. A country, due to its peculiar history and circumstances, might jump more or less straight from an agriculturally dominated society to the internet age, going from cows to Facebook without smokestacks in between. We can in a similar vein spare the philosophy of laws some toil by leapfrogging to the present situation in meta-ethics.

In ethics, Ideal Observer theories and projectivism/expressivism inched closer and closer to each other for decades. Pushed by the demands of making sense of the more "objective" aspects of our moral language, the former adopted rules of use that mirrored the language of moral realism. Meanwhile, pushed by internalist aspects of morality, Ideal Observer theory became subjectivist and in some cases even relativist (in response to Brandt), becoming about Ideal Advisors. Throw in minimalism about truth, representation, facts, and properties, and now few can tell the difference between realists and irrealists anymore.

This same story is unfolding with laws of nature. Suppose that you are in a physics lab doing a spin measurement on some neutrons. The projectivist makes a recommendation: for you, given your goals (prediction, explanation) and circumstances, I recommend using quantum mechanics. The systems theorist likewise says: the Ideal Advisor, who takes into account your goals (prediction, explanation) and circumstances, advises you to use quantum mechanics. The only substantial difference between the two theories lies in their semantic properties, and in particular, one being non-cognitivist and the other cognitivist. Now add minimalism about the relevant metaphysical and semantic notions to something like Ward's projectivism and mix. The projectivist can then say that it's a fact that Schrödinger's equation is a law, that Schrödinger's equation is a law is true, that the proposition possesses the property of lawfulness, and so on. Just as Blackburn can agree with Boyd on the central tenets of moral realism, so Ward can agree with Lewis and Armstrong on features of laws. The Problem of Creeping equally affects laws of nature as it does moral language. Concentrating on Humean views,

---

summaries of the actual world but describe worlds that aren't actual. We need this modal latitude to use laws. However, many of those lawful possibilities are such that, if they were actual, the most elegant summary of that world wouldn't be what the system theory dubs as lawful. Newton's equations have solutions—say a single particle always traveling inertially—whose simplest and strongest summary are not Newton's equations. The projectivist has no such problem. Suppose, for comparison, that ghosts played an important role in navigating through life. Ghosts almost by definition don't supervene on the natural world. A "projectivist" treatment in which the overly credulous paint the world with such spirits doesn't demand that we identify anything actual with ghosts. It allows a naturalistic understanding of ghosts without an implausible supervenience thesis, e.g., identifying ghosts with creaky noises in the dark. Same with laws. If I endorse using $F = ma$ in making predictions, that doesn't entail that its content supervenes on the actual. See Ward (2002) and Ismael (2015). In what follows we'll assume that the nomic Ideal Advisor view can overcome this challenge.

it is now very hard to tell the difference between a major form of Humean nomic realism and a major form of Humean nomic irrealism.

We didn't see this problem coming in the case of laws, I suspect, because the internalist, motivational aspects of morality were always front and center, whereas these aspects of laws of nature were systematically downplayed. The challenges to systems theory force us to face these features. We just don't care about alien laws. A good recommendation needs to take the audience into account and alien laws don't do this. We expect laws to help us navigate through life. Alien laws don't. As a result, they don't motivate us in any way. In retrospect, we in the metaphysics of science ought to have paid more attention to this aspect of laws. The laws have always been suspiciously kind to us.

1.8 The New Landscape

We need to sharply distinguish the two issues I've raised.

The first issue is that in meta-ethics, for good reasons, expressivism and Ideal Observer theory each made moves that brought them very close to one another. I've pointed out that, for similar good reasons, projectivism and systems theory about laws of nature have also made moves that leave them almost indistinguishable. This point is like Parfit's famous claim that the apparently deep disagreements amongst Kantians, contractualists, and consequentialists are not in fact so profound. He views them as "climbing the same mountain on different sides" (Parfit 2011, p. 385). The first point is similar, that projectivists and systems theorists are climbing the same mountain, differing only in certain semantic features.

The second issue, the Problem of Creeping, is that minimalism about truth and other semantic properties allows this remaining difference to vanish. This realization happened in meta-ethics and it is a problem for more than only expressivism and Ideal Observer theory. It is also a challenge one meets in distinguishing a Moorean realism from expressivism, for instance. I've argued that the same Problem of Creeping affects the conceptual landscape in laws of nature as much as in morality. This news will be received as no news to global expressivists such as Blackburn and Price, but it should be news to Humeans and others about laws of nature. The problem of alien properties suggested shifts that make it apparent that systems theorists and expressivists are climbing the same mountain from different sides. Yet one could still distinguish the views due to their different semantic properties. The Problem of Creeping removes that ability. Keeping the mountain metaphor, the Problem of Creeping places us at the summit where the climbers merge into one.

How should we respond to these two issues?

Taking the second first, note that I have *not* argued that we *should* adopt minimalism about truth. I am not saying that we are at the summit just described. I have only pointed out that *if* one is a minimalist, then philosophy of laws must face the same problem that meta-ethics currently does.

In meta-ethics it's not so clear how to respond to the Problem of Creeping. Naturally, one reaction has been for meta-ethicists to find new ways to draw the line between realism and irrealism. Even if moral realists and irrealists can say the same things, moral realists may demand types of explanation unacceptable to moral irrealists. Or perhaps inference patterns will reveal a difference. The literature contains a proliferation of explanationist and inferentialist ways of distinguishing the two views (see Dreier 2018 for a survey and references). Unfortunately, the different ways of drawing a line between realism and irrealism tend not to agree with one another. None seems obviously best. Dreier (2018, p. 532) argues that there is no One True Distinction, and he adopts an "irenic and pragmatist perspective, allowing that different ways of drawing the line are best for different purposes."

If I'm right that the analogy is strong, then we can expect philosophers of science keen to maintain a difference between projectivism and systems theory to also find new lines of differentiation between the two. If the parallels hold up, as I think they will, however, we should not expect to find the One True Distinction in laws of nature either.

Another response is to "move on" past cognitivism and non-cognitivism, representationalism and non-representationalism. I don't know precisely how to characterize this path due to its diversity, but the result seems to be articles in meta-ethics with titles that sound like self-help guides to bad break-ups. One other reaction in this neighborhood would be to embrace a global expressivism such as that held by Blackburn and Price. None of these moves will deliver a clear difference between realism and irrealism in ethics, nor can they be expected to do so between Humean realism and irrealism.

Of course, one can deny minimalism about truth and avoid the Problem of Creeping. As I said, I have not argued that we ought to embrace minimalism, so this move is perfectly fine for all I have said. However, it does leave us with the first problem. And it means that the principal difference between realism and irrealism hangs on the correct theory of truth and representation, not anything specifically about laws. That is a very unwelcome result for someone who felt that projectivism or systems theory was correct due to reflection on science and its laws of nature. The moves toward pragmatism among systems theorists and the moves toward norms of law-talk among projectivists took the views so close that only semantic differences like truth-evaluability remained, differences that Creeping potentially obliterates. Rejecting minimalism about truth doesn't resolve the first problem raised by this chapter, namely, learning that projectivists and systems theorists are climbing the same mountain.

Like leapfrogging to the internet age, it's not clear that we've landed in a good place. We can take solace in skipping the painful birth of many sophisticated new positions and directions, yet it's not clear how Humeans should now understand laws of nature.

1.9 Laws as Negotiated Settlements

Philosophy papers are often convincing when they deliver bad news and go awry when they try something positive. I don't know how to respond to the Problem of Creeping—so this chapter will end safely negative on this point. Risking the fate of going positive, in the space remaining I want to suggest a way of thinking about the "two sides of the same mountain" problem. I think we can gain insight on the "two sides" problem by re-examining some issues that arise in Hume's work—by going back to the Really Good Old Days. When we do, we'll see that drawing lines between Humean theories of laws and declaring a winner seem less interesting than they were before. We can also see some moves made that will help us in our current predicament.

Hume's theory of natural necessity is very rich. He is famous for his skeptical attack on necessity. Necessity is found, he says—or most commentators say he says—in us and not the world. Hume does not rest with skepticism. He also provides a great origin story of why creatures like us would manufacture (natural) modality in a world lacking naked modal facts. Hume's insight is that modal reasoning is highly adaptive or "fitness"-enhancing given the predicament we're in. We live in an uncertain and risky world. We're cognitively and perceptually limited creatures who receive no information from the future. We don't know what will happen next, but if we're to survive and thrive we'd better have some guidance. We dearly need to predict what will happen next, prepare for it, and possibly intervene. For this the actual world is not enough. Our epistemic limitations and practical deliberative contexts require us to have theories about what is non-actual, just in case what we think is non-actual turns out to be actual (for attempts to begin to spell this out, see Ismael 2015, Strevens 2007).

Hume famously provides a theory of how all this works with causation. The source of necessity is a psychological faculty, the feeling of expectation. When we meet an event of type F, we come to expect one of type G because we've witnessed events like G follow events like F many times before. The feeling of expectation plays a key role: it gives us the ability to predict, prepare for, and possibly intervene on G by making F obtain. The expectation enhances our "fitness." And he also tells us how this works, claiming that the mind employs certain "rules" for judging cause and effect, e.g., that like causes regularly precede spatiotemporally contiguous-like effects.

Notice that this Humean theory can be updated to contemporary times. In outline, he proposes that we have a psychological faculty that follows some rules in taking as input some non-nomic facts (i.e., regularities) and yields as output a type of mental state (i.e., expectations) that is justified by its usefulness. Recent work in causal learning theory and developmental psychology offers psychological faculties quite different than Hume's state of expectation. One might, for example, replace the feeling of expectation with a special type of cognitive map, namely, a causal map, which is a representational system that maintains an updated representation of the causal structure of one's environment (Carey 1985; Gopnik et al 2004; Gopnik and Meltzoff 1997; Wellman 1990). Work in causal methodology might substitute Hume's "rules" with more sophisticated theories such as structural causal models (Pearl 2000). But the original rationale is still very much the same: these rules and faculties help us get by. The general picture is one wherein the brain is a prediction engine. To play its role, given the knowledge asymmetry (that we know "more" about the past than future), it creates various models of the future and runs them forward using rules it has found useful in the past.

This core Humean picture is very compelling. There is a beautiful consilience between theory and empirical work in it, and it has the advantage of parsimony because it needn't attribute necessary connections to the world itself. But where do laws of nature fit into this?

Hume doesn't really focus much on laws as opposed to causes (except when discussing miracles).[8] For Hume they are empirical generalizations formed from the patterns of covariation we encounter, but not a lot is said. I want to suggest a way of thinking of laws in Hume's picture that falls out from a problem his theory of causation faces, a problem that is basically the same as the one we've just encountered. The problem is one of mismatches between the deliverances of our psychological system—understood as feelings of expectation or as modern causal maps—and our considered judgements about causation.

To succeed in the world, we need to be able to change our causal maps. When we gather new data or come to better understand the rules by which causation is derived, we need to be able to learn and update our maps. Alignment with others is also important. In a social environment, it's crucial that we share expectations about what will happen. It becomes important to bring my causal map in line with yours. Early hunters could only bring down prey if they shared similar causal maps. But not all causal maps are equal. There are many pressures to

---

[8] Hume does recognize that "the utmost effort of human reason is, to reduce the principles, productive of natural phaenomena, to a greater simplicity, and to resolve the many particular effects into a few general causes, by means of reasonings from analogy, experience, and observation" (E 4.12). So he sees value, like modern Humeans, in devising generalizations from which one can get a lot from a little.

change one's causal map. Some are better than others. What does "better" mean for the Humean?

Hume faces this question throughout his corpus, and especially in his theories of causation, aesthetics, and morality. Causation is an expectation formed by seeing constant conjunction for Hume. But no one person is likely to observe all instances of F's and G's. Maybe they're constantly conjoined only in my set of observables. Maybe they're not even conjoined there but due to perceptual error it seems like they are. Nonetheless the psychological feeling of expectation may arise. We agree that some can be experts in taste. Nonetheless, it might be that some low art pleases me, e.g., zombie films. And my moral sentiments vary with whether people are near or far, family or not, even though when we make moral judgements we agree that "virtue in rags is still virtue" (1978, T 584). In all of these cases our psychological faculties may deliver responses that depart from some standard.

To deal with the mismatches between the deliverances of our psychological faculty (e.g., expectation, approval, pleasure) and our judgements, Hume often appeals to a "general point of view." In resolving the mismatch between our steady moral discriminations and our less fixed actual sentiments, Hume writes:

> Our situation, with regard both to persons and things, is in continual fluctuation; and a man, that lies at a distance from us, may, in a little time, become a familiar acquaintance. Besides, every particular man has a peculiar position with regard to others; and 'tis impossible we cou'd ever converse together on any reasonable terms, were each of us to consider characters and persons, only as they appear from his peculiar point of view. In order, therefore, to prevent those continual contradictions, and arrive at a more stable judgment of things, we fix on some steady and general points of view.... (1978, T 581–2)

A general point of view is a "method of correcting our sentiments, or at least of correcting our language, where the sentiments are more stubborn and inalterable" (1978, T 582). Elsewhere he writes that we need "to correct these inequalities [the above fluctuations] by reflection, and retain a general standard of vice and virtue, founded chiefly on general usefulness (1975, E 229, n. 1).

The "general point of view" motivates some commentators (e.g., Rawls 1971, pp. 183–92) to interpret Hume as espousing a kind of Ideal Observer theory (Radcliffe 1994). Moral and aesthetic judgements are based on the sentiments of someone occupying the general point of view. For causation, Hume defines causation as

> an object precedent and contiguous to another, and so united with it, that the idea of the one determines the mind to form the idea of the other, and the impression of the one to form a more lively idea of the other. (1978, T 1.3.14.31/170)

Garrett (1997) points out that we can give "the mind" in the above a subjective or an "idealized" spectator interpretation. If the latter, we would be identifying "true" causation with the expectation that occurs in a hypothetical mind, a mind that accurately perceived all the relevant constant conjunctions and perfectly followed the rules. Put in terms of our contemporary example, we can make sense of "better" in terms of an objectively best causal map. This map portrays the links between variables that one would draw if one perfectly followed the rules (say, Pearl's structural causal models) and had all the facts. Making sense of a 'best causal map' leads us to speak of hypothetical beings who are fully informed and rational. The best causal map is the map that such a being has in her head.

Note that this question—whether there is in the limit a best causal map—is a very esoteric one. It is essentially the question of whether there is a limit to science. I don't know of good arguments for thinking there is or isn't such a limit. In any case we've seen where this route will take us. These hypothetical "in-the-limit" causal maps will have little or nothing to do with psychological faculties in individuals and or the modifications made to them through learning. And since both are used in our predictions, preparations, and interventions, the hypothetical causal maps will have little to do with helping us navigate through life. They will be alien.

Sayre-McCord (1994) urges us not to interpret Hume's general point of view in morality and aesthetics as the perspective of an idealized observer. He understands Hume as envisioning a kind of Hobbesian jungle of diverse sentiments and the general point of view as a kind of negotiated settlement that smooths away inconsistencies amongst them. For this to work, the standard provided by the general point of view must be salient, mutually accessible, and tend toward stable consensus (p. 217). By salient, he means that the standard must engage the sentiments. I can bring myself to see that a person from outside my narrow circle is virtuous by imagining their act done in my circle. When I do, my sentiments are triggered. Mutual accessibility means that you can do the same. And if together we can form a consensus that selfless sacrifice for one in need is virtuous abstracted away from one's circle, then we will have a standard that irons out the bumps in the peculiarities of our varying sentiments.

Transferring this picture to natural modality, we can agree with Gopnik (2003) when she writes that "Science simply puts these universal and natural capacities [to detect causation] to work in a socially organized and institutionalized way" (p. 241). Science becomes a kind of Hobbesian civil society (minus dictator) that avoids a war of all against all. It tries to reconcile the many conflicting and changing causal maps we all have, settling on ones that serve our goals—of the lab, of the field, of the public, and so on. On this picture there is no final correct causal map, no guarantee that science settles on the "right" one. Yet there are ongoing negotiated standards for how to reconcile conflict—look to experiment, and if experiment can't decide, turn to theoretical virtues like simplicity, fruitfulness, and unification.

If we adopt this picture—which I wish I had space to develop—then what do we say about laws of nature? Laws of nature are the product of this social organization and institutionalization, particularly compact and powerful ways of saying a lot about the consensus-best causal maps in the relevant fields. They are the projectible generalizations that emerge from this Hobbesian jungle of vying causal maps. They play an important role, for they are the recommendations on which our sciences have achieved consensus. But metaphysically laws are not so interesting: they are the somewhat imprecise results of negotiation among our individual causal maps. As with the products of any messy social negotiation, they are bound to be somewhat loose and contested. This is the reason why science doesn't fuss too much about anointing one package as nomic when there are competing systems—for instance, choosing between Schrödinger's wave mechanics and Heisenberg's matrix mechanics, or figuring out precisely which propositions are the laws of evolutionary biology or general relativity. As with legal laws, consensus is difficult to achieve, so further negotiation is only warranted when it matters a great deal to achieving our goals.

We can now appreciate what both Humean theories got right. Focusing on our psychological capacities, e.g., expectations, causal maps, we can understand why projectivism is tempting. The source of Humean modality lay in our psychological faculties, and the laws are what science has achieved consensus on as its best recommendation. Projectivism nicely captures these aspects of laws. But focusing on the later improvements, organization and consensus building regarding our causal maps, we can see why a systems theory is appealing. The best system competition represents the Hobbesian battle to determine what causal map is best. Both theories get something right. Does the metaphysician or philosopher of science really need to decide between the two? No, for neither tells the full story. Each is only climbing half of the mountain.

Stepping back, in meta-ethics whether moral judgements are literally true or false might seem (or have seemed) like a life-and-death matter. In the laws debate the situation has never been like that. What's important is the origin and purpose of modal discourse: why do we engage in modal discourse if the world is fundamentally amodal, just one thing after another? Given this focus, it seems that philosophers of science should be less worried than meta-ethicists about creepiness and about from which side to climb the mountain. The Problem of Creeping perhaps demonstrates that it is even less worth having this debate than it initially seemed. It might not matter whether realism or irrealism about (Humean) laws is true. What matters is developing the origin and purpose stories for modal discourse. Laws of nature are then a kind of almost optional late-stage wrapping up of this development, one connecting the story to scientific practice. Here it turns out that the two camps (realist and irrealist Humeans) have so far approached things in slightly different ways, but the two approaches in fact complement each other nicely.[9]

References

Albert, D. (2015). *After Physics*. Cambridge, MA: Harvard University Press.

Armstrong, D. (1983). *What Is a Law of Nature?* Cambridge: Cambridge University Press.

Ayer, A. J. (1956). 'What is a Law of Nature?', *Revue Internationale de Philosophie*, 10, pp. 144–65.

Beebee, H. (2016). 'Hume and the Problem of Causation', in Russell, P. (ed.), *The Oxford Handbook of Hume*. New York, NY: Oxford University Press, pp. 228–48.

Blackburn, S. (1986). 'Morals and Modals', in Macdonald, G., and Wright, C. (eds.), *Fact, Science and Morality*. Oxford: Basil Blackwell, pp. 52–74.

Blackburn, S. (1993). *Essays in Quasi-Realism*. Oxford: Oxford University Press.

Blackburn, S. (2015). 'Blessed are the Peacemakers', *Philosophical Studies*, 172, pp. 843–53.

Boyd, R. (1988). 'How to Be a Moral Realist', in Sayre-McCord, G. (ed.), *Essays on Moral Realism*. Ithaca, NY: Cornell University Press, pp. 181–228.

Brandt, R. (1955). 'The Definition of an 'Ideal Observer' Theory in Ethics', *Philosophy and Phenomenological Research*, 15(3), pp. 407–13.

Callender, C. (2017). *What Makes Time Special?* Oxford: Oxford University Press.

Callender, C., and Cohen, J. (2010). 'Special Sciences, Conspiracy and the Better Best System Account of Lawhood', *Erkenntnis*, 73, pp. 427–47.

Carey, S. (1985). *Conceptual Change in Childhood*. Cambridge, MA: MIT Press.

Cohen, J., and Callender, C. (2009). 'A Better Best System Account of Lawhood', *Philosophical Studies*, 145, pp. 1–34.

Descartes, R. (1644). *Principles of Philosophy*, in Cottingham, J., Stoothoff, R., and Murdoch, D. (eds. and trans.), *The Philosophical Writings of Descartes (vol. 1)*. Cambridge: Cambridge University Press, 1985.

Doppelt, G. (1978). 'Kuhn's Epistemological Relativism: An Interpretation and Defense', *Inquiry*, 21, pp. 33–86.

Dorst, Christopher. (2019). Toward a Best Predictive System Account of Laws of Nature', *British Journal for the Philosophy of Science* 70(3), pp. 877–900.

[9] My longtime colleague Nancy Cartwright, reflecting on my work on the best system, said to me (I paraphrase), "Craig, for someone who doesn't really believe in them, you care a lot about laws." In some ways this chapter is me finally working through this tension. Many thanks for comments to Nancy Cartwright, Eddy Chen, Jonathan Cohen, Jamie Dreier, Michael Hicks, Carl Hoefer, Siegfried Jaag, Barry Loewer, Elizabeth Miller, Markus Schrenk, Elliott Sober, the philosophers at Ranch Metaphysics 2020, and especially Elanor Cranor and Christian Loew.

Dreier, J. (2004). 'Meta-Ethics and The Problem of Creeping Minimalism', *Philosophical Perspectives*, 18, pp. 23–44.

Dreier, J. (2018). 'The Real and the Quasi-real: Problems of Distinction', *Canadian Journal of Philosophy*, 48, 532–47.

Earman, J. (1984). 'Laws of Nature: The Empiricist Challenge', in Bogdan, R. (ed.), *D. M. Armstrong*. Dordrecht: D. Reidel Publishing Company, pp. 191–223.

Eddon, M., and Meacham, C. J. G. (2015). 'No Work For a Theory of Universals', in Schaffer, J., and Loewer, B. (eds.), *A Companion to David Lewis*. Chichester: Wiley-Blackwell, pp. 116–37.

Feyerabend, P. (1975). *Against Method: Outline of an Anarchistic Theory of Knowledge*. London: Verso.

Firth, R. (1952). 'Ethical Absolutism and the Ideal Observer', *Philosophy and Phenomenological Research*, 12, pp. 317–45.

Garrett, D. (1997). *Cognition and Commitment in Hume's Philosophy*. Oxford: Oxford University Press.

Gibbard, A. (1990). *Wise Choices, Apt Feelings: A Theory of Normative Judgment*. Cambridge: Harvard University Press.

Gibbard, A. (2003). *Thinking How to Live*. London: Harvard University Press.

Gopnik, A. (2003). 'The Theory Theory as an Alternative to the Innateness Hypothesis', in Antony, L. M., and Hornstein, N. (eds.), *Chomsky and His Critics*. Malden, MA: Blackwell Publishing, pp. 238–54.

Gopnik, A., Glymour, C., Sobel, D., Schulz, L., Kushnir, T., and Danks, D. (2004). 'A Theory of Causal Learning in Children: Causal Maps and Bayes Nets', *Psychological Review*, 111, pp. 3–32.

Gopnik, A., and Meltzoff, A. N. (1997). *Words, Thoughts, and Theories*. Cambridge, MA: MIT Press.

Hall, N. (2015). 'Humean Reductionism about Laws of Nature', in Loewer, B., and Schaffer, J. (eds.), *The Blackwell Companion to David Lewis*. Oxford: Blackwell, pp. 262–77.

Halpin, J. F. (2003). 'Scientific law: A perspectival account', *Erkenntnis*, 58, pp. 137–68.

Hare, R. M. (1952). *The Language of Morals*. Oxford: Clarendon Press.

Hempel, C. (1965). *Aspects of Scientific Explanation and Other Essays in the Philosophy of Science*. New York: The Free Press.

Hicks, M. T. (2018). 'Dynamic Humeanism', *British Journal for the Philosophy of Science*, 69(4), pp. 983–1007.

Hume, D. (1975). *Enquiries Concerning Human Understanding and Concerning the Principles of Morals*, ed. L. A. Selby-Bigge, 3rd ed. revised by P. H. Nidditch. Oxford: Clarendon Press.

Hume, D. (1978). *A Treatise of Human Nature*, ed. L. A. Selby-Bigge, 2nd ed. Oxford: Clarendon Press.

Ismael, J. (2015). 'How to be Humean', in Loewer, B., and Schaffer, J. (eds.), *A Companion to David Lewis*. Oxford: John Wiley & Sons, pp. 188–205.

Jaag, S., and Loew, C. (2020). 'Making best systems best for us', *Synthese*, 197, pp. 2525–50.

Lewis, D. K. (1973). *Counterfactuals*. Oxford: Blackwell.

Lewis, D. K. (1983). 'New Work for a Theory of Universals', *Australasian Journal of Philosophy*, 61, pp. 343–77.

Loewer, B. (2007). 'Laws and Natural Properties', *Philosophical Topics*, 35(1/2), pp. 313–28.

Mill, J. S. (1843). *A System of Logic, Ratiocinative and Inductive*. Cambridge: Cambridge University Press.

Moore, G. E. (1903). *Principia Ethica*. Cambridge: Cambridge University Press.

Parfit, D. (2011). *On What Matters*. Oxford: Oxford University Press.

Pearl, J. (2000). *Causality: Models, Reasoning, and Inference*. Cambridge: Cambridge University Press.

Price, H. (2015). 'From Quasi-Realism to Global Expressivism – And Back Again?', in Johnson, R. N., and Smith, M. (eds.), *Passions and Projections*. Oxford: Oxford University Press, pp. 134–52.

Radcliffe, E. (1994). 'Hume on Motivating Sentiments, the General Point of View, and the Inculcation of "Morality"', *Hume Studies*, 20, pp. 37–58.

Railton, P. (1986). 'Facts and Values', *Philosophical Topics*, 14, pp. 5–31.

Ramsey, F. P. (1929/1990). *Philosophical Papers*, ed. D. H. Mellor. Cambridge: Cambridge University Press [includes the 1928 note "Universals of Law and of Fact" and the 1929 "General Propositions and Causality"].

Rawls, J. (1971). *A Theory of Justice*. Cambridge, MA: Harvard University Press.

Rosati, C. (1995). 'Persons, Perspectives, and Full Information Accounts of the Good' *Ethics*, 105, pp. 296–325.

Sayre-McCord, G. (1994). 'On Why Hume's "General Point of View" Isn't Ideal–and Shouldn't Be', *Social Philosophy and Policy*, 11, pp. 202–28.

Schmidt, M., and Lipson, H. (2009). 'Distilling Free-Form Natural Laws from Experimental Data', *Science*, 324, pp. 81–5.

Sidgwick, H. (1907). *Methods of Ethics*. 7th ed. London: Macmillan.

Sobel, D. (1994). 'Full Information Accounts of Well-Being', *Ethics*, 104, pp. 784–810.

Stevenson, C. (1944). *Ethics and Language*. New Haven: Yale University Press.

Strevens, M. (2007). 'Why Represent Causal Relations?' in Gopnik, A., and Schulz, L. (eds.), *Causal Learning: Psychology, Philosophy, Computation*. New York: Oxford University Press, pp. 245–60.

Taylor, B. (1993). 'On Natural Properties in Metaphysics', *Mind*, 102(405), pp. 81–100. (Reprinted in *The Philosopher's Annual*, XVI, pp. 185–204, 1993).

Van Fraassen, B. (1989). *Laws and Symmetry*. Oxford: Oxford University Press.

Ward, B. (2002). 'Humeanism without Humean Supervenience: A Projectivist Account of Laws and Possibilities', *Philosophical Studies*, 107, pp. 191–218.

Ward, B. (2003). 'Sometimes the World is not Enough: The Pursuit of Explanatory Laws in a Humean World', *Pacific Philosophical Quarterly*, 84, pp. 175–97.

Wellman, H. (1990). *The Child's Theory of Mind*. Cambridge, MA: MIT Press.

Woodward, J. (2014). 'Simplicity in the Best Systems Account of Laws of Nature', *British Journal for the Philosophy of Science*, 65, pp. 91–123.

2

Humean Disillusion

Jenann Ismael

2.1 Introduction: the Case for Humeanism

In philosophy, there tend to be different stages in a debate. One starts with a question or a problem (e.g., what are the laws of nature? Are there objective ethical truths? What is the most just form of society?). In initial stages as positions are being developed, arguments are properly addressed from one side of the debate to the other as objections are used to help refine positions. Once the debate has matured and we have a number of well-developed positions (i.e., where the choice is no longer so much one of consistency or coherence, where people agree that the positions are coherent and it is really a matter of drawing out implications), then I think arguments should be addressed not so much to those on the other side of the debate (the standard should not be 'will you convince your opponents?') but to those that have not made up their minds. This is the stage where we have well-defined and well-understood positions, and things have hardened into different viewpoints that typically organize a whole cluster of commitments. That is the phase that the debate over Humeanism has reached.

In its early years, the initially attractive aspects of Humeanism were in the foreground. It assumed the existence only of local matters of particular fact. And because it had an epistemology that depended only on the knowledge of such facts, it was tailor-made for the empiricist instincts of science-based metaphysics. The methodology, moreover, was supposed to recapitulate the methods that scientists invoked in choosing between theories. The idea was that science gathered a large and wide-ranging body of information about local matters of particular fact and systematized that body of fact using the methods that scientists actually use. The laws and chances were statements that appeared in a certain role in the systematization. The account provided a reductive, non-metaphysical account of laws and chances that captured the main insights of the simpler regularity and frequency-based accounts of yore, but because it united laws and chances into a single package and allowed systematization to operate on the package, it avoided simple counterexamples to those accounts (Ismael 2015). It was an absolute breath of fresh air for those who wanted a science-based metaphysics, i.e., who wanted to believe in the modal commitments of science without scholastic metaphysics. There were no relations among universals, no irreducible modal forces or

Jenann Ismael, *Humean Disillusion* In: *Humean Laws for Human Agents*. Edited by: Michael Townsen Hicks, Siegfried Jaag, and Christian Loew, Oxford University Press. © Oxford University Press 2023. DOI: 10.1093/oso/9780192893819.003.0003

anything added to the Humean mosaic to enforce laws. It was all about system-atizing bodies of local matters of particular fact in a way that was itself modelled on science.

The epistemology was particularly important for David Lewis, who introduced the Humean view into the literature, and for other empiricists. The non-reductive accounts of what laws and probability are that were on offer in the philosophical literature separated them from the local matters of particular fact that provided evidence for them, rendering them unknowable. Lewis's account had its own objectionable aspects, but those were gradually shed. Later Humeans discarded the 'natural properties' that Lewis appealed to (Loewer 1996), and others devel-oped the account to apply to special science laws (Cohen and Callender 2009). Both of these reforms brought it into closer alignment with science. Some aspects of the account remained schematic; criteria of simplicity, strength, and fit proved difficult to characterize explicitly, but that was because they were meant to desig-nate scientific criteria for choosing among systems that were themselves difficult to characterize. Since its beginnings, moreover, Humeanism has had the advan-tage of exceptionally eloquent and charismatic defenders: Lewis himself, of course, and since then Ned Hall, Barry Loewer, and David Albert. For the scien-tifically minded metaphysician or philosopher of science, for a while it seemed there was simply no other game in town.[1]

2.2 Disillusion

What has happened in the years since, however, is that Humeanism's faults have begun to surface. For at least some Humeans, the shoe has begun to pinch. It's a familiar story: you leave an old job or an old lover for a new one with none of its faults, and then over time its own faults begin to surface with increasing clarity.

I'll speak of Humeanism in the tradition stemming from Lewis, through Albert and Loewer, and argue that the Humean doesn't have an epistemology that makes sense for embedded agents. The indefinite extendibility of the Humean mosaic, the fact that chances are determined by the pattern over the *whole*, and the explicit commitment to a combinatorial principle for determining what patterns are pos-sible mean that there is no way for a Humean agent to use information about local matters of particular fact to arrive at beliefs about the chances, short of assuming a restriction on priors that is patently at odds with their own metaphysics.

Here's how the discussion will go: first, I'll run through the problem that Lewis himself articulated in the article that launched the philosophical literature on Humean chances: the Undermining Problem. The Undermining Problem was

---

[1] Maudlin has been a vocal and persistent critic of Humeanism and you will see some affinities with his position here. See Maudlin (2007).

that chances contain information from the future and so they are incompatible with certain ways that the future could go. That meant that to assert a theory of chance is to rule out certain ways the future could go, but if we use the chances themselves to guide credence about those futures, they will generally assign a non-zero probability, so we have a contradiction. Then I'll introduce the fix. Undermining teaches us that we need to temper knowledge of the chances to correct for ignorance of the future. If we do that, I will show that the contradiction disappears in a natural way, but now a new problem emerges: no matter how much information you have about local matters of particular past fact, an indefinite portion of the chance-making pattern lies in the future. That means that the idea that observation gives us information about the chances at all was mistaken. Without a boundary condition in the size of the mosaic, you never get any closer to knowing what the chances are.

The argument is simple: a combinatorial principle for which Humean mosaics are possible, the claim that laws and chances are determined by a global criterion applied to the mosaic, and the recognition that the mosaic is indefinitely extendible together mean that conditionalizing on local matters of particular fact brings us no closer to knowing what the chances are. If you are inclined to think that it is not in general a problem if your metaphysics of X makes X unknowable to embedded agents, Lewis himself recognized that it is devastating to an account of chance, because the role chances play is to guide belief in the face of ignorance about the future. I'll discuss a response suggested (in conversation) by Albert and Loewer that I argue doesn't work.

2.3 The Undermining Problem

The problem of undermining and the immense amount of energy and work that went into sorting it out brought a great deal of analytical clarity into discussion of the relationship between a base ontology of local facts and a vocabulary that might contain disguised information about distributed patterns in it.

Lewis started out by asking what chances could be, and he introduced the Principal Principle (PP) as an implicit definition of chance that identified chances by the role they play guiding belief. What the Principle said in its original formulation was that one should adjust one's credence to the chances no matter what other information one has, except in the presence of inadmissible information:

PP: $cr(A/\langle ch_t(A) = x \rangle E) = x$, provided that E is admissible with respect to $\langle ch_t(A) = x \rangle$

Where $cr(A)$ is one's credence in A at some time t and $ch_t(A)$ is the chance of A at t. The restriction to admissible information was needed to discount cases where

PP clearly becomes inapplicable; e.g., when one possesses information from the future of the sort one might get from a crystal ball or a privileged communication from God.[2]

The problem that Lewis noticed that this poses for a Humean account of chance has to do with the possibility of what he called 'undermining futures'.

Undermining futures are futures that are incompatible with the chances being what they are. We know that such futures have to exist on a Humean account of chance to the extent that the correct theory of chance depends on how the future goes. So long as there is some dependence of what the chances are on how the future goes, no matter how small, and so long as a theory of chance assigns some non-zero probability to any future that is nomologically compatible with the past, that was enough to show that there would be some futures assigned a non-zero probability by the chances and yet that were metaphysically incompatible with the chances being what they were.

The problem can be put in a nutshell. It is that, on the one hand, chances are those things that play the role of chance in PP, so whatever we assign as reference, it had better be able to play that role. But, on the other hand, no Humean truth-maker *could* play that role because any Humean truthmaker introduces inadmissible information and undermines the applicability of PP.

That is the central difficulty of a Humean account. If laws and chances are to be identified with (some function of) the whole pattern of facts, then beliefs about the laws and chances have to be tempered by ignorance of the facts. If presumed knowledge of the chances outruns knowledge of the facts, then we aren't going to have advance knowledge of them in a way that allows us to use them to guide credence about the future. The reason that seems like disaster is that chances are *there* to serve the epistemic purpose of guiding belief in the face of ignorance of the future.

This is how Lewis put the problem, looking back in "Humean Supervenience Debugged":

> If I'd seen more clearly, I could have put the core of my reduction like this. According to the best-system analysis, information about present chances is inadmissible, because it reveals future history. But this information is not inadmissible, as witness the way it figures in everyday reasoning about chance and credence.[3]

---

[2] The formula that Lewis wrote down was more complicated than what he said in prose. I'm sticking with the prose formulation, which comes closest to capturing the pre-theoretic idea of how information about chance guides belief. If you prefer the more complex principle, substitute it here.

[3] Lewis (1994, p. 486). Ned Hall and Mike Thau were instrumental in getting him to see it this way.

The insight that Lewis took from Hall and Thau is that (i) it is true that if what the chances are depended too sensitively on some particular event e, then we can't use beliefs about the chance of e to guide expectation about e (so, for example, if which theory of chance were true depended sensitively on the outcome of the particular toss of a given coin, then we couldn't use the chances to guide belief about that toss), but (ii) the correct theory of chances doesn't depend sensitively on the particular events that we typically use them to guide expectation about.

Consider a world consisting of nothing else but a sequence of tosses of a single coin. The chances supervene on *total* histories: the bearing of a single toss on the chances ought to be proportional to the length of the sequence. In a world the length of our world, a single toss isn't going to shift probabilities for which theory of chance is true in any appreciable way, so we can use the chances to guide belief about the outcomes of particular tosses.

And that was enough to reconcile the everyday role of chance with the Humean commitment to the existence of undermining futures. The quantitative disparity between the information contained in a theory of chance about some particular event and the information that that event carries about which theory of chance is correct means that we can ignore the latter. (An analogy: although in principle every object exerts gravitational attraction on every other, the gravitational influence the earth exerts on a piece of dust so outweighs the effect the dust exerts on the earth that we can ignore the latter in calculating the former.) We can get a quantitative measure of that degree of dependence that we might think of as a measure of the degree of inadmissibility of information about chance and we reformulate the PP to reflect that it is applicable only when the degree of inadmissible information is low. It is low in the case of coin tosses and everyday events, making chance information admissible, and high in the case of undermining futures, making it inadmissible. Contradiction dispelled, and consistency restored.

And what we learn from all of this is that the reason that it pays for creatures like us (i.e., creatures that have information about the past, but whose information about the future is always derivative of what they know about the past) to think about laws and chances has to do with the 'balance of information'. The *balance* of information is such that if we have a large enough body of information about the past, we can use that to stabilize beliefs about which theory of chance is correct, and then use our theory of chance to guide credence about everyday events. So even though in *principle* which theory of chance is correct depends on *every* event, *in fact* our theories are largely indifferent to (cannot be undermined by) beliefs about the particular events we want to use them to guide belief about. If the world was too simple, or our actions and the events we are interested in predicting are not highly localized relative to the chance and law-making patterns in the Humean mosaic, it wouldn't work.

All of that seemed to make good sense and recover confidence in the Humean account of what chances are. A very big part of what recommends that account is that it gives us this very natural epistemology. The account of what the chances are connects them both to what we count as evidence for them, and also to what they guide belief about.

2.4 Ignorance about Chances

Undermining taught us that beliefs about what the correct theory of chance is, and hence about *what the current chances are*, on a Humean account, are hostage to the outcome of future observations so that any application of PP has to be tempered to reflect our ignorance about what the chances are. That means that the simple PP, which tells us what to do if we know what the chances are, needs to be supplemented with a principle that says how to let beliefs about chance guide credence where one *doesn't* know what the chances are. On this way of understanding it, undermining simply makes explicit the need for some principle about how to form credences in the absence of certainty about the chances.

In principle, you could write down any number of these. One could, for example, adopt as credences the chances assigned by the theory of chance assigned highest credence, or divide credence evenly among the top four, or ... you can think of any number of them.[4] The simplest suggestion is that one should create a weighted mixture of all theories of chance metaphysically compatible with history so far.

GPP: $cr(A) := \Sigma cr(C_{chi})ch_i(A)$, where ch_i is the chance assigned to A by epistemically possible theory of chance ch_i.[5]

Where should the weights come from? They should reflect current credences in the theories of chance in question and ultimately have to be rooted in priors. One can interpret the priors in a Bayesian way or impose additional constraints. But we want to impose the requirement that you can't assign a zero probability to any of the theories of chance that is metaphysically compatible with history so far. If you respect that requirement, you are not going to be in the problematic position—i.e., of plunking for a theory of chance that is metaphysically incompatible with the future going a certain way, and accepting the theory's recommendation to assigning it a non-zero credence—because you are never

---

[4] Not all of these are going to behave well diachronically under all conditions; see Pettigrew (2012). But think about what these kinds of arguments show: that there are some ways that the Humean mosaic could turn out to be where such a principle would lead to reasoning that would lead to non-Bayesian conclusions. But the Humean can respond here as below by saying that the success of our methods rests on the mosaic not being 'pathological' in certain ways. If that response is legitimate below in conjunction with induction, it is legitimate here.

[5] This is the one I proposed (Ismael 2008). I've suppressed the temporal parameters to avoid making the expression unpleasantly wieldy. I now prefer Pettigrew's Aggregate Principle: see Pettigrew (2016).

going to be plunking for a single theory of chance at all. At any given moment in any given history, there is going to be lots of different ways the future could go, and you are going to be mixing chances drawn from theories of chances corresponding to different histories. (That's true even if the world is in fact finite. If the world ends tomorrow, the credences I form on the eve of destruction are going to incorporate chances from lots of theories corresponding to histories in which it continues.) So we are golden.

This was a way of recognizing that if theories of chance supervene on *total* histories, we should be as ignorant of which theory of chance is correct as we are of how the future will go. And it seemed to me that it was a point in favor of Humeanism that as soon as you recognize this very natural idea, undermining problems go away. All of this is just getting increasingly precise and explicit about exactly what using information about local matters of particular fact to update beliefs about the laws looks like in a Humean world, assuming no non-Humean necessary connections. Since the laws supervene on the whole mosaic and specifically encode information about global features, the epistemology turns into a matter of using information about local matters of particular fact to form beliefs about global properties of a four-dimensional manifold of such fact.[6]

One of the primary arguments *for* Humeanism was that it gave us the connection between chances, on the one hand, and the local matters of particular fact that (i) provide evidence for them and (ii) they guide belief about, on the other. Lewis was quite explicit that whatever analysis one gave of what chances are, it ought to make sense that they should guide credence in accord with PP. So, for example, if the chance of e was the degree to which Angela Merkel preferred e to occur, it wouldn't make sense for the chance of e to guide your credence in it. The complaint he made about non-Humean theories was that they fail this test.

> Be my guest—posit all the primitive unHumean whatnots you like. [...] But play fair in naming your whatnots. Don't call any alleged feature of reality "chance" unless you've already shown that you have something, knowledge of which could constrain rational credence. [...] I don't begin to see, for instance, how knowledge that two universals stand in a certain special relation N*N* could constrain rational credence about the future coinstantiation of those universals.
>
> (Lewis 1994, p. 484)

For Lewis and for many of us who liked Humeanism, it was just as much because the epistemology was so clean and straightforward. Patterns in the Humean

---

[6] This generalizes to the deterministic case where chances become 1 or 0. In that case, we are using information about local matters of particular fact to guess the laws and then using the laws (in conjunction with the past) to derive predictions for the future.

mosaic were exactly the kinds of thing local matters of particular fact could give one information about. It made perfect sense not only that chances should be what guide credence in the face of uncertainty about local matters of particular fact but that *evidence for what the chances were* would come from information about such things.

2.5 The New Problem

But now a new problem comes into focus. Undermining showed us that we have to temper our knowledge of the chances to reflect our ignorance of the future. When we correct for our ignorance of the future by tempering our beliefs in which theory of chance is correct, the chances themselves are unavailable for the purposes for which they were designed. We cripple Humean chances for the purposes of guiding belief in the face of ignorance.

We can see this by thinking about how the Humean epistemology would work in a bounded universe of definite size. We are going through the world, picking up information about local matters of particular fact. Each observation brings us incrementally closer to knowledge of the whole mosaic. Since the chances are determined by a compressibility criterion applied to the mosaic as a whole, we also get incrementally closer to knowledge of the chances. Of course it's true that we won't have certainty about the chances until the end of time. And indeed, it was crucial to the dissolution of the undermining problem that we never attain certainty. But every time we learn a local matter of particular fact, we fill in one of the tiles in the mosaic. That means: cross off all of the potential ways the world could be incompatible with what we have learned and redistribute our credences over the rest. Over time, since there is a smaller and smaller number of ways the mosaic could be, we get closer and closer to knowledge of the correct theory of chance.

As soon as we remove the boundaries that define the mosaic, nothing we observe brings us closer to knowing what the total pattern is and nothing brings us any closer to knowing the chances. Literally, *nothing* that we observe tells us *anything* about the chances.

Here's the argument. We start with three premises:

(i) The set of possible mosaics is obtained by a combinatorial principle; any assignment of physical quantities to spacetime points represents a possible mosaic;

(ii) The laws and chances are determined by a global criterion applied to the mosaic; and

(iii) The mosaic is indefinitely extendible.

Indefinite extendibility means just what it sounds like. It means that the Humean mosaic is open-ended; it stretches indefinitely into the future. Note that it doesn't entail that the Humean mosaic is infinite. It just means that there is no particular finite size that it is constrained to be. Any finite history can be extended indefinitely into the future.

Why think the Humean mosaic is indefinitely extendible? There are two reasons. From a Humean perspective, to deny indefinite extendibility would be to hold that the existence of any collection of events was incompatible with the existence of some other. And that would be to deny Humeanism, because Humeanism was precisely the denial that there was any necessary connection between distinct existences. Lewis used to introduce what it meant to be distinct existences by saying the existence of one placed no restriction on the existence of the other.[7] The reason that indefinite extendibility matters is that the laws and chances supervene on the Humean mosaic as a whole, not on any part of it. And any submanifold, no matter what the spread of events over that submanifold is, can be embedded in indefinitely many mosaics whose total spread of events—judged by the criteria of overall fit—supports any chosen theory of chance that you like. This should be intuitively obvious. Fill in any patch P of observations up until a time t, and choose any theory of chance T_w, and it is easy to find a mosaic that would embed P and whose overall pattern would support T_w. And no observation or set of observations could make one theory of chance more probable than another. A nice fit with the pattern in one patch of spacetime can be undone by another and overridden by any larger patch of the same size. Best fit with the pattern over some finite patch has no bearing on best fit over the manifold as a whole, in a manifold of indefinite size. In an indefinitely extendible universe, nothing that you learn from observation will get you any closer to knowing what the best systematization of the whole mosaic will ultimately be, and that means that there is no way of learning from experience what the current chances are.

It is important to understand how misleading it is to think of 'the mosaic' as though it is a definite four-dimensional structure of known size. In a finite universe of known size, every tile in the mosaic gives us *some* information about the global pattern. We start out with a finite set of ways the universe could be; for the Humean these are obtained by recombination on assignments of values of physical quantities to spacetime points. Every observation allows us to rule out some of these, and we redistribute credence over the rest. Every observation brings

---

[7] Since indefinite extendibility is a modal claim, it comes in different modal strengths: the 'can' of metaphysical possibility and the 'can' of physical possibility. Only the weaker is needed for the argument here, but according to our best current physics, the stronger also holds. In Newtonian mechanics and SR spacetime is infinite in every direction. In GR, the global structure of spacetime depends on the matter distribution a theorem of John Manchak's (2009) shows that any model of General Relativity (finite or infinite) can be extended indefinitely by interspersing volumes of spacetime in a way that preserves the truth of the field equations.

us closer to knowledge of the pattern of events over the mosaic as a whole, and hence closer to knowledge of the chances. Once indefinite extendibility is taken into account, in neither of the senses above do we get closer to knowledge of the chances. Conditionalizing on individual local matters of particular fact doesn't get us any closer to making the *global* assessments of fit that determine the chances. So if the Humean mosaic is indefinitely extendible, i.e., if it spreads along every dimension without limit and is not metaphysically constrained to be some particular finite length, and chances are determined by overall fit with the whole spread of events, we have no intelligible story from the inside of how to go from a body of observed fact to hypotheses about what the chances are.

Consider a numerical analogy. You are given a small patch of a two-dimensional array of integers (positive and negative whole numbers) and told that they form part of an array of indefinite extent. (See Figure 2.1.)

Now suppose that I ask about global properties of the array as a whole: What do *all* of the numbers sum to? What is the probability that the sum of the top row is greater than 7893? What is the probability that overall the diagonal contains more 5's than 4's? What is the probability that the array as a whole sums to 667?[8]

If there are answers to these questions, how does what you've been given here (this section of the mosaic) count as *evidence* for them? And what do you *learn* when another number is revealed? What does an additional bit of the array tell you about the totals in question? How are you supposed to update your probabilities?

If the mosaic were bounded and of known size, it doesn't matter what your priors are, you learn from what you see. Conditionalizing on new observations takes you closer to knowing the pattern over the total mosaic, and hence closer to

3	−6	−16	−4	16	6	1
5	9	−14	−10	14	7	−11
15	13	19	−20	17	−23	−21
−2	−8	−18	0	18	8	2
−15	−13	−17	20	−19	23	21
−5	−7	24	10	−24	−9	11
−1	−12	22	4	−22	−12	−3

Figure 2.1 A numerical analogy

[8] It would be fair to ask, what is meant by 'the array' here? Just being given an initial segment like this is not enough to specify an array. If reference to 'the array' is to be well defined, some principle must be provided for specifying the array (a recursive function, for example), and under some conditions that principle can be used to fix answers about global properties like sums or lower or higher bounds. In the case of the Humean mosaic, there is no recursive principle or other means of demarcation. We are told only that it includes all events—past, present, and future.

knowing the correct theory of chance. The indefinite extendibility of the mosaic, however, means that nothing that you see takes you any closer to knowing what the full mosaic looks like; nothing that you see makes one theory of chance more likely than another.[9] Here's a quick hermeneutic argument: Humean credence for what the chances are has to be distributed across all epistemically possible mosaics. Divide the set of possible mosaics into equivalence classes according to which theory of chance they support. Every time we conditionalize on an observation, we cross off mosaics in each of these equivalence classes, but we never reduce the number of classes nor (on any natural way of counting) diminish their relative sizes.

If part of what made Humeanism attractive was that it gave us a connection between chances, the local matters of particular fact that we treat as evidence for them, and the future events that they guide credence about, indefinite extendibility severs that connection. By linking the Best System specifically to the *totality* of facts, Humeanism cuts the probabilistic/evidential connection between local matters of particular fact and the correct theory of chance. In a finite manifold of known size, it didn't matter what your priors were, conditionalization on local matters of particular fact would modify your beliefs about chance in prescribed ways. The combination of indefinite extendibility, a criterion for determining laws and chances that applies at the global level, and the absence of constraints on the relationship between one event, or one submanifold, and another means that conditionalization on local matters of particular fact does not generally have any impact on beliefs about the chances. No amount of looking at the world will bring you any closer to knowing what the chances are. There is simply not enough structure on the probability space to tell you how to learn from experience.

At least two of these things seem non-negotiable. The defining metaphysical dogma of Humeanism—that there are no necessary connections between local matters of particular fact—entails both the absence of constraints on the relationship between contents of one submanifold and the next, and indefinite extendibility. The remaining one—the global criterion for determining what the chances are—is just the BSA. So it is difficult to know how to resolve the issue. For the Humean, theorizing about chances is theorizing about the global properties of an indefinitely extendible manifold in which every local matter of particular fact is metaphysically independent of every other, and there is no connection between the pattern over some initial segment and the pattern over the whole. Any leaning towards this or that total pattern in a submosaic could be followed by another in which all trends were reversed or overwhelmed.[10]

---

[9] The likelihood of a hypothesis is the probability of the evidence conditional on the hypothesis.
[10] The vocabulary of submosaics is more accurate relativistic language. Indefinite extendibility applies to the spatial dimensions as well as the temporal, and any submosaic could be embedded in a larger in which all observed trends were reversed or overwhelmed.

Identifying laws and chances with distributed patterns in the mosaic of fact works really well when dealing with a finite mosaic of known size. It gets the connection with past facts that made it possible to use observed facts as information about the laws and chances, and the connection with future facts that made them good guides for belief in the face of ignorance. Every local matter of particular fact takes you incrementally closer to knowing the pattern over the total mosaic and so it is inherently information about the chances. Indefinite extendibility severs the connection between local matters of particular fact and the chances.

Global supervenience works well for many things. It is plausible that facts about beauty or value supervene globally on the physical facts. In those cases, reductions are to quell any worry that the facts in question introduce something metaphysically strange or something that can't be fit into a naturalistic view of the world. But it does *not* work well for chance since (as Lewis painfully pointed out in connection with undermining) the epistemic role of chance is to *guide belief in the face of uncertainty of the future*. Whatever chances are, they have to be the kinds of things that could play the epistemic role of chances, which means that they have to be the kinds of thing that situated agents could know about and use to guide expectation.[11] But once indefinite extendibility is taken into account, the whole neat epistemology that I described in the first half of the chapter, in which every local matter of particular fact takes you quantifiably closer to knowing what the chances are, while strategically avoiding undermining by not committing itself to the non-existence of undermining futures, falls apart. The problem is a tension between the Humean account of what the chances *are* (distributed patterns in the manifold) and what they *do* (guide belief about the future in the face of uncertainty).[12] Agents need to be able to learn about the chances, and to form reliable beliefs about them in the situation in which they need them to guide belief about the future. Unless there is a sensible account of how agents can have reliable beliefs about Humean chances in advance, Humean chances can't be the sorts of things we use to guide belief. It's really the same problem that Lewis saw behind the worry about undermining.

Everything I've said up until now should be uncontentious. It is the epistemological upshot of yoking the Humean account to the pattern over the whole Humean mosaic where the mosaic is indefinitely extendible and intrinsically

---

[11] We can be unfussy and non-committal about what 'know' amounts to here, beyond saying that it falls short of certainty and involves the ability to learn from experience. Any way of firming up the notion that captures its central philosophical uses and falls within this range will do.

[12] Did Lewis consider indefinite extendibility? I don't think so. Lewis did not note that the step from 'all that there is is one damn thing and then another' to 'there is a totality of local matters of particular fact' is a substantive one. As far as I know, the only place that Lewis discussed it is in "Postscript to 'Things Qua Truth-makers': Negative Existentials," written with Gideon Rosen (Lewis and Rosen 2003). They argue in that piece that the world should be treated as a concrete particular and provide the truthmaker for negative existentials. This addresses truthmaker problems associated with indefinite extendibility but doesn't help with the epistemic considerations here.

unconnected. The situation for the Humean agent trying to form beliefs about chances is the precise analogue of someone given what they are told is the initial segment of an indefinitely extendible string of integers in which each is stipulated to be independent of the others, and they are forced to make judgments about the sum. In that setting, what they have been given tells them nothing about the sum, and the revelation of additional numbers tells them nothing more: it brings one no closer, rules nothing out, narrows nothing down.[13]

And it is worth pointing out that indefinite extendibility not only problematizes the epistemology of chances but also problematizes the role chances play guiding belief, and for something like the same reason. Chances defined by global fit with an indefinitely extendible manifold have no definite, quantifiable bearing on what happens next in the here and now.

2.6 A Humean Response

Here is a response suggested by Loewer and Albert.[14] They say that the problem I've pointed out is just the problem of induction. The solution is to put constraints on priors that favor worlds which are inductively hospitable. What that has to mean in this setting is that the laws and chances that would be derived from systematization from a large enough initial submosaic are reflective of those in the whole. They go on to say that everybody has the problem of induction, so this isn't specific to Humeanism. And they add that the problem is double for non–Humeans. Humeans need induction to generalize from knowledge of part of the manifold to the whole, but then they are done. Non-Humeans, by contrast, need induction to go from part of the manifold to the whole, and then on top of that they have an unbridgeable gap between the pattern of events in the manifold and the laws. Since non-Humeanism is a non-reductive view, even knowing the whole Humean mosaic won't fix the chances.

Let's look a little more carefully at what this response amounts to. In probabilistic terms, it amounts to recognizing that there is no link internal to the probability space between beliefs about local matters of particular fact and beliefs about chances. The logical structure of the space leaves beliefs about chances unaffected by conditionalization on local matters of particular fact. The proposal here is to impose a link *externally* by constraining priors so that they correlate the pattern of events over the submosaic from which observation is drawn to the pattern over the whole. It is non-trivial to make this precise in a way that would

[13] Integers, of course, are positive and negative whole numbers, so one can take away with the next number what one added with the one before, or take away with the next string what one added with the string preceding. Indefinite extendibility is compatible with the universe being *actually* finite, but credence for what the chances are has to be distributed across all epistemically possible mosaics.
[14] In conversation.

achieve the desired effect, but let's suppose that it could be done. Let's review the situation: When you have a finite mosaic of known size there is a link between local matters of particular fact and facts about the total mosaic that is built into the probability space so that, when you conditionalize on observation, you automatically update your credences for laws and chances. In that setting, it doesn't matter what priors you start with, conditionalizing on local matters of particular fact will take you incrementally towards knowledge of the chances because it will take you incrementally closer to knowing what the whole mosaic is like. That link is severed in an indefinitely extendible setting. Conditionalizing on particular observations has no intrinsic effect on the global pattern that determines the chances. The Humean suggests that we supplement Humeanism with constraints on priors that tie the systematization of the past to the systematization of the whole. The availability of such priors tells us nothing. One can *always* choose priors that will set up a link between facts that have no intrinsic link. I could choose priors that link the value of Genentech stock to the price of tea in China. I could choose priors that link the color of my socks on any given day to the color of the Queen's underwear. That doesn't mean that I can learn about the Queen's underwear by looking at my socks.

In physical terms, constraints on priors embody assumptions that one makes about the way the universe is *before* you have any evidence and they guide how you update on what you see. The constraints that the Humean would need to impose in order to make scientific practice as it stands a way of learning about chances would be the assumption that the laws and chances of the best systematization of the patch of the universe that falls within our own past reflects those laws and chances that one would get from systematizing the whole. The problem with this for the Humean is that putting constraints on priors that heavily discount mosaics where systematizations of the past do not reflect systematizations of the whole is discounting the bulk by far of what they regard as possible mosaics. Consider, for example, the full set of random strings of integers of some particular length—say 20,000 integers long—obtained by a combinatorial principle. The overwhelming majority of those strings will not be ones that satisfy this condition. So the Humean is committed to pairing a combinatorial principle for obtaining future possibilities with an epistemic principle that says: even though there are as many ways the future could go as there are ways of assigning physical quantities to spacetime points, in forming beliefs about the future heavily discount all but a very small sliver of those.

If we follow this line, for the Humean induction is a heavy bet that is *pre-empirical*, *essential* to being able to draw any conclusions from experience about the chances, and at odds with a metaphysics that is explicitly committed to recombination at the level of local matters of particular fact. It is also arbitrary, since we could just as well, and with no less justification, adopt priors that assume that the global pattern will form an American flag or that it will surround our little patch of spacetime in a sea of bland uniformity. The intuitive naturalness of the assumptions embodied in these constraints should not mask how unnatural they are from a Humean point of view. Think about the connection between Humean chances and the kind of regularity that is being assumed to support induction. The existence of this kind of regularity makes summarizing *possible*, but few summarizable mosaics would be supported by the Humean epistemology with the inductive premise. There are innumerable ways that the Humean mosaic could be that would make it easy to summarize but not hospitable to induction. If the mosaic formed a giant pattern of the American flag, for example, or if after a brief period of apparent complexity everything turned into a simple uniform gas, or if it followed a plan laid out in two pages of the Book of Mormon. It makes sense if you live in a world with the sort of structure that supports inductive practices that you would exploit it in summarizing, but it makes no sense to *assume* it in a Humean world looking forward. Indeed, it is at odds with the defining metaphysical dogma of Humeanism. Humeanism is defined by the denial that there are necessary connections between local matters of particular fact. It is defined by the belief that no matter what has happened up until this moment, moving forward there are as many ways the future could go as there are ways of assigning events to spacetime that lie in the future. Betting heavily on priors that favor induction for a Humean is like me saying: "look, I know the color of the Queen's underwear is metaphysically independent of the color of my socks. I know, that is to say, that any combination of colors for my socks and her underwear is possible. But I can see my socks, so I'm going to assume priors that heavily discount all combinations except those in which they are the same and use observations of my sock to update my credences for the color of her underwear." If you are a Humean, any regularities that emerge over the course of history can be exploited after the fact to give a compact description, but there is nothing that rationalizes assuming this sort of regularity looking forward.

Dustin Lazarovici (2020) has recently argued that typical Humean manifolds won't permit systematization. Exactly the same argument will show that *even if we restrict attention to those manifolds that permit systematization*, the assumption that Loewer and Albert are recommending we build into the pre-empirical constraints will typically fail and fail quite badly. Following the Humean epistemology in a Humean universe in which there *are* Humean chances and laws will typically lead us *away* from (not towards) them.

This proposed supplement or revision to Humeanism comes at considerable cost in the elegance of the view. Canonical Humeanism says: looking forward, there are in fact as many ways the world could be as you would get by extending history without constraint. You wait until the mosaic is complete and then you systematize the full body of fact. Systematization is just informational compression. It exploits whatever regularities the manifold possesses and combines it with whatever assumptions of fact pack enough predictive punch to warrant their inclusion. The laws and chances are just axioms in this systematization. There is no more metaphysics than that. The epistemology is very simple: God solves for the laws and chances and delivers them to agents to use as guides to belief.[15]

Here's the new view. It's the same metaphysics as above: looking forward from any point in history, there are as many ways the world could be as we would get by assigning values of physical quantities to spacetime points in the future. The divine epistemology and metaphorical sit-down with God are discarded. It is now agents who have to discover the chances. Even though the metaphysics says that looking forward from any point in history, there are as many ways the world could be as we would get by assigning values of physical quantities to spacetime points in the future, the epistemology says that you must take as a pre-empirical assumption that the laws and chances derived from any large enough submanifold would reflect the laws and chances derived from a global systematization. This amounts heavily weighting your priors to ignore all but a small sliver of Humeanly possible completions of the mosaic. Since the metaphysics is explicitly committed to combinatorial possibilities for the future, the only thing that keeps this from being flat-out inconsistent is that one reserves *nominal possibility* that the future might be among the vast majority of worlds whose overall systematization is different from that of the initial segment.

2.7 What's the Alternative?

At this stage, the Humean will ask: what's the alternative? He will say that one has to assume induction works for scientific practice to make sense. Nobody has a solution to the problem of induction. There is no metaphysical guarantee that the future will be like the past. And if one assumes that induction works, the Humean account is still the best game in town.

There are two things to say about this. The first is that if it's a question of choosing priors that make sense of scientific practice, why not start with priors that expect no correlations unless there are connections? If those are the priors that you start with, everything in your experience will be telling you that the world is *not* an intrinsically unconnected pattern of fact. You will find a wide body of correlations—some local and temporary, some deep and pervasive—and

---

[15] Readers of this literature will be familiar with the primordial myth motivating the Humean account in which David Albert relates a sit-down with God where you ask God to tell you about the world and He begins to list every event one by one. Albert says,

"[Y]ou explain to God that you're actually a bit pressed for time, that this is not all you have to do today, that you are not going to be in a position to hear out the whole story. And you ask if maybe there's something meaty and pithy and helpful and informative and short that He might be able to tell you about the world which (you understand) would not amount to everything, or nearly everything, but would nonetheless still somehow amount to a lot. Something that will serve you well, or reasonably well, or as well as possible, in making your way about in the world" (Albert 2015, p. 23).

scientific theorizing will take the form of systematizing those connections. Once those connections are systematized, they provide the basis for prediction, intervention, design.[16] In the old days the assumption was that the 'connections' would ultimately take the form of deterministic laws. Although most theorizing is probabilistic (we have large bodies of data, look for correlations among the values of measured quantities, and assess likelihoods of hypotheses about the structure of the source), the presumption was that stochastic regularities would be underwritten by deterministic laws and eliminated in the fundamental theory. Since quantum mechanics we have at least become comfortable with fundamentally stochastic laws, and statistical mechanics has made the role of probability even in deterministic settings clear. In either case the information that provides the basis for theorizing is probabilistic, and theories that are the product of systematization will involve probabilities in various guises.

What is there to recommend these priors over the Humean priors? I'm Bayesian enough to think that priors are priors. There are no rational constraints on priors and no *a priori* argument for one set of priors over another. Descriptively, this is close to the set of priors that we employ in everyday reasoning. It is a generalization of what the child does when approached with a new toy or an opaque box with handles and buttons (Gopnik et al. 2004), and I suspect that it is close to the unreflective, default view of most scientists.[17] It does a better job hermeneutically of making sense of scientific practice. Observed correlations suggest connections.

---

[16] Saying that they provide the basis for intervention and design is a way of saying that we don't just rely on them to make predictions about the future; we rely on them in the kind of hypothetical reasoning and assessment of possibilities that is characteristic of those contexts. That is all that their modal force comes to. The remarks above were focused on chances since there is a well-developed discussion that puts the epistemology of chance front and center, but I switch now to talking about laws and chances together. These are the joint product of global systematization; together they provide the fundamental modal outputs of theorizing.

[17] Consider an analogy. Unlike people, countries can last indefinitely long into the future. Credit ratings are assigned to countries, as they are to people, to assess creditworthiness and assist lenders in deciding whether to issue loans. Suppose I asked you: what is the credit rating of a country? Is it a summary of the borrowing behavior over the course of its existence? If it was, you'd never be in a position to so much as offer a guess at the credit rating of a country. It would always depend on how things go, and since countries can continue indefinitely into the future, you wouldn't even be able to say whether what you knew from past behavior had any probative value and what that value was. Credit ratings weren't meant to summarize behavior over complete history (past, present, and future) but to extract from past behavior information relevant to future behavior. They are needed to tell banks whether to lend money in the here and now. Credit ratings take into account payment history, amount owed, length of history, types of credit used, things like that. And the reason that credit ratings can play the role they play, i.e., the reason that they are a good predictor of future behavior, is that there are stable features of the system to which they are assigned that is manifested in their past history and is a good indicator of future behavior. The whole practice of assigning credit ratings is predicated on the idea that there is some kind of regularity that guides what people do that we can separate for all of the contingencies of their situation and that will provide some guidance about how they'll behave in the future. In the case of credit ratings, this is all quite informal and seat of the pants, but it is not different in kind from what we do in scientific settings. The whole business of looking for laws and chances is predicated on the idea that there is some kind of regularity that guides the behavior of physical systems which we can separate from all of the contingencies of their situation and that will provide some guidance about future behavior and behavior across a range of hypothetical situations.

Connections are analyzed and tested. Once established, the connections are relied on in prediction, design, and intervention.[18]

What are these 'connections'? Formally they are the symbolic expressions that can restrict the joint values of parameters, or the probabilities of values (e.g., f = ma, Maxwell's Equations, Born's Rule). They often take the form of laws of temporal evolution (Schrödinger's equation, Hamilton's equations). Physics offers no 'analysis' of what these connections are. Philosophers became interested because certain ways of thinking about laws made them seem problematic. The way we present physical theories makes it sound like we start with a space of possibilities and pare them down by adding laws. This gives us a sense that laws act as restraints that *keep things* from happening. And the notion of law, of course, is borrowed from the human domain and suggests restrictions on freedom. Hume problematized the idea of physical necessity with his critique of causation and forced a reckoning with the idea that there is anything in the world over and above events and their regularities. It became common among empiricists to hold that laws were regularities and to deny that there is anything in the world that enforces them.

I want to suggest that this was the wrong way to think from the beginning. Notions of physical necessity are dual to notions of physical possibility and we should focus instead on the notion of possibility. Physical theories give us a notion of physical possibility that comes from a kind of analysis and synthesis that is familiar since Newton. We are accustomed to thinking in terms of analysis of complex systems into spatial components, but the real basis of systematization is analysis of complex *motions* or behaviors into simple ones and then recombination of those simple motions to derive the possibilities at the higher levels.[19] The result of this process is that there's a radical reduction of degrees of freedom looking forward from any point in spacetime relative to what we would get if we followed the Humean prescription of treating every local matter of particular fact as an independent degree of freedom. If we view possibility as the more basic notion and treat it in this constructive way, the idea of laws as restraints goes away. Theories are an attempt to discern latent possibilities in the world. It is about

<hr/>

[18] The fact that they are relied on in design and intervention is what makes them modal. In design and intervention we are assessing not only what will happen but what would happen under a range of merely possible conditions. Laws are meant as much to capture the full modal latitude inherent in the world as to generate predictions. See Hicks (2018) for a nice discussion of the difficulty that traditional Humeanism has accounting for this aspect of practice. Hicks's article shares much of my own sense that understanding Humeanism means abandoning the atemporal God's-eye view epistemology and giving a situated account that gets the role of laws and chances from an embedded perspective right. See also Jaag and Loew (2020).

[19] 'Motions', here, doesn't mean simple change of position, but change of state. Although Newton was primarily interested in change of position, nowadays we recognize internal properties alongside position; the state of a system is represented by a point in a phase space, and 'motion' means change of state represented by movement through phase space. In the philosophical literature what I've been calling 'ranges of motion' have sometimes been called powers. See Demarest (2017).

looking at the phenomena with an eye to separating the dimensions along which the phenomena can vary. Laws are expressions of restrictions not in the sense of 'restraints' but in the sense of 'boundaries'.

The problem with Humeanism is that it starts with a metaphysics that is explicitly committed to recombination at the level of local matters of particular fact. If we start from a more neutral standpoint with no pre-empirical commitment to how many ways the world could be and think of science as in the business of discovering the immanent possibilities in the world, Humeanism will look like it is overreaching. The more conservative view will turn out to be the one that recognizes only such immanent possibilities as are implicit in the phenomena. Through a combination of theory, observation, and experiment designed to push nature to exhibit her full capacity for independent variation, the scientist is trying to establish the limits of what is possible. That still leaves the notion of possibility, but it is a notion that has every claim to be conceptually rock-bottom independent of any connection to science. A view like this involves a reworking of the way that we think about physical modality. It has the virtue of involving none of the extraneous metaphysical machinery invoked by traditional anti-Humean accounts,[20] and I think it is closest to the one that most working scientists adopt. If I had to give it a name, I'd call it the 'natural nomological attitude' (echoing Arthur Fine's 'natural ontological attitude' perhaps only in name).

The second thing to say is that I agree that there's no solution to the problem of induction in the form of a rational or metaphysical guarantee that induction works. But precisely because we can easily imagine worlds in which those practices fail, the success of the inductive practices embodied in science is a datum or a clue that we can use in understanding how our world is structured. Instead of grafting an inductive assumption onto a metaphysics for science that is at odds with it, we should lead by asking *why* science works. And we should be looking for an understanding of how the world is structured that makes the inductive methods embodied in our scientific practices sensible and non-arbitrary.

---

[20] Anti-Humean views are typically modalist, in a sense characterized nicely by Michael Hicks: "The regularity theory holds that laws of nature are merely generalizations.... Modalist views are less metaphysically perspicuous than the regularity theory because they claim that to be laws, a generalization must be backed, made true, or associated with a relation between properties (Dretske [1977]; Tooley [1977]; Armstrong [1983]), the essences of properties Shoemaker [1980]; Ellis [2001]; Bird [2007]), *sui generis* facts about production (Maudlin [2007]), or irreducible counterfactuals (Lange [2009])....Although each of these metaphysical machines is distinct, they are united in holding that facts about nomic necessity, or facts with modal implications of some sort, are fundamental. In contrast to these views, the regularity theorist holds that the laws are not backed or made true by anything beyond their instances and they are made laws by nothing more than the sum of non-nomic facts at a world" (Hicks 2018, p. 984). These kinds of metaphysical machines, as Hicks calls them, are foreign to science, and Humeans rightly oppose them. Laws need no metaphysical backing. The modal force of laws just means that we rely on them not simply in predicting what *will* happen but in the kind of hypothetical reasoning we rely on in decision, design, intervention, and control. See my Ismael (2017), and John Norton (2021).

We don't have to look very far to find such an understanding. Here are some unguarded descriptions from physicists about how theorizing works in the kind of fundamental theory for which the BSA was originally designed. The first is from James Hartle:

> Identifying and explaining the regularities of nature is the goal of science. Physics, like other sciences, is concerned with the regularities exhibited by particular systems. Stars, atoms, fluid flows, high temperature super- conductors, black holes, and the elementary particles are just some of the many examples. Studies of these specific systems define the various subfields of physics — astrophysics, atomic physics, fluid mechanics, and so forth. But beyond the regularities exhibited by specific systems, physics has a special charge. This is to find the laws that govern the regularities that are exhibited by all physical systems — without exception, without qualification, and without approximation. The equality of gravitational accelerations of different things is an example. These are usually called the fundamental laws of physics. Taken together they are called informally a "theory of everything".   (Hartle 2002, p. 4)

The second is from Gerard T'Hooft, a very different type of physicist:

> What is a "Theory of Everything"? When physicists use this term,...we have a deductive chain of exposition in mind, implying that there are 'fundamental' laws describing space, time, matter, forces and dynamics at the tiniest conceiv-able distance scale. Using advanced mathematics, these laws prescribe how elementary particles behave, how they exchange energy, momentum and charges, and how they bind together to form larger structures, such as atoms, molecules, solids, liquids and gases. The laws have the potential to explain the basic features of nuclear physics, of astrophysics, cosmology and material sci-ences. With statistical methods they explain the basis of thermodynamics and more. Further logical chains of reasoning connect this knowledge to chemistry, the life sciences and so on.   (T'Hooft 2017, p. 2)

These are relatively generic. One can find similar remarks in Feynman, Wigner, Hawking. The sort of picture is guiding the imagination when they are looking for theories is very different from the Humean one. The scientist doesn't think of herself as in the business of summarizing a global mosaic of intrinsically uncon-nected fact. She sees herself as in the business of trying to uncover an immanent substructure in the composition of material systems. The idea that is guiding the practice is that if you look closely enough at the fine-scale structure of the matter, you will find that systems everywhere are made of the same components with a limited range of motion. Observed regularities are used as guides to this sub-structure. The laws describing components are identified and then combined with facts to yield predictions for systems built from the same components. Scientific theories involve analysis to isolate the fundamental components of matter and the laws that describe their behavior, and then synthesis to proscribe the bounds of what is possible. Those boundaries are then viewed as genuine constraints on what can happen, not just retrospective summaries of what did.

That is important because it addresses another source of discomfort about Humeanism. On the view here, the components of a complex system have modal profiles that it is the task of theory to discern and experiment to test. They can be arranged into arbitrary configurations and we can ask about such configurations how they would behave under conditions that may or may not be actual. In so doing, we are drawing conclusions that are rooted in the intrinsic structure of the system, and that we rely on not simply to *predict* but to *engineer* the future: to design and steer and avoid and forestall. One can see how the pieces of this view fit together to make sense of that practice. Not everybody shares this discomfort, but I have never been able to arrive at a way of thinking about modality on a Humean view that seems halfway adequate to its practical role. For the Humean, systematization is about informational compression. The things that end up as laws, and that govern our beliefs about what can and can't happen, are just the axioms in the system that achieves the best global fit. Ideally, the laws of the Best System would be so strong that they predict everything that happens. As it is, some things get squeezed out in the process of compression and are undetermined by the laws, leaving us with a range of possibilities. These appear as simply the regrettable overflow of systematizing a complex domain. But when an engineer is designing an airplane or we are exploring new ways of producing nuclear energy, one wants to think of the laws as genuine constraints on what *can* happen looking forward.

None of this is offered as a philosophical *analysis* or *metaphysics* of laws. It really just amounts to taking science at face value and declining to see physical modality as problematic. Humeanism gets a lot of traction by opposing certain philosophical views of what the laws are, but Humeanism doesn't have enough structure on the domain to make it possible to learn from data without wheeling in constraints on priors that are at odds with its own metaphysics. The scientist relies on the idea that there are laws that can be discovered by investigating the material substructure of the part of the universe they have access to and projected into other parts of the universe, on the defeasible assumption that they are configurations of the same components. Where the Humean sees induction as separate from and prior to the discovery of laws (and chances),[21] she sees it as the whole point of theorizing. If one wants a modern empiricist position tailored to science, why not simply accept the world of science at face value? It is not a world

---

[21] I'll omit the '(and chances)' below for ease of expression.

of intrinsically unconnected matters of particular fact arranged in a mosaic that—if nature is kind—can be retrospectively summarized. It is a world of components with limited ranges of motion that combine into larger configurations to form the objects we see around us and whose behavior is derivative of that of their parts.

Is this an anti-Humean position?[22] The laws, so understood, don't govern or guide events; they are not transcendent relations between universals; they don't *produce* their instances or enforce behavior; and they aren't 'backed' by special modal truthmakers, so it is nothing like the familiar anti-Humean positions. On the other hand, it does recognize immanent constraints in the manifold of events. On this view, laws aren't retrospective summaries of what *did* happen; they are genuine constraints on what *can* happen looking forward. I'm torn between calling it neo-anti-Humean and anti-neo-Humean. The latter seems more appropriate because it is not a descendant of anti-Humeans in the philosophical tradition that Humeans oppose.[23] It is defined, rather, in opposition to Humeanism in the style that it takes in the hands of its most influential proponents.

2.8 Conclusion

Humeanism is an attractive view if one looks from a God's-eye view and one asks, 'what kinds of thing' would it make sense for God to recommend as guides for the beliefs of creatures like us (limited creatures with no specific information about the future)? But when one demands an epistemology that embedded agents can employ, these three things together leave us without a sensible story about how to learn about what the chances are from what you observe:

(i) Chances are determined by a global criterion applied to the mosaic as a whole;
(ii) There are as many ways that the mosaic could be as there are assignments of local quantities to spacetime points; and
(iii) The Humean mosaic is indefinitely extendible.

A fix for the problem suggested by Albert and Loewer was to presuppose that induction works. I argued that this was an unstable position for the Humean because it says effectively: even though the defining doctrine of Humeanism is that there are as many ways the future could go as there are assignments of

---

[22] There are different ways to elaborate the metaphysics of a view like this. We could take laws or powers as primitive, for example. I don't myself have a preference.
[23] See Carroll (2016), also Armstrong (1983) and Schaffer (2016).

physical quantities to spacetime points, one should nevertheless assign nominal probability to all but a tiny sliver of them.

The Humean can, and I suspect will, stick to his guns. But I think that when the Humean is forced to be explicit about how embedded agents are supposed to learn about the chances, some of the sheen comes off Humeanism, and empiricists might be persuaded to look elsewhere.[24]

References

Albert, D. (2015). *After Physics*. Cambridge, MA: Harvard University Press.

Armstrong, D. M. (1983). *What Is a Law of Nature?* Cambridge: Cambridge University Press.

Bird, A. (2007). *Nature's Metaphysics*. Oxford: Oxford University Press.

Carroll, J. W. (2016). 'Laws of Nature', *The Stanford Encyclopedia of Philosophy* (Fall 2016 Edition), ed. Edward N. Zalta, https://plato.stanford.edu/archives/fall2016/entries/laws-of-nature/.

Cohen, J., and Callender, C. (2009). 'A Better Best System Account of Lawhood', *Philosophical Studies*, 145(1), pp. 1–34.

Demarest, H. (2017). 'Powerful Properties, Powerless Laws', in Jacobs, J. (ed.), *Causal Powers*. Oxford: Oxford University Press, pp. 38–53.

Gopnik, A., Glymour, C., Sobel, D. M., Schulz, L. E., Kushnir, T., and Danks, D. (2004). 'A Theory of Causal Learning in Children: Causal Maps and Bayes Nets', *Psychological Review*, 111(1), pp. 3–32.

Hartle, J. (2002). 'Theories of Everything and Hawking's Wave Function of the Universe', https://arxiv.org/abs/gr-qc/0209047 (accessed 2 Mar. 2021).

Hicks, M. (2018). 'Dynamic Humeanism', *British Journal for the Philosophy of Science* 69(4), pp. 983–1007.

Ismael, J. (2008). 'Raid! Dissolving the Big, Bad Bug', *Noûs*, 42(2), pp. 292–307.

Ismael J. (2015). 'How to be Humean', in Loewer, B., and Schaffer, J. (eds.) *A Companion to David Lewis*. Oxford: John Wiley & Sons, pp. 188–205.

Ismael, J. (2017). 'An Empiricist Guide to Objective Modality', in Slater, M., and Yudell, Z. (eds.), *Metaphysics and the Philosophy of Science: New Essays*. New York: Oxford University Press, pp. 109–25.

Jaag, S., and Loew, C. (2020). 'Making Best Systems Best for Us', *Synthese*, 197, pp. 2525–50.

[24] This paper was originally presented at a conference organized by Barry Loewer. I want to thank him for the invitation and thank him and David Albert for really helpful discussion at the conference. I am also indebted to Christian Loew for the invitation to include my paper here and to him and Siegfried Jaag for very helpful comments.

Lazarovici, D. (2020). 'Typical Humean Worlds Have No Laws', ahttp://philsci-archive. pitt.edu/17469/ (accessed 2 Mar. 2021).

Lewis, D. (1994). 'Humean Supervenience Debugged', *Mind*, 103(412), pp. 473–90.

Lewis, D., and Rosen, G. (2003). 'Postscript to 'Things Qua Truth-makers': Negative Existentials', in Lillehammer, H., and Rodriguez-Pereyra, G. (eds.), *Real Metaphysics: Essays in Honour of D. H. Mellor*. London: Routledge, pp. 39–41.

Loewer, B. (1996). 'Humean Supervenience', *Philosophical Topics*, 24(1), pp. 101–27.

Manchak, J. B. (2009). 'Can We Know the Global Structure of Spacetime?', *Studies in History and Philosophy of Modern Physics*, 40(1), pp. 53–56.

Maudlin, T. (2007). *The Metaphysics Within Physics*. Oxford: Oxford University Press.

Norton, J. (2021). 'How to Make Possibility Safe for Empiricists', https://www.pitt. edu/~jdnorton/papers/Empiricist_possibility.pdf (accessed 2 Mar. 2021).

Pettigrew, R. (2012). 'Accuracy, Chance, and the Principal Principle', *Philosophical Review*, 121(2), pp. 241–75.

Pettigrew, R. (2016). *Accuracy and the Laws of Credence*. Oxford: Oxford University Press.

Schaffer, J. (2016). 'It is the Business of Laws to Govern', *Dialectica*, 70(4), pp. 577–88.

T'Hooft, G. (2017). 'Free Will in the Theory of Everything', https://arxiv.org/ abs/1709.02874 (accessed 2 Mar. 2021).

3

Knowing the Powers

Wolfgang Schwarz

3.1 Introduction

If the world contains primitive modal elements—irreducible laws, powers, potentialities, or propensities—how could we know about them? Humeans have long
worried that we could not. Put bluntly, the worry is that observation and experiment only tell us what *does* happen; they don't directly reveal what *might* or *must*
or *would* happen under non-actual circumstances. If modal phenomena are
reducible to facts about occurrent events, then it is no surprise that observing
occurrent events can give us information about modality. By contrast, knowledge
of *primitive* modality seems to require an inexplicable leap from observations of
one kind of fact to conclusions about an entirely different kind of fact.

Let's call this the *access problem* for non-Humean accounts of (natural)
modality. It is a subproblem of a more general problem, often highlighted by
Lewis. As Lewis pointed out, it is not enough to posit 'unHumean whatnots'
(Lewis 1994, p. 239). We also need a credible story of how these whatnots could
play the familiar roles of laws, powers, potentialities, or propensities. Part of
that role concerns the methods by which these things can be discovered.

In its blunt formulation, the access problem is easy to resist. Arguably, some
modal facts *can* be directly observed: we can see that a surface is slippery, or that
its slipperiness causes a fall. In other cases, inference to the best explanation
might be hoped to get us from, say, an observation of frequencies to hypotheses
about chance.

Like many Humeans, I am not convinced by these replies. I don't think they get
to the heart of the access problem. My aim in this essay is to explain why.

In a nutshell, I will argue that if the world has primitive modal elements, then
there are different *a priori* conceivable ways in which these elements might be
arranged, many of which are compatible with our total history of perceptual
experience. If, for example, certain objects in our environment have some primitive
power F, then there are conceivable (although perhaps metaphysically impossible)
scenarios in which the objects display the same observable behaviour even
though they have a different power F^*. Since the two kinds of scenario deserve
equal *a priori* credence, and are equally compatible with our perceptual experiences, we can't know that we are in an F scenario.

Wolfgang Schwarz, *Knowing the Powers* In: *Humean Laws for Human Agents.* Edited by: Michael Townsen Hicks,
Siegfried Jaag, and Christian Loew, Oxford University Press. © Oxford University Press 2023.
DOI: 10.1093/oso/9780192893819.003.0004

I will outline three lines of response. One is to concede that we can never know which un-Humean whatnots are present in our surroundings—except in a certain shallow sense of 'know'. Another is to posit implausible epistemic norms that would allow us to favour F scenarios over F^* scenarios without any relevant evidence. A third option, I will argue, is to reconstrue the anti-Humean position as a doctrine about ideology rather than ontology. This helps with some forms of the access problem, but others remain. For example, difficult questions would still arise for our knowledge of objective chance.

3.2 The Access Problem

Let me try to explain what I see as the heart of the access problem.

I will assume without argument that we have abundant knowledge of natural modality. We know that nothing can travel faster than the speed of light, that electrons are disposed to repel one another, that ordinary wine glasses are fragile, and that a properly minted coin has a 50% chance of landing heads when tossed.

I will also assume that we know (at least roughly) how we know these facts. We may discover that a glass is fragile by observing its delicate build, or by watching it crack under pressure. We may discover that nothing can travel faster than the speed of light by noticing that this is implied by a theory which in turn is supported by a host of theoretical and empirical considerations.

Our question is therefore not how—by what methods—we can discover modal facts. Rather, the question is whether a certain metaphysics of modality can make sense of these methods.

To illustrate, consider a crazy view which identifies fragility with the property of having been touched by an angel. A glass's fragility, on this view, does not imply anything about its physical composition or about whether it will break when struck. (There is nothing incoherent about a scenario in which, say, a glass breaks under light pressure even though it has never been touched by an angel.) As a consequence, it becomes mysterious how observing the glass's build, or its behaviour under stress, could reveal anything about its fragility—that is, about whether it has been touched by an angel.

Suppose we throw the glass against the wall and see it shatter. To get from this observation to the conclusion that the glass has been touched by an angel, we would need 'bridge principles' according to which, say, objects that shatter have usually been touched by an angel. But how could we have discovered these principles without some other means of knowing whether something has been touched by an angel?

It is not a response to say that, on the crazy view in question, having been touched by an angel is identical to fragility, and that surely we can find out that a glass is (or was) fragile by watching it shatter. The problem is that having been touched by an angel *cannot be* identical to fragility, because the assumed identity would make a mystery of the methods by which we discover fragility. To assume that fragility equals past angelic touch is to presuppose what is in question: that past angelic touch can play the role of fragility.

Now consider a (still somewhat crazy) anti-Humean view on which fragility is a metaphysically primitive property—a special un-Humean property with a 'dispositional essence'.[1] Again, we may ask why observation of a glass's shattering should tell us anything about this primitive property. We would, it seems, have to assume a bridge principle linking the primitive property with the observed events. But how could we have discovered that principle?

Again, it is not a response to insist that the posited primitive property is identical to fragility, thereby presupposing what is in question. For the vast majority of properties, you can't find out whether a glass has them by watching it shatter. An identification of fragility with some property X is only plausible if there are *independent* reasons to believe that the way we find out about fragility is a way to find out about X, without already assuming that X is fragility. If X is a primitive whatnot, it is hard to see what these independent reasons might be.

The problem arises not just for friends of primitive powers. Whenever anti-Humeans identify a modal phenomenon with the presence of some un-Humean whatnot (an irreducible law, a higher-order universal, or whatever), they owe us an explanation of how our methods for identifying the phenomenon could serve as methods for identifying the un-Humean whatnot.

3.3 Permutations

You may think such an explanation is not hard to find. Suppose we inspect a glass and come to believe that it is fragile. If all goes well, our judgement is true, and appropriately caused by the glass's fragility. If so, we *see* that the glass is fragile, and thereby come to know that it is fragile. What is needed to ensure our knowledge, then, is that whatever property is identified with fragility stands in the appropriate causal relation to our judgement. And why shouldn't a primitive power satisfy that condition? What's the problem?

There are, in fact, several problems. It will take a while to sort them all out.

Let's briefly set aside modality and think about diagnostics. Imagine a patient consulting a doctor, with symptoms of arthritis. An X-ray confirms this suspicion. Looking at the X-ray image, the doctor can *see* that the patient has arthritis.

---

[1] This is still somewhat crazy because fragility is plausibly not fundamental, even on anti-Humean accounts. However, the issues I will discuss would also arise for more plausible candidates for primitive dispositions—negative unit charge, perhaps—so I will stick with the familiar case of fragility.

But this does not mean that any remaining epistemological questions about her knowledge can be delegated to the physiology of visual perception.

For one thing, the doctor's training and background knowledge play a role. An amateur like me could not see that the patient has arthritis by looking at the X-ray. Moreover, the X-ray does not provide *conclusive* evidence for the diagnosis. It is easy to think of scenarios in which the patient does not have arthritis, even though the doctor's perceptual evidence is just the same: the patient might have an unknown disease that looks like arthritis in an X-ray image; the X-ray technician might have accidentally swapped the patient's X-ray image with that of an earlier patient; and so on.

We can nonetheless describe the doctor as knowing (and seeing) that the patient has arthritis because—roughly speaking—these 'error scenarios' (in which the patient does not have arthritis but the X-ray image looks the same) are (a) in some sense remote, (b) the doctor gives them little credence, and (c) she is right to give them little credence.

If I were to look at the X-ray image, condition (a) might still be satisfied, but the corresponding conditions (b) and (c) would fail. I would not be confident that the patient has arthritis and, even if I were, my confidence would be irrational. That's why I couldn't know (or see) that the patient has arthritis just by looking at the X-ray image.

Now back to the view that fragility is some primitive property F. Let's grant that the presence of F (partially) causes our perceptual experience when we inspect the glass, and thereby our judgement that the glass is fragile. However, the experience will not be conclusive evidence for the presence of F. We can distinguish different kinds of error scenarios.

First, there are scenarios we would intuitively describe as situations in which it appears to us as if we are inspecting a fragile glass, but in reality something else is going on. Perhaps we are being deceived by an evil demon. Or perhaps we are inspecting a special glass that looks fragile but is actually unbreakable. Even if we've thrown the glass against the wall and saw it shatter, the glass might only have become fragile as it hit the wall, so that we would be wrong to conclude that it was fragile before. In normal situations, scenarios like these can be regarded as remote, and don't stand in the way of knowledge. Let's set them aside.

Here is a different kind of error scenario. Consider a world that is much like the actual world except that the primitive property F has been replaced (in all its instances) by a different primitive property F^*—a property that (let's say) nothing has in the actual world. Anything that is, in our world, caused by the glass being F is here caused by the glass being F^*. In particular, it is F^*, not F, that (partially) causes our experiences as we inspect the glass.

By assumption, the two kinds of scenario do not differ in the light waves that arrive at our retina, in our subsequent brain states, or in the phenomenology of our experience. Our experience therefore does not put us in a position to rule out the F^* hypothesis.[2]

We also can't dismiss the F^* scenarios as far-fetched. In an F scenario, the glass has one primitive property; in the corresponding F^* scenarios, the glass has a different primitive property. *A priori*, the two scenarios are on a par. Any *a priori* preference in favour of either scenario would be arbitrary and irrational.

But if there is no *a priori* reason to favour one scenario over another, and our perceptual experiences can't distinguish between the two scenarios—insofar as we have the same (kind of) experiences in both scenarios—then both scenarios deserve roughly equal (posterior) credence. So we can't know that we're in an F scenario.

I find this argument persuasive. If you agree, feel free to skip the next section, in which I will investigate an argument for the opposite conclusion.

3.4 Deep Knowledge and Shallow Knowledge

The argument for the opposite conclusion begins with a semantic (or conceptual) ascent. Arguably, the word (or concept) 'fragile' functions in a semi-demonstrative way to pick out a certain property in the world around us—a property that causes things to break when struck.[3] We can dismiss as far-fetched scenarios in which the term fails to pick out anything. So we can be confident that 'fragile' applies to some things in the world around us. That is, we can be confident that some things around us are fragile.

Now, if 'fragile' picks out F, then to be confident that some things are fragile is to be confident that some things are F. Any scenario in which F is replaced by an alien property F^* is a scenario in which nothing is fragile. So we can be confident that we are not in an F^* scenario.

Let's call this argument *the verbal trick*. To see what's wrong with it, let's return to our doctor. While we were talking about fragility, another patient has arrived whose symptoms are compatible with both arthritis and fibromyalgia. This time, the X-ray image comes back blurry, and the doctor remains unsure if the patient has arthritis or fibromyalgia.

Assume the patient actually has arthritis. The doctor might then use the following trick to identify his ailment. First, she introduces a new name, say

---

[2] You might object that if F is a more plausible candidate for a primitive property (see note 2 above), then swapping F by F^* will make a difference to our brain state and thereby to our experiences. Alternatively, you might suggest that our experience is directly sensitive to what primitive properties are present in our environment. But relations of causal or metaphysical dependence aren't pertinent here. Suppose we were initially unsure about whether our environment contains instances of F or F^*. The question is whether our perceptual experience alone would then put us in a position to conclusively rule out the F^* hypothesis. The answer is no.

[3] Remember that 'fragile' is a placeholder for something like 'negative unit charge'.

'julitis', for the patient's disease. Having paid attention to this act of baptism, she is rationally confident that the patient has julitis. She can ignore scenarios in which the patient does not have julitis. But 'julitis' denotes arthritis (and not fibromyalgia): julitis and arthritis are the same disease. Any scenario in which the patient has fibromyalgia is a scenario in which the patient does not have julitis. So the doctor can be confident that her patient has arthritis, and not fibromyalgia.

We are dealing with a 'Frege case'. Arthritis is known to the doctor under two modes of presentation: as the illness called 'arthritis' about which she learned in medical school, and as the illness called 'julitis' that causes her present patient's symptoms.

Fregeans hold that these modes of presentation affect the truth-conditions of attitude reports: to know (or be confident) that the patient has *arthritis*, the doctor would have to know (or be confident) that the patient has the illness called 'arthritis' about which she learned in medical school. Evidently, the doctor did not acquire any such knowledge through her introduction of the name 'julitis'. Fregeans would conclude that the doctor may know that the patient has julitis, but not that he has arthritis.

Fregeans will spot the same mistake in the above verbal trick. We may be confident that some things in the world are fragile, but it doesn't follow that we can be confident that some things are F, even on the hypothesis that 'fragile' picks out F.

The Fregean account of attitude reports is controversial. Let's assume that it is false, so that we can truly report the doctor as knowing that her patient has arthritis.

Nonetheless, it should be uncontroversial that the doctor has not gained any useful information through her introduction of a new name. She has made no genuine epistemic progress with respect to her patient's disease. She is in no better position to prescribe a treatment. Medical textbooks rightly do not mention the introduction of new names as a diagnostic method.

I will say that the doctor has *shallow knowledge* that the patient has arthritis, but lacks *deep knowledge* of the same fact. Perhaps the English word 'knowledge' means shallow knowledge. I doubt it, but it doesn't matter. I'm interested in deep knowledge—a kind of knowledge that tracks genuine epistemic progress.

Deep knowledge of an empirical hypothesis H requires a history of perceptual experience that favours H over its alternatives—in the sense that, among scenarios in which you have these experiences, H scenarios have (significantly) greater *a priori* credibility than $\neg H$ scenarios.

For example, consider your past and present experiences related to your hands. There are scenarios in which you have the same[4] history of perceptual experiences even though you don't have hands. (You might be a brain in a vat.) But these scenarios deserve much lower *a priori* credence than scenarios in which

---

[4] What does it mean that your experiences in two scenarios are 'the same'? Good question. Roughly, the experiences should have the same phenomenology. The full answer, I think, is complicated—see Schwarz (2018).

you have the experiences and you also have hands. Provided that the other conditions for knowledge are satisfied, you may therefore have deep knowledge that you have hands.

By comparison, the doctor's history of perceptual experiences does not (significantly) favour scenarios in which her patient has arthritis over scenarios in which he has fibromyalgia. There are scenarios in which she has all the same experiences (her experience of listening to the patient's description of his symptoms, the blurry X-ray image, her introduction of the name 'julitis', etc.), but in which the patient has fibromyalgia. These scenarios do not deserve much lower credence than corresponding scenarios in which the patient has arthritis. So the doctor does not have deep knowledge that the patient has arthritis.

When it comes to distinguishing F and F^*, we are in the same position as the doctor with arthritis and fibromyalgia. We may be able to pick out F demonstratively, and we may have a word for it, but we don't have deep knowledge that anything around us is F.

3.5 Humean Knowledge

My permutation argument from Section 3.3 resembles a well-known argument purporting to show that we could never discover the identity of fundamental (categorical) properties in a Lewisian metaphysics—for how could we tell apart scenarios that merely differ by swapping these properties? Some have, in effect, responded that we might still have shallow knowledge about fundamental properties (e.g. Langton 2004; Schaffer 2005). But I think Lewis (2009) was right to accept the argument for deep knowledge.[5] Like Lewis, I do not find the conclusion especially problematic. It's not like we all thought we knew the relevant facts, and then Lewis tells us that we don't. On Lewis's view, we can't even state or entertain the propositions of which we are ignorant (see Kelly 2013; Dasgupta 2015).

Anti-Humeans might adopt a similar response to my permutation argument. They might concede that we can't have deep knowledge of whether a glass (or electron) has a specific power F. But this is not how anti-Humeans usually present their view. The irreducibly modal facts they posit are supposed to be familiar facts about dispositions, potentialities, or laws—facts that aren't beyond our epistemic reach.

Anti-Humeans might also complain that my constraints on 'deep knowledge' are too demanding. Can *Humeans* explain our deep knowledge of the glass's fragility?

Let's see. On a typical Humean analysis, a glass is fragile iff (roughly) it has a material structure which, together with the laws of nature, entails that it is likely

---

[5] Lewis held that our ordinary concept of knowledge is a concept of deep knowledge (see Lewis 1996).

to break under moderate stress, where the relevant laws are certain regularities in the history of the universe. One might reasonably wonder how we could find out that a glass satisfies this condition (of having a material structure etc.) simply by watching it shatter, given that the condition requires suitable regularities in the entire history of the universe.

What would the relevant error scenarios look like? They would be scenarios in which the glass shatters upon being thrown at the wall, but in which this event is not an instance of any general regularity in the history of the world.

Now we have other experiences to draw on. These other experiences suggest that the dynamics of physical systems, at least to the extent that we have observed them, is fairly predictable. Not only have we often seen delicately built objects break under stress, we have noticed that similar physical systems of all sorts generally respond in similar ways when put in similar conditions. It is still *possible* to have all these experiences in a world without relevant dynamical regularities. There are, for instance, scenarios in which the dynamics of the systems we have observed is regular, but the dynamics of unobserved systems is entirely irregular. But scenarios like these deserve little *a priori* credence.[6]

On the Humean account, non-trivial knowledge of natural modality always involves an element of conjecture. But at least our experiences often favour some modal hypotheses over others—assuming that we may treat some scenarios as *a priori* more credible than others. If we can't assume that the regularities in the observed part of the world still hold in the unobserved part,[7] the Humean epistemology is doomed (as Ismael, in Chapter 2 of this volume, notes)—but so is everyone else's. I will return to this point in Section 3.10.

3.6 Ontology and Ideology

It is time to revisit a premise of my argument in Section 3.2. A friend of primitive fragility, I claimed, would need 'bridge principles' to connect the hypothesis that a glass is F with the observation that it shattered upon being thrown against the wall. Similarly, in Section 3.3, I claimed that when we see the glass shatter, our experience is neutral between scenarios in which the glass is F and scenarios in which it has a different disposition F*.

This may seem strange. According to our (imaginary) friend of primitive fragility, F is, by its very essence, a disposition to break under stress. Dispositions

---

[6] I am not suggesting that this story mirrors some kind of inference we are supposed to make. On a cognitive level, it may well be that we come to believe that the glass is fragile because we trust our experience, which 'presents it' as fragile. See Beebee (2003) for how Humeans might account for this kind of presentation.

[7] To make this assumption precise, we would need a criterion for distinguishing gruesome regularities from genuine regularities. Humeans disagree on what that criterion should look like.

are not independent of their manifestations. Why, then, should we need bridge principles connecting the disposition with its manifestation? Why should we allow for scenarios in which a different disposition F^*—a disposition to *glow* under stress, perhaps?—has traded places with fragility, causing things to break under stress? Such a scenario makes no sense.

I agree that we can't arbitrarily swap an object's dispositions while holding fixed the manifestations. But our question is whether dispositions can be identified with primitive properties. We should not presuppose a positive answer by assuming that these primitive properties can be referred to as 'the disposition to break under stress' or 'the disposition to glow under stress'. That's why I have used the neutral names 'F' and 'F^*'.

Different versions of the permutation problem arise for different types of anti-Humeanism. On some views (e.g. Heil 2010; Williams 2019), fundamental powers have both a dispositional character and a non-dispositional, qualitative character. A relevant F^* scenario for the permutation argument might then be a scenario in which the qualitative character of F is swapped with the qualitative character of some other power. Scenarios like these may be deemed metaphysically impossible: the pairing between qualitative and dispositional characters is supposed to be metaphysically necessary. But I never said that the F^* scenario is metaphysically possible. I only assumed that it can't be ruled out *a priori*.

Others (e.g. Bird 2007; Mumford and Anjum 2011) deny that fundamental powers have a qualitative aspect. Here the posited fundamental properties are assumed to have an entirely modal essence, an essence that grounds counterfactuals and other modal truths. A relevant F^* scenario might then be a scenario in which a different fundamental property grounds the same modal truths. Perhaps it is a scenario in which a different property F^* shares F's essence. (Again, we need not assume that such scenarios are metaphysically possible, as long as they can't be ruled out *a priori*.) Alternatively, it could be a scenario in which F is swapped for a property with an entirely different modal essence—an essence, perhaps, that grounds counterfactuals about glowing when struck.

I admit that I have trouble understanding talk about essence and grounding. But I have been told that grounding is an objective relation that need not be epistemically transparent: X can ground Y even if there is no *a priori* connection between X and Y. This suggests that if F^* is a property that grounds counterfactuals about glowing when struck, there can still be *a priori* conceivable scenarios in which F^* is present even though the relevant object wouldn't glow (but rather break) when struck.

To escape the permutation argument, we would need not just a metaphysical but an *a priori* connection between the presence of F and counterfactuals about breaking. Ideally, we would have a connection that goes both ways, so that the observed breaking of a glass is evidence for the presence of F. This can be made to work, but it might require rethinking the metaphysics of powers.

Remember the Humean project of analysing counterfactuals in ultimately non-modal terms. The project has not been a resounding success. One might reasonably hold that it will never succeed. More strongly, one might hold that there is no way to determine the truth-value of (arbitrary) counterfactuals from suitably different propositions.

Even if counterfactuals are in this sense primitive, the truth-value of a counterfactual may still be *constrained* by other propositions. For example, it is widely held that a counterfactual $A > C$ entails the corresponding material conditional $A \supset C$ (and so the falsity of $A \supset C$ entails the falsity of $A > C$). Accepting this entailment does not commit us to the Humean reductive project. $A > C$ can entail $A \supset C$ even if $A > C$ is primitive.

When I say that $A > C$ entails $A \supset C$, I don't mean that some opaque metaphysical relation holds between $A > C$ facts and $A \supset C$ facts. I mean that *a priori* reasoning is enough to rule out any putative scenario in which $A > C$ is true and $A \supset C$ false. We don't need empirical bridge principles to infer $A \supset C$ from $A > C$.

Now return to the shattering glass. Let's assume (as before) that when we see the glass shatter, we gain deep knowledge that the glass shatters and that it has been thrown: scenarios in which the glass doesn't shatter or hasn't been thrown can be ignored.[8] The question is how we can get from here to knowledge of any irreducibly modal facts about the glass.

If counterfactuals are primitive, then one such fact is *Thrown > Shatter*. And it is not hard to see how we could get from *Thrown ∧ Shatter* to *Thrown > Shatter*.

The inference might be a simple matter of deduction. Some hold that $A \wedge C$ logically entails $A > C$. But the inference might be justified even without that controversial assumption. Suppose, before we saw the glass shatter, the live possibilities divided into *Thrown > Shatter* scenarios and *Thrown > ¬ Shatter* scenarios, with any remaining scenarios deserving little credence. Since *Thrown > ¬ Shatter* entails *Thrown ⊃ ¬ Shatter*, and *Thrown ∧ Shatter* is logically incompatible with *Thrown ⊃ ¬ Shatter*, observation of *Thrown ∧ Shatter* then allows us to rule out all *Thrown > ¬ Shatter* scenarios, leaving most of our credence on *Thrown > Shatter* scenarios.

So the problems I raised for our knowledge of F do not arise for our knowledge of (supposedly) primitive counterfactuals likely *Thrown > Shatter*—nor do

---

[8] This assumption may be too generous. Friends of powers often hold that the glass's shattering is itself just a collection of dispositions. We must then ask how we could tell that *these* dispositions are instantiated. Let's say the shattering of the glass involves, among other things, counterfactuals about how the bits of glass would affect a sheet of paper held in their way. Since no paper is actually held in the way, how do we know that these counterfactuals are true? The answer I'm about to give for our knowledge of *Thrown > Shatter* does not carry over. This problem is especially acute for 'holist' views on which the fundamental powers are individuated by the total nomic profile of all powers. For the glass to have a particular power then requires the truth of many non-trivial counterfactuals about the entire nomic structure of the world.

they arise for our knowledge of fragility if that is analysed in terms of such counterfactuals. Watching the glass shatter might give us deep knowledge that it is (or was) fragile.[9]

The key to this solution is that the primitive counterfactuals are not posited as *ontological* (or *typological*, see Busse 2018) primitives. They are primitive *ideology*. We have assumed (on behalf of the anti-Humean) that the truth-value of (some) counterfactuals is not settled by any non-counterfactual truths. Informally, if God wanted to give a complete description of reality from which one could in principle infer all truths, she would have to explicitly include counterfactuals. In that sense, the counterfactual operator '>' is a piece of primitive ideology: it has to be used in any complete description of the world.

Treating '>' as primitive ideology is not the same as positing a primitive piece of ontology. We don't have to assume that counterfactuals are made true by the presence of a special entity. Indeed, we should not, since that would bring back the permutation problem.

3.7 Generalizing

When I stated the access problem, I focused on a particular un-Humean whatnot: primitive fragility. I also focused on a particular error possibility: scenarios in which that property is replaced by another primitive power. It might be useful to state the problem in more abstract and general terms.

Anti-Humeans commonly assume that there are metaphysically primitive facts about the presence and distribution of un-Humean whatnots. Let M be the totality of these facts. Let H be the totality of all fundamental Humean facts (if any). Finally, let E be the complete truth about the character of our perceptual experience (past, present, and future).

The first premise in my sceptical argument is that $E \wedge H$ is *a priori* consistent with alternative hypotheses about the presence and distribution of un-Humean whatnots. If M specifies that some things in our environment have a primitive power F, then $E \wedge H$ is *a priori* compatible with scenarios in which instead these things have F^*. If M specifies that a certain higher-order universal N relates F and G (see Armstrong 1983), then $E \wedge H$ is compatible with scenarios in which N instead relates G and H.

---

[9] I say 'might' because other problems remain. The problems from note 9, for example. Also, how do we know that the glass is irreducibly such that it would break under stress as opposed to, say, such that it would break *under stress at room temperature* (but not at other temperatures), or such that it would break *under stress or under UV light*? And how do we know that it does not have an irreducible propensity to randomly break, irrespective of any stress? These alternatives are all compatible with the observed *Thrown* ∧ *Shatter*.

Anti-Humeans will typically regard many of these scenarios as metaphysically impossible. If F has an 'essence' that 'grounds' the truth of counterfactuals like *Thrown > Shatter*, they might say, then any scenario in which a glass has F, is thrown, and yet fails to shatter is metaphysically impossible. Similarly, necessitarians about laws will say that scenarios in which N relates G and H even though some G fails to be H are metaphysically impossible. I do not assume otherwise. I only assume that the relevant scenarios can't be ruled out *a priori*.

My second premise is that scenarios that differ merely by shuffling around un-Humean whatnots sometimes deserve equal *a priori* credence. For example, two scenarios that only differ by swapping fundamental powers deserve the same (or roughly the same) *a priori* credence. More specifically, there are many scenarios that agree with respect to E and H, differ substantially in the distribution of un-Humean whatnots, but have roughly equal *a priori* credibility.

From these two premises,[10] it follows that we lack (and will never acquire) deep knowledge about the distribution of un-Humean whatnots.

A perhaps familiar special case of this argument involves scenarios in which the un-Humean whatnots have all been removed, while holding fixed E and H. The argument claims that we have no reason to favour worlds with un-Humean whatnots over corresponding 'Hume worlds' with all un-Humean elements removed (see e.g. Earman and Roberts 2005). I have instead focused on worlds where the un-Humean elements are swapped around, because I suspect that friends of primitive powers will maintain that Hume worlds really should be given little *a priori* credence. I disagree, but I don't know how to argue the point. I hope that few will be tempted to maintain that scenarios in which certain things around us have some primitive property F are *a priori* more credible than scenarios in which they have a different primitive property F^*.

Instead of rejecting one of the premises, anti-Humeans might accept the conclusion and admit that we can never know (or even have evidence about) which un-Humean whatnots are present in our environment—except in the cheap and shallow sense in which we can know a patient's ailment by giving it a name. I don't think this kind of 'modal humility' would be a fatal flaw, but (as I mentioned earlier) it does not fit how anti-Humean views are usually presented.

Another response is to not posit any un-Humean whatnots and merely insist that certain modal truths are conceptually basic.[11] This blocks the argument because there can be *a priori* connections between modal and non-modal

---

[10] In fact, from the second alone, but it is useful to state the weaker premise 1 as a separate assumption.

[11] Vetter (2021) also argues that anti-Humeanism is best understood not as an ontological thesis but as the view that some modal truths cannot be explained in ultimately non-modal terms. However, Vetter's 'explanation' relation is metaphysical. To meet the epistemic challenge, we would need an epistemic relation.

hypotheses, even if the former are not reducible to the latter. A primitive counter-factual might still entail the corresponding material conditional. Similarly, 'it is a law that all Gs are Hs' might entail 'all Gs are Hs' even if the concept of a law is irreducible.

Admittedly, the move to ideology also has its costs. For example, it appears to violate an attractive 'truthmaker principle' according to which all truths are made true by what fundamental things there are and what fundamental properties and relations they instantiate. Friends of powers, in particular, often endorse the intu-ition that facts about what a glass would do when struck are grounded in more fundamental facts about the glass's intrinsic properties. With the move to ideol-ogy, they would have to give up this intuition.

One might also argue that inexplicable *a priori* connections are mysterious. On typical Humean accounts, we can explain why, say, a counterfactual entails the corresponding material conditional. No such explanation can be given if counter-factuals are ideologically primitive.[12]

Special problems also remain for our knowledge of chance.

3.8 Propensities

So far, we have looked at 'deterministic' powers. Almost all powers with which we are familiar in science and everyday life are non-deterministic. A fragile glass is not guaranteed to break when thrown against the wall—not even if we specify its exact microstate, along with the precise manner of throwing and the state of the rest of the universe. The evolution of the universe's wave function will give rise to branches on which the glass survives unscathed. Without fixing the microstate of the universe, even statistical mechanics tells us that the glass might survive.

I will use 'propensity' to refer to non-deterministic dispositions. The canonical expression of a propensity involves objective probability or chance: object o has probability x to O under C. Sometimes we don't have a precise numerical representation. We may understand qualitative propensities as coarse-grained versions of numerical propensities. We can also allow for cases where the circum-stances C are trivial. Perhaps a radium atom has a certain propensity to decay within a certain time span, under any circumstances whatsoever.

In Section 3.6 I argued that ideological anti-Humeans might explain our knowledge of counterfactuals (and dispositions) by exploiting the entailment between counterfactuals and the corresponding material conditionals. With probabilistic counterfactuals (and dispositions), this move no longer works,

---

[12] Thanks here to Siegfried Jaag.

because probabilistic counterfactuals do not entail corresponding material conditionals.

Informally, a counterfactual $A > C$ is not 'completely modal'. The counterfactual has implications not just for what might or would or could happen but also for what does happen: $A > C$ entails that either not-A or C. But expressions of non-trivial propensity have no such implications. The fact that a coin has a 50% chance of landing heads when tossed logically entails nothing about actual outcomes. So how can we discover the propensity by observing outcomes?

As before, this is not quite the right way to ask the question. We know how to discover propensities. The question is whether an anti-Humean interpretation of propensities can make sense of these methods.

Let's look at an example. We want to find out whether a coin is fair or biased towards tails. A good approach is to make a large number of tosses. If the coin lands heads about half the time, it is safe to conclude that it is fair. If we get more tails than heads, we may conclude that it is biased.

We could model this whole process as an inference to the best explanation, following the frequentist approach to statistical inference. But we can get a clearer picture by adopting a Bayesian perspective.

The Bayesian treatment invokes another kind of probability, which earlier I've called 'credibility'. Credibility is an epistemic notion—something like rational degree of belief. (How exactly credibility should be understood doesn't matter for our purposes.) Intuitively, if you keep getting more tails than heads in repeated tosses of a coin, the more *probable* it becomes that the coin is biased. Here 'probable' means credible.

Let's suppose, for simplicity, that there are only two live possibilities for our coin: either the coin is fair (H_F) or it is biased 2:1 towards tails (H_B). Initially, the two possibilities have equal credibility. Then we start tossing. The first four outcomes are heads, tails, heads, and heads. Intuitively, this supports the hypothesis that the coin is fair, since the observed outcomes are more probable given H_F than given H_B. More precisely, conditional on H_F, the probability of getting the observed outcomes is $1/2^4 = 1/16$. Conditional on H_F it is $2/81$. By Bayes's Theorem, the credibility of H_F therefore rises from 0.5 to about 0.72.

The crucial step here is the move from an assumption about the coin's propensity to the credibility of certain outcomes. On the assumption that the objective probability of heads on each toss is $1/3$, for example, the credibility of the observed outcomes is $2/81$.

Whether this is plausible depends on what we mean by 'objective probability'. Suppose we identify a coin's objective probability of landing heads with the relative frequency with which an angel has touched that side of the coin. On this interpretation, assumptions about objective probability arguably tell us nothing about how the coin will land. If I told you that an angel has touched the heads side of a coin once and the tails side twice, would you feel rationally compelled to assign credence 2 / 81 to the hypothesis that the coin will land heads, tails, heads, heads in the next four tosses? Hardly. The angelic touch theory does not fit our epistemic practice.

Lewis famously argued that the same problem affects anti-Humean interpretations of chance:

> I haven't the faintest notion how it might be rational to conform my credences about outcomes to my credences about some mysterious unHumean magnitude. Don't try to take the mystery away by saying that this unHumean magnitude is none other than *chance*! I say that I haven't the faintest notion how an unHumean magnitude can possibly do what it must do to deserve that name.
>
> (Lewis 1986, pp. xv–xvi)

Lewis here assumes that the anti-Humean identifies chance with an ontologically primitive 'unHumean magnitude'. As in the case of deterministic dispositions, it helps to instead treat chance (or propensity) as primitive ideology. On this view, truths about propensities are not entailed by suitably different truths; a complete description of the world would have to explicitly specify the propensities. (We have a somewhat odd primitive of the form '...has probability...of...under....') Just as the connection to truth might be built into the classical concept of lawhood, so the connection to credibility might be built into the concept of a propensity.

This time, however, the price to pay is not just a violation of the truthmaker principle. We don't just get inexplicable modal facts. We also get inexplicable epistemic norms. There is no hope of finding an informative explanation of *why* rational credence should be guided by assumptions about propensities. The bridge principle linking propensities to (rational credence about) outcomes— what Lewis (1980) called the 'Principal Principle'—must be accepted as a basic norm of rationality.

I have no objection to basic norms of rationality. I have already committed to basic norms that license inferences from the observed to the unobserved. But the fewer basic norms, the better. We all face the problem of induction: we can't refute the inductive sceptic. Primitivists about propensity face an additional sceptical problem: they can't refute a 'propensity sceptic' who denies that we have any reason to be confident that something is going to happen if we know that it has a high chance.

In effect, by assuming a primitive norm linking propensity and credence, the anti-Humean concedes that she has no explanation of why the standard methods of discovering propensities work. In that sense, the access problem has not been solved—although the move to ideology at least makes the position intelligible.

Once again Humeans do not face this problem. On popular Humean interpretations of objective probability, the Principal Principle can be derived from independently plausible epistemic norms (see Schwarz 2014). Humeans can refute the propensity sceptic (unless the propensity sceptic is also an inductive sceptic).

3.9 Resiliency

A noteworthy feature of propensities is that if a system has a non-trivial propensity to O under C, then it is hard to predict whether O will come about in any given instance of C. It is hard to predict when a radium atom will decay, or how a fair coin will land when tossed.

More precisely, on the assumption that a coin is fair, the credibility of getting heads is $1/2$ even if we take into account all kinds of other information that we could easily obtain—say, about the current time, the weather, the result of previous tosses, and so on. Information about propensities tends to 'screen off' other available information. This feature of propensity hypotheses is closely related to what Skyrms (1980) called *resiliency*, and it plays an important role in our epistemic practice.[13]

We might think of resiliency as a further link between chance and credence, stating that chance hypotheses screen off a wide range of other propositions. But Lewis arguably intended this to be part of his Principal Principle.

In our terminology, Lewis's Principle says that the *a priori* credibility of an outcome, conditional on the hypothesis that its objective probability is x, is also x, and it remains x when conditionalizing on further 'admissible information'. Some have suggested that 'admissible information' should be understood as any information that is screened off (from the outcome) by information about objective probability. But this would make the admissibility clause in the Principle redundant. I prefer a more substantive reading of the clause, on which it captures the resiliency aspect of chance hypotheses. On this reading, 'admissible information' is a placeholder for the domain of resiliency. Lewis tentatively suggested that information about the past is usually admissible. But the details depend on the relevant chance hypothesis and its theoretical context. In quantum mechanics, information about a system's present microstate is admissible; in statistical mechanics it is not.

---

[13] We may have already seen resiliency in action. In the previous section, I claimed that on the assumption that the chance of heads on each toss is $1/2$, the (epistemic) probability of heads, tails, heads, and heads is $1/2^4 = 1/16$. This assumes that the individual tosses are (probabilistically) independent. Where did that assumption come from? One might suggest that a fair coin has not only a 50% propensity to land heads when tossed but also a 25% propensity to land heads and then tails when tossed twice, and so on. I prefer to keep the propensities simple: the coin only has a 50% propensity to land heads when tossed. The independence assumption is then an instance of resiliency: given that the chance of heads on each toss is $1/2$, the credibility of heads on the second toss is $1/2$, even after taking into account the previous outcome.

If propensities are primitive, I see no way of deriving the Principal Principle from more basic norms, including the part of it that captures resiliency. But whatever we think about the other part of the Principle, is hard to believe that the resiliency part is primitive. Surely it is not a brute fact of epistemic normativity that probabilistic hypotheses in quantum mechanics screen off information about microstates, but those of statistical mechanics do not!

The best response I can think of for anti-Humeans is to hold that truly primitive propensities only exist in an indeterministic universe, and that (by conceptual necessity) their domain of resiliency is fixed to include every proposition about the past, or every proposition that carries no information about what happens causally downstream of the relevant circumstances. The propensities associated with coin tosses or the diffusion of gases are somehow derived, in a way that explains their varying domain of resiliency.

I would feel uneasy about this response. It appears unduly opinionated about the domain of resiliency for fundamental propensities, which to me looks like an empirical question. (Indeed, shouldn't we replace 'in the past' by 'in the past light cone'?) Also, what is the proposed derivation of higher-level propensities? Should we understand them along Humean lines, perhaps as best-system probabilities? They can hardly be derived from fundamental propensities: it is an open question whether our world is deterministic, but it is not an open question whether there are fair coins. Perhaps the best option for anti-Humeans is to base higher-level propensities on properties of the lower-level dynamics, using the 'method of arbitrary functions' (see, e.g., Strevens 2003). But the details would need to be spelled out carefully, and I have not seen that done.

3.10 Conclusion

If we assume that reality contains facts of a certain kind, we may ask how we could know about these facts. When it comes to facts about natural modality, I have assumed that this is an empirical matter: somehow, our perceptual experiences are supposed to shed light on the world's modal character. But our experiences don't simply reveal the modal facts, settling any doubts or questions one may have had about the presence and distribution of un-Humean whatnots.

By this measure, our experiences directly reveal almost nothing. They don't settle all doubts about whether we have hands, or whether the sun will rise tomorrow. In each case, we need bridge principles linking our experiences with hypotheses about the external world. Some of these principles can be learned, but some must be *a priori*. Most of our empirical knowledge rests on the *a priori* assumption that our surroundings are more or less as they perceptually appear to be, and that certain regularities in the observed part of the world continue to hold in the unobserved part.

Anti-Humeans sometimes claim that we are only justified to make this last assumption, about the 'uniformity of nature', if we believe in un-Humean whatnots, whose presence guarantees the world's regularity. I disagree. I don't understand how un-Humean whatnots could guarantee or explain the world's regularity in the first place. Moreover, worlds with un-Humean whatnots are not automatically regular. We would need a further *a priori* assumption that the un-Humean whatnots themselves are simple, unchanging, and come in regular patterns. There are conceivable scenarios in which things constantly change their powers, or in which things are constantly replaced by other things with different powers (see Tugby 2017). There are also scenarios in which the primitive powers are utterly gerrymandered—in which electrons have a disposition to repel one another on Tuesday afternoons in the vicinity of apple trees, but to attract one another on Wednesday evenings at the beach. Scenarios like these will have to be deemed *a priori* improbable. So we all need *a priori* assumptions about the simplicity and uniformity of nature.

Standard forms of anti-Humeanism seem to require further assumptions, and much stranger ones. Take any complete hypothesis about the presence and distribution of un-Humean whatnots. The hypothesis might specify which things have a certain fundamental power F, which first-order universals are related by the necessitation relation N, or something along these lines. If this hypothesis—this scenario—is compatible with our total history of perceptual experience, then so are many other scenarios in which the un-Humean whatnots have been rearranged: F has been swapped with F^*, N relates different first-order universals, and so on. Some of these permuted scenarios may be too gerrymandered to deserve significant credence. But many are just as simple and uniform as the original scenario. If we are meant to have deep knowledge of natural modality, we would need *a priori* principles favouring some of these scenarios over equally simple permutations. We might need a principle according to which, in the absence of any relevant evidence, you should be confident that your surroundings contain F rather than F^*.

I fear some anti-Humeans would be willing to go down that route. But they don't have to. I have outlined two alternatives.

One is to concede that we can't have deep knowledge about the presence and distribution of un-Humean whatnots. It would, I think, be a serious cost to say that we can't have deep knowledge of whether a glass is fragile or whether radium-226 can decay into radon. But perhaps these propositions could be somehow analysed as 'structural', so that their truth does not depend on the identity of any un-Humean whatnot (just as it does not depend on the identity of fundamental properties in Lewis's metaphysics).

The other alternative is to do away with un-Humean whatnots and construe the anti-Humean position as one about ideology rather than ontology (or typology). Many of the troublesome permutations then become *a priori* impossible.

Either way, the access problem is still not fully resolved. In particular, something like the Principal Principle, with its curious restriction to 'admissible' information, will still be needed as a basic norm.[14]

References

Armstrong, D. M. (1983). *What Is a Law of Nature?* Cambridge: Cambridge University Press.

Beebee, H. (2003). 'Seeing Causing', *Proceedings of the Aristotelian Society*, 103(1), pp. 257–80.

Bird, A. (2007). *Nature's Metaphysics*. Oxford: Oxford University Press.

Busse, R. (2018). 'The Adequacy of Resemblance Nominalism About Perfect Naturalness', *Philosophy and Phenomenological Research*, 96(2), pp. 443–69.

Dasgupta, S. (2015). 'Inexpressible Ignorance', *Philosophical Review*, 124(4), pp. 441–80.

Earman, J., and Roberts, J. T. (2005). 'Contact with the Nomic: A Challenge for Deniers of Humean Supervenience About Laws of Nature Part II: The Epistemological Argument for Humean Supervenience', *Philosophy and Phenomenological Research*, 71, pp. 253–86.

Heil, J. (2010). 'Powerful Qualities', in Marmodoro, A. (ed.), *The Metaphysics of Powers: Their Grounding and Their Manifestations*. London: Routledge.

Kelly, A. (2013). 'Ramseyan Humility, Scepticism and Grasp', *Philosophical Studies*, 164(3): 705–26.

Langton, R. (2004). 'Elusive Knowledge of Things in Themselves', *Australasian Journal of Philosophy*, 82(1), pp. 129–36.

Lewis, D. K. (1980). 'A Subjectivist's Guide to Objective Chance', in Jeffrey, R. C. (ed.), *Studies in inductive logic and probability*, Vol. II. Berkeley: University of California Press, pp. 263–93.

Lewis, D. K. (1986). *Philosophical papers*, Vol. II. Oxford: Oxford University Press.

Lewis, D. K. (1994). 'Humean Supervenience Debugged', *Mind*, 103(412), pp. 473–90.

Lewis, D. K. (1996). 'Elusive Knowledge', *Australasian Journal of Philosophy*, 74, pp. 549–67.

Lewis, D. K. (2009). 'Ramseyan Humility', in Braddon-Mitchell, D., and Nola, R. (eds.), *Conceptual Analysis and Philosophical Naturalism*. Cambridge, MA: MIT Press, pp. 203–22.

Mumford, S., and Anjum, R. L. (2011). *Getting Causes from Powers*. Oxford: Oxford University Press.

Schaffer, J. (2005). 'Quiddistic Knowledge', *Philosophical Studies*, 123(1–2), pp. 1–32.

[14] Thanks to Jenann Ismael, Siegfried Jaag, Barbara Vetter, and members of the Kollegforschungsgruppe 'Human Abilities' at HU Berlin for helpful comments on an earlier version.

Schwarz, W. (2014). 'Proving the Principal Principle', in Wilson, A. (ed.), *Chance and Temporal Asymmetry*. Oxford: Oxford University Press, pp. 81–99.

Schwarz, W. (2018). 'Imaginary Foundations', *Ergo*, 29, pp. 764–89.

Skyrms, B. (1980). Causal Necessity. *A Pragmatic Investigation of the Necessity of Laws*. New Haven and London: Yale University Press.

Strevens, M. (2003). *Bigger Than Chaos*. Cambridge, MA: Harvard University Press.

Tugby, M. (2017). 'The Problem of Retention', *Synthese*, 194, pp. 2053–75.

Vetter, B. (2021). 'Explanatory Dispositionalism,' *Synthese*, 199, pp. 2051–75.

Williams, N. E. (2019). *The Powers Metaphysic*. Oxford: Oxford University Press.

4

Naturalism, Functionalism, and Chance

Not a Best Fit for the Humean

Alison Fernandes

4.1 Introduction

How should we give accounts of scientific modal relations, such as laws and
chances? According to the Humean, we should do so by reducing these
relations to parts of non-modal actuality: typically, patterns in actual events,
where the relevant events do not metaphysically necessitate each other or 'build
in' facts about modality.[1] Modal relations are nothing 'over and above' the
non-modal.

Here are three motivations for being Humean. Firstly, one might be worried
about admitting 'mysterious' elements into one's ontology (Loewer 2012, p. 121).
Modal relations are strange. Humean accounts reduce modal relations to the
non-modal. They provide a straightforward account of what modal relations are
and their relation to the non-modal. If we take modal relations as primitive, by
contrast, we seem saddled with strange entities, and have to explain their connection to actual events.

Secondly, one might be motivated by a kind of functionalism. One may wish to
account for modal relations by considering the *role* such relations play in our lives
and scientific theorizing. Perhaps chance, for example, should be accounted for
by identifying something that plays its role of guiding credences, and so forth. It
might seem that Humean accounts are particularly well suited to meet this aim,
since they can use the role of modal relations to specify what non-modal relations
scientific modal relations reduce to. We'll see some examples below. The association between Humeanism and functionalism has become so strong that even
non-Humeans take it that Humeans are uniquely interested in showing why
modal relations are fit to play their roles (Hall 2015), and that Humean

---

[1] I use 'reduction' to refer to whatever metaphysical dependency relation the Humean adopts.
I won't discuss how the non-modal 'Humean base' should be characterized—see Maudlin (2007, ch. 2),
Miller (2014), and Bhogal (2017, pp. 457–9) for discussion.

Alison Fernandes, *Naturalism, Functionalism, and Chance: Not a Best Fit for the Humean* In: *Humean Laws for Human
Agents*. Edited by: Michael Townsen Hicks, Siegfried Jaag, and Christian Loew, Oxford University Press.
© Oxford University Press 2023. DOI: 10.1093/oso/9780192893819.003.0005

explanations are of a kind that even non-Humeans should adopt (Ismael 2015). Recent Humean accounts have been particularly explicit in their functionalist motivations. While Lewis (1983, 1994) was largely content to appeal to broad criteria such as generality and simplicity in his analysis of laws, recent accounts (Hicks 2018, Dorst 2019, Jaag and Loew 2020) have focused on refining these criteria and justifying their relevance by arguing that we need laws to be simple, general, and to satisfy other criteria, if they're to be useful to us. If one has functionalist motivations, and Humeanism is required to meet those, one has strong reason to be Humean.

Thirdly, one might be motivated towards Humeanism by naturalism of a kind of that envisages a close connection between science and metaphysics. The general thought is that metaphysics shouldn't attempt to replace science or revise it in its image—"[metaphysics] should wherever possible prefer scientific explanations over metaphysical postulation" (Loewer 2012, p. 136). Instead, metaphysics is constrained by science and provides accounts of the kinds of relations and entities that science is concerned with. I'll develop this idea further below. It may seem that, by being functionalist, Humean accounts can deliver the modal relations used in science—those that feature in scientific derivations and explanation. If so, Humeanism may seem like a good choice for those wanting a naturalistic metaphysics.

The argument of this chapter is that two of these motivations do *not* count in favour of Humeanism. If one is motivated by functionalism, one has no reason to adopt Humeanism over its rivals. If one is motivated by a naturalist connection between science and metaphysics, one has reason to *reject* Humeanism. Motivations of the first kind (finding modal entities strange) may still lead one to Humeanism. But for those more concerned with the function of modal relations and the fit between science and metaphysics, one should look elsewhere.

To make this argument, I will focus on the case of chance: objective probabilities that apply in the single case. In Section 4.2 I discuss the positive claims made by Humeans: that Humean accounts fit well with science, and that *only* Humean chances can be shown to play the role of chance. In Section 4.3 I examine recent attempts to show that Humean chances satisfy the chance role and show how they rely on indifference reasoning. In Section 4.4 I argue that, notwithstanding this concern, Humeans have no special advantage when it comes to showing that chances are fit to play the chance role. In Section 4.5 I consider whether the Humean can respond by advocating a revision of science—and argue this response fits ill with naturalism. In Section 4.6 I argue that there is a deeper tension: Humeanism implies a disunity between science and metaphysics of a kind that naturalists should reject. In Section 4.7 I offer a brief sketch of an alternative naturalist justification.

4.2 The Humean's Claimed Advantages

Humeans claim their accounts can recover the chances used in science—they can recover the particular *values* of these chances, and their features, such as their *objectivity*. Regarding the first, Loewer (2001, 2004), Albert (2015, ch. 1), Frigg and Hoefer (2015), Hoefer (2019, ch. 7), and Schwarz (2016) claim that Humean chances will have values matching those used in science, including those of classical statistical mechanics. While my arguments don't rest on there being chances if the laws are deterministic, classical statistical mechanics will serve as a useful example.[2] In Boltzmannian classical statistical mechanics, as explicated by Albert (2000), even though the laws are deterministic, probabilities play an essential role in scientific derivations of macroscopic behaviour. Loewer claims that only Humeans can account for probabilities in these deterministic settings (2001, p. 619). They may do so, for example, by using a 'Best Systems' account, according to which chances derive from the true axiomatic system that best balances *simplicity* (in its number of axioms and their complexity) against *strength* and *fit* (how much information the system provides about actual events) (Lewis 1986; Loewer 2001). In the case of statistical mechanics, including a simple probability measure over initial states allows one to derive a range of macroscopic behaviour that wouldn't otherwise be derivable. So, the probability measure is plausibly included in the Best System.

Humeans also claim that because patterns in actual events are *objective*, and not mere recommendations for belief, Humean chances can feature in scientific explanations and derivations (Albert 2000, p. 64; Loewer 2001, pp. 611–12). Humean accounts deliver objective statistical-mechanical chances that can explain macroscopic behaviour and the second law of thermodynamics.

Another major motivation for Humeanism is functionalism. Humeans have argued that, in order to give an adequate account of chance, one must be able to show why chances, understood in those terms, play the *role* of chance—why they satisfy chance–credence principles like the Principal Principle, for example (Lewis 1986). While the exact form of the chance–credence principle is controversial (Section 4.5), there is broad agreement that we should align our credences in some way to what we take the chances to be (perhaps conditional on further information), given we have no evidence that overrides our use of chance-based reasoning. According to Lewis's Principal Principle (PP) (1986), for example, we should align our credence that an event occurs, $Cr(A)$, to its chance of occurring,

---

[2] For arguments we need statistical-mechanical chances, see Loewer (2001), Glynn (2010), Handfield and Wilson (2014), Emery (2015, 2016). For a non-Humean account, see also Demarest (2016).

Ch(A), as follows (where X is the proposition that Ch(A) = x, and E is any 'admissible' evidence):[3]

PP: $Cr(A|XE) = x$

Our credence in an event, conditional on what we take its chance to be (and any other admissible evidence) should equal what we take its chance to be. Evidence counts as admissible if it provides information about A only by providing information about A's chance. Assuming the chance theory of a world (T) is admissible, and a particular specification of what further information is admissible (G):

PP: $Cr(A|ET) = Ch_G(A)$

Some take the task of giving an account of chance to be to show that chance–credence principles are *justified*, given what chance is (Strevens 1999, p. 256; Hoefer 2007, 2019). Others take chance–credence principles as primitive. The task is then to show that chance can play the role specified by these principles (Lewis 1986; Loewer 2004; Ismael 2011; Schwarz 2014). I will refer to the criterion in either case as showing that the chance role is *satisfied* or that the account of chance is *justified*.

Humeans have claimed that Humean chances are *uniquely* well placed to satisfy the chance role (Lewis 1994, p. 484; Loewer 2004, pp. 1121–3; Hoefer 2007, p. 595). Lewis (1994, p. 484) thinks he sees 'dimly but well enough' how a Humean justification might go, but sees no prospects for other accounts. Loewer (2004, p. 1123) claims that any attempted justifications by propensity theorists will be illicitly question-begging. Hoefer (2007, p. 595) claims *only* Humeans chances can be shown to "deserve to guide action under circumstances of ignorance." Humeans claim what I call a '*special advantage*': Humean chances can be shown to be fit to play the chance role in a way not available to other accounts. The usual contrast is with propensity accounts. Because Humean chances reduce to patterns in events, and because we can be shown to be at least *reasonable* in aligning our credences to (appropriately related) patterns in actual events, Humean chances can be shown to be suitable for guiding credences. Moreover, *only* Humean accounts can show that chance satisfies chance–credence principles, since only they reduce chance to patterns in actual events.

If the above claims are right, those motivated by naturalism and functionalism have strong reason to adopt a Humean account of chance, independently of their prior metaphysical commitments.

---

[3] *A* is the proposition that a particular event occurs, but I also talk loosely of events having chances. Following Meacham (2005) and Handfield and Wilson (2014), I won't explicitly index chances to times.

4.3 Humean Justifications

In this section I show how Humean justifications rely on a) a reduction of chance to patterns in actual events, and b) indifference reasoning. While this is not my major argument, for those with qualms about indifference reasoning its role in these justifications should give them pause.

While Lewis suggests there will be rational constraints for belief based on knowledge of frequencies and symmetries (1994, p. 484), I'll focus on Schwarz's more explicit justification. Using indifference principles, Schwarz (2014) argues that agents are required to assign the same credence to event-sequences with the same proportion of event-types ('exchangeability'). This requirement implies that one's credence in an event of a given type at a particular point in the sequence should equal the proportion of events of that type in the whole sequence (the relative frequency). So, beliefs in relative frequencies rationally constrain credences. Provided Humean chances are appropriately related to relative frequencies, they constrain credences. Schwarz then argues that the *expected* value of the chance assigned by a Best System should equal the *expected* relative frequency (since we have no reason to believe a Best System would assign a higher or lower value)—a second application of indifference reasoning (Schwarz 2014, p. 98). Even if one accepts indifference reasoning in general, this second application is especially problematic—it would imply a random probability generator could equally well constrain rational credences. Arguably one needs a closer connection between Best Systems chances and relative frequencies.

Loewer (2001, 2004) aims to avoid indifference reasoning in justifying his Best Systems Account. But indifference reasoning turns out to be unavoidable. Loewer argues that agents are rationally required to align their credences to what they take the Best Systems chances to be, since a Best System, by definition, provides the best combination of simplicity and informativeness about actual events (2004 p. 1123; see also 2001 p. 617, n. 9). This argument depends on indifference reasoning. Firstly, informativeness of a system overall does not imply informativeness with respect to a *particular* event or chance setup. I may be rationally required to follow what I take to be informative about this very box of gas, but not what I take to be informative across all space–time. Moreover, a Best System aims for the *best* balance between simplicity and informativeness. A system could be more inform-ative than the Best System (while being simple enough to use), even though it doesn't deliver the best overall balance. Indifference principles might require agents to assign the same chance to events of a given type across space–time (similar to Schwarz's 'exchangeability'). But without indifference reasoning, agents do not have a compelling reason to follow what they take the Best Systems chances to be.

Hoefer (2007, 2019, ch. 4) might seem to avoid some of these problems, since he relates chances more closely to relative frequencies. Chances paradigmatically derive from relative frequencies in event sets that satisfy certain properties, such as 'looking chancy' and having stable distributions across time, space, and naturally selected subsets. (Otherwise, chances derive from relative frequencies from structurally similar chance setups.) Hoefer argues that agents are at least reasonable to align their credences to the chances, since, "at *most* places and times in world history," the relative frequency of outcomes in a short run will approximate the relative frequencies overall (the chances) (2007, p. 582; 2019 p. 102). Agents do as well as they reasonably can over these short runs when they align their credences to the relative frequencies since, absent very detailed information about the past or information about the future, both of which are inaccessible in ordinary situations, the best they can hope to do is to group events by type and align their credences to the relative frequency. The justification then applies in the single case: if setting one's credence to the chance is reasonable over a short run, it is reasonable at any step of which the short run is composed.

Hoefer does not invoke indifference principles directly. Instead, what plays a similar role in his justification is the limit on how well agents can be expected to predict events. Agents can't be expected to predict all events correctly, since chances that would allow them to aren't knowable. If agents can't usefully distinguish between events of a given type, it seems the best they can do is have a single credence in events of that type. Given this constraint, setting your credences to the relative frequencies is a better strategy than any other (at most times and places). There is still, however, a problematic use of indifference reasoning elsewhere. What would be a reasonable strategy across most times and places determines what is a reasonable strategy here and now. This use of indifference makes Hoefer's account vulnerable to the same kind of objection as above. On any single trial, or short run, an agent may do better (and may believe they'll do better) by aligning their credences to a significantly different single value. While Hoefer's statistical conditions limit the amount of variance, they don't exclude variance altogether. Unless we *require* credences to take the same single value over *all* short runs of a given type (indifference reasoning), Hoefer's justification fails.

While the details of these Humean approaches differ, there are general lessons to draw. Firstly, some form of indifference reasoning is unavoidable for Humean justifications. There needs to be some way of generalizing from what is rational across time and space to what is rational here and now. If one is wary about use of indifference principles, their use here is a serious concern. Secondly, Humean justifications rely on the reduction of chance to patterns in actual events. The reduction is what limits the potential divergence between relative frequencies and chances and is supposed to provide Humean accounts with their unique advantage. In the next section, I argue that considering the scientific use of chance undercuts this claimed advantage.[4]

---

[4] For other concerns about Humean justifications, see Hall (2004) and Strevens (1999).

4.4 The Undercutting Argument

One way in which Humeans might be thought to have no unique advantage
in showing the role of chance is satisfied is that chance–credence principles are
epistemic principles. They tell us how we should align our credences, given
what we take the chances to be—not what the chances actually are. Different
accounts can agree on how we reason to chances, so none has any advantage.
But I don't think this fully answers the Humean. The Humean will say that only
they have a good *metaphysical* account of how chances are ultimately evi-
denced by the relative frequencies—because they reduce chances to patterns in
actual events (see, for example, Schwarz, Chapter 3 in this volume). Still, the
epistemic response reminds us of something important—the Humean justification cannot rely merely on how we ordinarily reason using chances, absent
the story about reduction.

Instead, my argument makes use of 'undermining worlds'—worlds that have a
chance of occurring, but, were they actual, the chances would be different at those
worlds (given a Humean account). I explore the relation to more standard
undermining worries below (Section 4.5). To begin, consider how we reason
using chances. Say you've constructed an appropriate chance setup involving tossing a coin. You toss the coin a large number of times. The relative frequency of
heads is 0.5, and the sequence 'looks random'. You reason that the chance of heads
is 0.5. You're then told you will be offered a series of bets on a finite series of subsequent tosses. If you have no inadmissible evidence, and reason using the
Principal Principle, your credence in heads will be 0.5. You're now asked, "Will
you do well to align your credence to what the chances are, or what you take them
to be?" You reason as follows. Say you're right about the chances. Then there is a
high chance that you'll do better to align your credences to the chances than
another single value. But it is by no means *guaranteed* you will. Every coin toss
may come up tails. A similar point holds for what you take the chances to be—
there is a high chance that the relative frequencies in a sufficiently long run
indicate the chances, but no guarantee. If this use of chance-based reasoning is
correct, the Humean is wrong to look for a guarantee that agents will do well to
align their credences to the chances (or what they take them to be)—at best there
is a high chance of doing well.

For simplicity, the case above used a local finite chance setup. The Humean
might argue it is only at the global level, for *most times and places*, that Humean
chances are guaranteed to play the chance role. Let's construct a global case. Say
the conditional chances of our world are those of classical statistical mechanics
(using Albert (2015) and Loewer's (2004) 'Mentaculus' account of chance), or a
collapse version of quantum mechanics (say GRW). Using these chances, conditional on the initial macrostate of the universe and the macrostate of an appropriate chance setup, there is a non-trivial chance that the chances and relative

frequencies diverge to arbitrarily high degrees.[5] There is, for example, a non-trivial chance that, at most times and places, agents will *not* do well to align their credences to what the chances are. Whether chances are determined by a Best Systems account (Schwarz, Loewer), or tied more closely to relative frequencies (Hoefer), there is a non-trivial chance that *all* agents at *all* times and places do poorly.

The Humean may point out that agents will still do well to align their credences to what they *would take* the chances to be in worlds where the frequencies and chances diverge. But, recall, the Humean justification can't simply rest on how we ordinarily reason to chances if they're to claim an advantage over their rivals.

So far, I have only claimed that how we reason using chances, particularly in scientific contexts, allows for the frequencies and chances to diverge. I take this claim to be a non-controversial feature of our reasoning with chance. Ismael describes potential divergence as "part of the logic of those [modal] concepts" (2015, p. 190). In Section 4.5 I'll consider how the Humean may recommend a revision of this practice. For now, I want to consider how this use of chance reasoning creates trouble for the Humean.[6]

The concern for the Humean is as follows. Humeans aim to deliver chances that a) reduce to patterns in actual events, and b) are *used* in science. In attempting to show the chance role is satisfied, these two aims come into conflict. The Humean needs the reduction to limit the divergence between chances and frequencies. But, if Humean accounts deliver chances as they are used in science, then the above 'scientific route' becomes available for considering whether agents will do well if they align their credences to the actual chances. In this use of chance reasoning, there is *no guarantee that agents will do well*—there is a merely a high chance they will. Divergences between the actual chances and possible relative frequencies must be allowed when reasoning *using* chances. My argument is that this 'scientific' use of chance *undercuts* the metaphysical guarantee claimed by the Humean.

For example, say we take Schwarz's exchangeability approach (for simplicity, applied to a relative frequency account). Even if principles of rationality require agents to treat certain event sequences as exchangeable, and thereby align their credences with what they take the relative frequencies to be, there is a non-trivial chance that they'll do poorly by doing so. Say we take Loewer's informativeness approach. Even if agents take the Best System at their world to be informative, there is a chance that what is the Best System at our world *won't be* very

---

[5] By non-trivial, I mean such outcomes are not ruled out. If the universe is infinite, and one employs probabilities of 0, the highest degree of divergence may have chance 0. But then chance 0 would not rule out an outcome, since the actual history of the universe would also have chance 0.

[6] My argument shares features in common with Strevens's (1999, p. 255). But it avoids Hoefer's (2007, p. 583) response, since it relies on naturalism and the fact that Humean accounts attempt to deliver chances as used in science.

informative at all. Say we take Hoefer's approach, which relies on the claim that, at most times and places, the Humean strategy will do as well as any other single credence strategy over short runs. There is a chance that, at most times and places, the Humean strategy does *worse* than a single credence strategy over short runs. In each of these cases, if we align our credences to the chances in the way the Principal Principle requires, we'll have a positive credence that agents will do poorly when they follow the Principal Principle. So, whether we reason externally (Hoefer) about the actual chances, or internally (Schwarz, Loewer) about what we believe about the actual chances, *when we reason using chances, there is no guarantee that we will do well to follow the Principal Principle.*

My claim is that this *use* of chance reasoning undercuts the metaphysical guarantee provided by the Humean reduction. Once we acknowledge the chance-based possibility of agents doing poorly, we have no reason to be moved by the Humean's metaphysical guarantee. The scientific perspective in which we *use* the chances trumps the metaphysical guarantee. Science and its use of chance reasoning is our best guide to what we should believe about empirical matters—such as whether agents will do well to align their credences to the chances. Our credences should be guided by these chances, in the way science suggests, and not by anything else. In claiming science has priority over metaphysics, I'm adopting a kind of naturalism that the Humean was aiming towards—metaphysics does not attempt to replace science. So, when metaphysical and scientific explanation potentially compete, it is the metaphysical explanation that has to give way.

More is needed to defend this argument. I'll begin with a few quick objections. Firstly, one might worry that an application of the Principal Principle is needed to reason from a belief in a *chance* that agents may do poorly to a *credence* that agents may do poorly. This is true. But the Principal Principle is not in question here—just a particular Humean route to showing the chance role is satisfied.

Secondly, the chance of the relative frequencies and chances diverging significantly is small. One might therefore argue that we can neglect this possibility. However, a chance being small doesn't license its neglect; small-probability events should sometimes be taken *very* seriously. Moreover, if small chances *could* be neglected, there would still be no unique advantage for the Humean over their rivals.

Thirdly, the Humean might complain that I am begging the question. The Humean justification relies on the Humean reduction of chance *ruling out* cases where the chances and frequencies diverge. By allowing the chances and frequencies to diverge, aren't I simply assuming the Humean reduction fails? No. The undercutting argument *does not* assume that the *actual* chances and *actual* frequencies can diverge in non-Humean ways. The argument was that by considering a strategy where one aligns one's credences to what the *actual* chances are, there is a chance-based possibility that the frequencies *may* diverge from what the actual chances *are*—not from what the chances *would be at those worlds*. I am not assuming the Humean metaphysics is false.

Surely, the Humean goes on, showing the role of chance is satisfied requires considering whether one would do well to align one's credences to what the chances *would be*—not what they *actually are*. I'm not so sure. As far as I can tell, the question of justification admits of both readings—one in which one aligns one's credences to the actual chances ('rigid reading'), and one in which one aligns one's credences to what the chances would be ('non-rigid'). I can see nothing so far to decide between the two. Neither reading is more general. Using the rigid reading, you can still consider how successful one would be, given different chances were actual. Moreover, even if the Humean insists on the non-rigid reading, trouble remains.

The undercutting argument can still be given, as soon as the Humean explains how Humean chances are 'held fixed' in scientific reasoning. In scientific contexts, we don't take the chances to vary as the relative frequencies do. Instead, we standardly hold the chances 'fixed' at their actual values. For example, on being told the chance of the coin landing heads is 0.5, you consider possibilities on which the coin always lands heads, and take it that the chance would still be 0.5 if this possibility were to obtain.[7] When reasoning 'metaphysically', and assuming the Humean reduction holds, you may reason that the chances would be different. But, in scientific contexts, the chances are held fixed.

Moreover, it is Humeans themselves who have sometimes recommended holding the chances fixed in this way. Humeans have responded to worries about how Humean laws and chances can explain—the 'explanatory objection'—by distinguishing between scientific and metaphysical explanation. According to the objection, Humean laws aren't sufficiently independent of the actual events so as to explain them (Dretske 1977, p. 267; Armstrong 1983; Maudlin 2007, ch. 6; Lange 2013; Shumener 2019). If the laws are explained by actual events (the Humean reduction), then the laws can't explain those same events—this would be circular. Various responses have been made. The most popular relies on distinguishing between metaphysical and scientific explanation (Loewer 2012; Miller 2015; Hicks and van Elswyk 2015; Bhogal 2020)—see Emery (2019) for discussion. The thought is that while actual events *metaphysically* explain the laws, "this metaphysical explanation doesn't preclude [Humean] laws playing the usual role of laws in scientific explanations... [Scientific and metaphysical explanation] are different enterprises" (Loewer 2012, p. 131). What has made this response appealing is, I suggest, the fact that the Humean typically aims to *recover* scientific practice and leave it untouched by their metaphysics. (I'll consider more revisionary responses below.) If Humean laws are, when reasoning scientifically, independent

---

[7] See Demarest and Miller (Chapter 5 in this volume) for formulations of the relevant counterfactuals, and discussion of how undermining worries may generalize. While we have similar aims, some of the differences between our views are: a) I rely more on the use of chances in scientific practice, and less on the intuitive truth of certain counterfactuals; b) I don't think the Humean has a particular problem with higher-level science generalities—I take these to derive from fundamental physics.

of the actual events so as to be able to explain them, they are 'held fixed', even as the actual events vary.

Humeans make a similar move in response to another standard objection: that Humean laws don't allow the initial conditions or other states to vary independently of the laws, as is standard in science (Tooley 1977; Carroll 1994; Maudlin 2007, p. 68; Bhogal 2020). Some, including Humeans, answer this concern by appealing to forms of contextualism or projectivism (Halpin 2003; Ward 2003; Roberts 2008, ch. 10; Loew and Jaag 2020; Dorst 2022). In its strongest form, *any* set of true axioms can be 'projected' onto the Humean mosaic, and taken as the laws, depending on our context and interests. In a weaker form, the actual Humean laws are held fixed as the initial conditions (or other states) are allowed to vary.

If either of these responses is right, Humean chances are 'held fixed' in scientific reasoning. Even under a non-rigid reading, there is a chance that agents will do poorly by aligning their credences to what the chances *would* be. So, Humean accounts are open to the undercutting argument that, when reasoning scientifically, there is no guarantee agents will do well to align their credences to what the chances would be. If the Humean recovers the scientific practice of holding the laws and chances fixed, there becomes a straightforward scientific possibility of agents doing poorly. The claim is that this scientific use of chance undercuts the Humean's metaphysical justification. The Humean loses their special advantage in showing that the chance role is satisfied.

The Humean might instead try to explain why it merely *seems* Humean chances are held fixed in scientific contexts, even though they're not. One suggestion is that, in large universes, the Humean chances aren't sensitive to changes in frequencies in small local systems. So, in local cases, we assume the relative frequencies can vary entirely independently of the chances (Loewer 1996, p. 117). But this strategy is not well suited to all contexts in which statistical-mechanical explanations are applied. Loewer (2012) and Albert (2000) apply the statistical-mechanical probability distribution to the initial macrostate of the whole universe. If standard statistical-mechanical explanations are to work, the chances must still be held fixed even though we're no longer considering changes to the relative frequencies in small local systems.

4.5 Revising Science

The undercutting argument above is related to standard undermining arguments against Humeans. Lewis notes that the possibility of chances being otherwise than they are creates trouble for Humean accounts when combined with the Principal Principle—the 'Big Bad Bug' (Lewis 1986, 1994). The worry is that the Principal Principle will give contradictory advice: the chances being what they are (and Humeanism being true) metaphysically rules out certain possibilities that the 'scientific' use of chance (the Principal Principle) rules in. My argument also exploits the gap between the Humean metaphysics and the scientific use of chance. But while I think standard solutions to the Big Bad Bug (Thau 1994; Lewis 1994; Hall 2004) are enough to remove the threat of contradiction, they don't help the Humean produce a satisfactory account of chance—or at least this is what I will now argue.

The Humean can respond to the Big Bad Bug by being revisionist about how we use chance-based reasoning. According to the revisionary Humean, the Humean reduction limits the possibility space used when reasoning *using* chances in ways that go beyond the limits provided by probability theory. Say you know the chance of the coin landing heads on each toss is 0.5, say this is the only thing going on in a universe, and I ask you (even in a scientific context) whether it is possible for the coin to land heads every time it is ever flipped. The answer is: No, it is not. The fact that the (Humean) chance is 0.5 rules this possibility out.

This 'revisionary' Humean response is successful against the Big Bad Bug. We were wrong to think we should align our credence in an event (conditional on what we take its chance to be and any other admissible evidence) to what we take its chance to be. Instead, we should align our credence in an event (conditional on the same) to what we take chance to be *conditional on the complete chance theory of a world* (T). According to the 'New Principle' (Lewis 1994; Thau 1994; Hall 1994, 2004):

NP: $Cr(A|ET) = Ch_G(A/T)$

By conditionalizing chances over the complete chance theory (T), the New Principle rules out cases where the chances are 'undermined': where events (that have some chance of occurring) would make the Humean chances otherwise than they are. Because the New Principle excludes the possibility of the chances and relative frequencies diverging greatly, even when we reason *using* chances, it excludes the possibilities used to undercut the Humean justification.

An alternative revision is to revise what the *chances* themselves are (Arntzenius and Hall 2003; Schaffer 2003). According to this revision, the chances are the Humean chances conditional on the complete theory of chance—thus conditionalizing out the possibility of undermining. Under either of these responses, there is only a small modification to the credence recommended (in local setups) compared to standard scientific practice. If so, perhaps this revision is of no great concern.

My response to these and other revisions (Hoefer 2019) is as follows. If the Humean is claiming a *unique* advantage in showing the chance role is satisfied, then the first revision is untenable. If the New Principle replaces the Principal Principle, we no longer have a theory-neutral criterion by which to judge accounts of chance. The New Principle implies that when we use chance to guide our beliefs, we should bracket out the possibility (one consistent with the Humean metaphysics) of the chances being otherwise than they are. But if the New Principle is designed as a fix for the Humean, it's not a theory-neutral criterion by which to judge accounts of chance. It doesn't matter if only Humean accounts can be shown to satisfy the New Principle.[8]

Hall (2004) argues the New Principle isn't simply a fix for the Humean—it is a principle we should all adopt. Chance, he thinks, is like an 'analyst-expert': an expert who is extremely good at evaluating the relevance of one proposition for another. When you ask such an analyst-expert for advice, you want her advice conditional not only on what evidence you happen to have but on whatever conditions would be necessary and sufficient for her to *count* as being an expert— evidence contained in T and evidence that she might not have herself (Hall (2004), p. 103). So, all should accept the New Principle. But Hall's reasoning relies on the possibility of there being evidence that is sufficient for someone to *count* as being a chance expert. The non-Humean has good reason to deny this: while there might be evidence for the chances from actual events, no evidence about actual events *guarantees* the truth of a chance theory. The non-Humean should keep to the Principal Principle.

Regardless of whether the Humean claims a unique advantage, there are two further concerns with these responses. The first concern is about simplicity—see also Schwarz (2016). One of the claims of the Best Systems Humean programme was that scientific modal relations were to derive from simple and general axiomatic systems. One of the hallmark features of laws (and chance functions) that the Humean account is supposed to deliver is that they are *simple*—simple in form, and simple to apply. But, while the above revisions to the Principal Principle or the chances may be quantifiably small in the difference in chance *values*, they add significantly to the *complexity* of the chance function or its employment. Consider again the case of employing the statistical mechanical distribution to the initial macrostates of the universe. According to the above response, the conditional chances that should guide our derivations and predictions aren't those given by the standard statistical mechanical distribution, but those conditionalized on whatever must be the case for those to be the Best Systems chances. The simplicity of probability assignments and their employments is gone.

The second concern is that a Humean that recommends revising scientific practice has to give up a strong claim to be *recovering* scientific practice. Even if the revision is quantifiably small, it's still a revision. I don't think scientific

---

[8] Lewis thought that Humean chances might be the closest thing that could play the role of chance given by the Principal Principle, so it wouldn't matter that they didn't strictly satisfy it (Lewis 1994). But, as Arntzenius and Hall (2003) argue, there are things that satisfy the Principal Principle perfectly—they just don't look like chances. See also Schaffer (2003).

practice is sacrosanct. There are cases where philosophical reflection on science leads to fruitful revision. But in this case, unlike others, scientific practice itself doesn't seem to be in trouble—it's not contradictory or incomplete. The Humean can still claim an ontological advantage—they reduce chance to something less mysterious. But this ontological advantage comes at a naturalistic cost. Naturalism and functionalism combined, moreover, provide no reason to adopt a Humean account.

4.6 The Deeper Tension

I have argued that Humeans have no advantage in showing chance satisfies its role. I'll now argue that undermining worries point to a deeper naturalistic tension between Humean metaphysics and science.

Following Loewer (2012, p. 131), I take the distinction between science and metaphysics to concern the explanatory relations used. The explanatory relations used in science are relations such as laws, chances, and causation. An explanation is partly scientific when these relations are *used* as explanatory relations. In metaphysics, the relevant explanatory relations are of other kinds—such constitution, (metaphysical) reduction, ground, nature, or essence. An explanation is partly metaphysical when these relations are *used* as explanatory relations. Scientific laws and chances can still feature in purely metaphysical explanations, provided they aren't appealed to *as* explanatory relations. The Humean, for example, can give a *metaphysical* explanation of why the chances don't diverge too much from the relative frequencies, because they don't employ chances as explanatory relations in this setting. I take it Loewer has this in mind when he claims that only scientific relations may be 'probabilistic' (2012, p. 131). He is not claiming probabilities can't feature in metaphysical explanations, but that the metaphysical explanatory relations won't themselves be probabilistic.

My concern with the Humean explanation of why chance satisfies its role is *not* that the explanation is metaphysical. My concern is that the Humean explanation *competes* with scientific explanation in a way that prevents us from combining metaphysics and science into a single unified explanatory practice.

One way to bring this worry out is to consider contrast cases. In other cases, metaphysical and scientific explanations can be fruitfully combined.[9] We explain the movement of a limb by considering how the movement of its *constitutive* parts is *caused*. We explain why a mercury column counts as a thermometer by considering what *constitutes* a mercury column and the *required role* (perhaps *nature*) of a thermometer and using scientific laws to explain how the mercury

---

[9] Bhogal (2020) discusses such cases under the label of 'chaining'.

column is able to function as a thermometer. A typical 'functional justification' is partly metaphysical (in virtue of using constitutive relations, and (perhaps) natures, or essences) and partly scientific (in virtue of using laws and other scientific relations).

But this is patently not what's going on in the case of the Humean justification. In the Humean justification, the laws and chances are not used as explanatory relations. The Humean justification can, in fact, proceed *entirely independently of what the laws and chances actually are*. All one needs to offer the Humean justification is the metaphysical reduction of chance to patterns in actual events. Nothing about the actual laws or chances is needed in order to generate the 'possibility space' within which the Humean justification operates. It can include worlds with laws and chances that are vastly different from our own, that have entirely different structural features from our own, and that are even too disordered to have chances (or laws)—and still the relative frequencies will approximately match the Humean chances (when they exist).

Some might see this lack of dependence on the actual chances and laws as an advantage. What I see it as doing is making clear that the actual Humean chances aren't used as explanatory relations within the Humean justification. This is a problem. At best what the Humean can do is *overlay* the Humean metaphysical justification on top of the scientific use of chance. But this doesn't make the chances required in the justification. They remain redundant. There is no deeper unity between the scientific and metaphysical parts of the explanation. What this means, in turn, is that chances, taking the values they do, *aren't* doing their usual work of guiding credences and explaining within the Humean justification. That work is done instead by the nature of chance, combined with principles of indifference.

There is something decidedly odd about all this. The Humean, having shown that chances (and laws) are fit to guide credences, and scientifically explain, abandons them when it comes to explaining how chances are fit to play their roles. This is not how metaphysics should operate. The metaphysical natures of things should combine coherently with science-based reasoning to offer explanations and justification.[10]

This objection does not depend on assumptions about explanation involving metaphysical dependency. While I agree with Emery (2019) on the need for science and metaphysics to combine, one doesn't need metaphysical grounding to make this claim. There is value in science and metaphysics cohering, independently of one's metaphysical commitments. The value of this coherence may even

---

[10] Other coherence problems arise if a metaphysics of chance builds in features that science should explain. Taking chances to be intrinsically temporally or causally asymmetric, for example, as under traditional propensity accounts, prevents any temporal or causal asymmetries in chance being explained in scientific terms.

be explicated in purely instrumental terms—it is *useful* to have practices that combine. For this reason, the Humean can't defend the lack of cohering by appealing to the fact that (scientific) explanation has only instrumental value, while metaphysical explanation aims at 'elucidating underlying structure' (Bhogal 2020). Ultimately there are both theoretical and instrumental advantages to explanatory practices that combine, reasons that metaphysics is answerable to.

The upshot is that Humean metaphysics introduces a worrying disconnect between metaphysics and science. This is the deep tension I alluded to earlier between recovering the scientific use of chance and providing a reductive account of modality. If the Humean recovers the scientific uses of chance, they recover something explanatorily disconnected from their own metaphysics. For those motivated by the form of naturalism I began with, this is a serious concern. While the Humean can recommend a revision of scientific practice, the worries I raised above (Section 4.5) still hold—such a revision fits ill with naturalism.

4.7 Science Justifies Science

To end, let me offer a brief sketch of how one might use scientific relations to show how chance is fit to play its role. One option is to mimic a Humean justification, and argue there is a *high chance* that the relative frequencies approximate the chances over long runs (the weak law of large numbers).[11] So, there is a high chance that, by aligning your credences to the chances, you will do reasonably well at predicting events over long runs.[12] This option uses indifference reasoning. But there is an alternative that avoids indifference reasoning.[13]

In any given case where you choose how to act based on credences, you have a *higher chance* of doing well if you align your credences to the chances rather than to any other value. Doing worse is possible, but unlikely. For example, say you know there is to be only one coin toss of a fair coin, and you adopt credences based on the chances (conditional on known information).[14] The coin toss is fair, in the sense that its chance of landing heads conditional on the macroscopic characterization of the chance setup is 0.5. If you don't know the exact microstate of the chance setup now (or have any other information about the outcome of the coin toss), the relevant chances are 0.5. If you align your credences to these

---

[11] One needs a version of the weak law that applies in the finite case.

[12] For a proof of this chancy justification of the principle, see Strevens (1999, Appendix B). For discussion of related attempts, see Mellor (1971, pp. 55–6) and Strevens (1999, p. 27, n. 11).

[13] For a related projectivist alternative using credences, see Ward (2005).

[14] Following Hall (2004), my preferred account of chance dispenses with admissibility. The conditional probabilities are still as objective as the laws—they take the value they do, independently of what anyone wants or thinks about the matter. Alternatively, one can assume one has no inadmissible evidence. See Elga (2007, pp. 114–15) for the physics behind such chance setups, and why we typically don't have such evidence. My thanks to Michael Hicks for discussion.

chances, you will only accept bets at odds of higher than 1:1 on heads, and odds of higher than 1:1 on tails—and will be indifferent to or reject all other bets.[15]

Adopting the strategy of aligning your credences to the chances (conditional on known information) implies that, for any bet that you will accept, there is a higher chance that you will gain money than that you will lose money. For any bet that you reject, there is a higher chance that you would lose money than that you would gain money. For any bet you are indifferent about, there is an equal chance of losing or gaining money. Provided the credences and chances are conditional on the same evidence each time, there is always a higher chance you gain money rather than lose it. Applying the Principal Principle, you should have a higher credence that you gain money rather than lose it. So, you are justified in aligning your credences to the chances, in both the 'internalist' and 'externalist' sense. The same kind of reasoning also works for what you take the chances to be. So, chance is fit to play the chance role.

Some have argued that chances can't be used to justify the Principal Principle, or show chance is fit to play its role (Strevens 1999, pp. 255–6; Hoefer 2007, p. 588; 2019, ch. 1). It might seem that, by using chance, *any* account of chance can be justified—making giving an adequate account of chance too easy. As Loewer puts it, "[w]ithout relying on the PP there is *no* non-question begging reason to think that setting one's degrees of belief by propensity chances will result in having high degrees of belief in truths and low degrees of belief in falsehoods" (2004, p. 1123).[16] The implication is one should *not* rely on the Principal Principle. Loewer goes on:

> ...since propositions about propensity chances are facts logically completely distinct from the propositions they assign chances to it is *utterly mysterious* why they should tell us anything about what degrees of belief to have in those propositions.   (Loewer 2004, p. 1123)

This kind of reasoning relies on chance being initially suspect: guilty until proven innocent. We're not entitled to appeal to chance reasoning, including assigning chances and using them to guide our credences, until we've a) said what chances are, and b) shown that they can appropriately guide our credences. Without the former, chance remains metaphysically mysterious. Without the latter, we either don't know the Principal Principle is justified or don't know that the account of chance is adequate.

---

[15] If you do have information about the exact microstate or other information about future outcomes, then the chance conditional on this information may be different. For example, if the information implies that the coin will land heads, the chance of its landing heads (conditional on that information) is 1—and you will accept all bets on heads, and no bets on tails.

[16] See also Hoefer (2019, p. 43). For similar concerns with respect to laws, see Lewis (1994, p. 484), and van Fraassen's (1989) inference and identification problems.

We should resist this stance. We don't come to the project of accounting for chance unsure of our entitlement to assign chances and use chance reasoning. We don't have to earn our entitlement to reason using the Principal Principle. While it's reasonable to wonder what might justify it, its actual use is not in question.

There would be greater concerns if the chance role were shown to be satisfied in a genuinely trivial way—a way that *anything* could satisfy. But this is not the case. The above justifications require there to be a *higher* chance that an agent will do well than otherwise. While this is a relatively weak constraint, it is still a constraint—one not met by many functions we might otherwise identify as chance functions. It is not merely by *identifying* something as chance that it is shown to play the chance role—as identifying someone as Armstrong might seem to imply they had large biceps. A chancy justification is not a case of 'premise circularity' where the conclusion is trivially derivable from a limited number of the premises. Instead it is analogous to a case of 'rule circularity'—chance reasoning is used in justifying chance reasoning.[17] The chancy justification shows there is coherence in our theories and beliefs about chance. This is not to say that the *source* of the justification is the coherence—that any coherent package would be equally well justified. But the way to give justifications is always to work from within the package of theories and beliefs we already accept. We revise these as required, when incoherencies are met.

The approach is naturalist, insofar as it takes science as our best guide to the empirical world. It is both functionalist and naturalist in that it uses science to explain how scientific relations are able to perform their functions. The programme shares much in common with Quinean naturalism, as explicated by Verhaegh (2018). We do not look to justify science 'from the outside'; instead, we use our developing knowledge of science to explain how we come to think scientifically and what function such thinking (and such relations) serve.

A variety of accounts of chance will be compatible with a chancy justification of chance, including metaphysically minimal accounts such as 'pragmatist' propensity accounts (Peirce 1910; Levi 1990) and Sober's 'no-theory theory' of chance (2010). These accounts rely on our use of chance reasoning in science to provide the only account of what chance is, and are my preferred option. I have concerns with attempts to provide a more robust metaphysical account of chance by appeal to causal and temporally properties—as standard propensity accounts do (see note 11). But these arguments are not my focus today. Instead I have been concerned to explicate a tension between the Humean's reductive metaphysics and the scientific use of chance—a tension that is independently problematic, and that undercuts the Humean's claim to provide a unique satisfaction of the chance role.

---

[17] Thanks to Alastair Wilson for the suggestion.

4.8 Conclusion

Humeans are right to be interested in showing how modal relations like chance satisfy their roles. But they go wrong in thinking this should be achieved by reducing modal relations to actual events. The scientific use of chance reasoning undercuts the metaphysical justification provided by the Humean. This undercutting argument points towards a deeper tension between the Humean reduction of chance and the use of chance reasoning in science—the Humean must either accept a strong disconnect between science and metaphysics or seek an unwarranted revision of science. None of this is to say one *cannot* be Humean—but these are heavy naturalistic costs to bear.

There is an alternative. One can show how chance is fit to play its role using chance-based reasoning—by reasoning *scientifically* about science. According to this naturalistic functionalist approach, one can use the resources of science to explain why scientific modal relations are fit to play their roles. While I have not pursued the details here, I take it this is a more promising approach to the metaphysics of scientific modal relations.[18,19]

References

Albert, D. Z. (2000). *Time and Chance*. Cambridge, MA: Harvard University Press.

Albert, D. Z. (2015). *After Physics*. Cambridge, MA: Harvard University Press.

Armstrong, D. (1983). *What Is a Law of Nature?* Cambridge: Cambridge University Press.

Arntzenius, F., and Hall, N. (2003). 'On What We Know about Chance', *British Journal for the Philosophy of Science*, 54(2), pp. 171–9.

Bhogal, H. (2017). 'Minimal Anti-Humeanism', *Australasian Journal of Philosophy*, 95(3), pp. 447–60.

Bhogal, H. (2020). 'Nomothetic Explanation and Humeanism about Laws of Nature', in *Oxford Studies in Metaphysics Volume 12*, https://oxford.universitypressscholarship.com/view/10.1093/oso/9780192893314.001.0001/oso-9780192893314-chapter-6 (accessed 19 Mar. 2021).

Carroll, J. (1994). *Laws of Nature*. Cambridge: Cambridge University Press.

Demarest, H. (2016). 'The Universe Had One Chance', *Philosophy of Science*, 83(2), pp. 248–64.

---

[18] See Fernandes (2017) for such an approach in the case of causation.
[19] My warm thanks to the following people for comments and discussion: Elizabeth Miller, Michael Hicks, Nina Emery, David Albert, Barry Loewer, Alastair Wilson, Michael Janssen, Catherine Kendig, Yann Benetreau-Dupin, Matthew Brown, Anjan Chakravartty, Edouard Machery, Sharon Crasnow, and Greg Frost-Arnold.

Dorst, C. (2019). Toward a Best Predictive System Account of Laws of Nature', *British Journal for the Philosophy of Science*, 70(3), pp. 877–900.

Dorst, C. (2022). 'Why Do the Laws Support Counterfactuals?', *Erkenntnis*, 87(2), pp. 545–66.

Dretske, F. (1977). 'Laws of Nature', *Philosophy of Science*, 44, pp. 248–68.

Elga, A. (2007). Isolation and Folk Physics. In Huw Price and Richard Corry *(eds.), Causation, Physics, and the Constitution of Reality, Oxford:* Oxford University Press. 106–119.

Emery, N. (2015). 'Chance, Possibility, and Explanation', *British Journal for the Philosophy of Science*, 66(1), pp. 95–120.

Emery, N. (2016). 'A Naturalist's Guide to Objective Chance', *Philosophy of Science*, 84(3), pp. 480–99.

Emery, N. (2019). 'Laws and their instances', *Philosophical Studies*, 176(6), pp. 1535–61.

Fernandes, A. (2017). 'A Deliberative Approach to Causation', *Philosophy and Phenomenological Research*, 95(3), pp. 686–708.

Frigg, R., and Hoefer, C. (2015). 'The Best Humean System for Statistical Mechanics', *Erkenntnis*, 80, pp. 551–74.

Glynn, L. (2010). 'Deterministic Chance', *British Journal for the Philosophy of Science*, 61(1), pp. 51–80.

Hall, N. (1994). 'Correcting the Guide to Objective Chance', *Mind*, 103(412), pp. 505–18.

Hall, N. (2004). 'Two Mistakes About Credence and Chance', *Australasian Journal of Philosophy*, 82(1), pp. 93–111.

Hall, N. (2015). 'Humean Reductionism about Laws of Nature', in Loewer, B., and Schaffer, J. (eds.), *The Blackwell Companion to David Lewis*. Oxford: Blackwell, pp. 262–77.

Halpin, J. F. (2003). 'Scientific law: A Perspectival Account', *Erkenntnis*, 58, pp. 137–68.

Handfield, T., and Wilson, A. (2014). 'Chance and Context', in Wilson, A. (ed.) *Asymmetries of Chance and Time*. Oxford: Oxford University Press, pp. 19–44.

Hicks, M. T. (2018). 'Dynamic Humeanism', *British Journal for the Philosophy of Science*, 69(4), pp. 983–1007.

Hicks, M. T., and van Elswyk, P. (2015). 'Humean Laws and Circular Explanation', *Philosophical Studies*, 172, pp. 433–43.

Hoefer, C. (2007). 'The Third Way on Objective Probability: A Skeptic's Guide to Objective Chance', *Mind*, 116(2), pp. 549–96.

Hoefer, C. (2019). *Chance in the World*. New York: Oxford University Press.

Ismael, J. (2011). 'A Modest Proposal About Chance', *Journal of Philosophy*, 108(8), pp. 416–42.

Ismael, J. (2015). 'How to be Humean', in Loewer, B., and Schaffer, J. (eds.), *A Companion to David Lewis*. Oxford: John Wiley & Sons, pp. 188–205.

Jaag, S., and Loew, C. (2020). 'Making best systems best for us', *Synthese*, 197, pp. 2525–50.

Lange, M. (2013). 'Grounding, Scientific Explanation, and Humean Laws', *Philosophical Studies*, 164, pp. 255–61.

Levi, I. (1990). 'Chance', *Philosophical Topics*, 18, pp. 117–49.

Lewis, D. K. (1983). 'New Work for a Theory of Universals', *Australasian Journal of Philosophy*, 61, pp. 343–77.

Lewis, D. K. (1986). 'A Subjectivist's Guide to Objective Chance', in *Philosophical Papers: Volume II*. Oxford: Oxford University Press.

Lewis, D. K. (1994). 'Humean Supervenience Debugged', *Mind* 103(412), pp. 473–90.

Loew, C., and Jaag, S. (2020). 'Humean Laws and (Nested) Counterfactuals', *Philosophical Quarterly*, 70(278), pp. 93–113.

Loewer, B. (1996). 'Humean Supervenience', *Philosophical Topics,* 24(1), pp. 101–27.

Loewer, B. (2001). 'Determinism and Chance', *Studies in History and Philosophy of Modern Physics*, 32, pp. 609–29.

Loewer, B. (2004). 'David Lewis's Humean Theory of Objective Chance', *Philosophy of Science*, 71(5), pp. 1115–25.

Loewer, B. (2012). 'Two Accounts of Laws and Time', *Philosophical Studies,* 160(1), pp. 115–37.

Maudlin, T. (2007). *The Metaphysics within Physics*. New York: Oxford University Press.

Meacham, C. (2005). 'Three Proposals Regarding a Theory of Chance', *Philosophical Perspectives*, 19, pp. 281–307.

Mellor, D. H. (1971). *The Matter of Chance*. Cambridge: Cambridge University Press.

Miller, E. (2014). 'Quantum Entanglement, Bohmian Mechanics, and Humean Supervenience', *Australasian Journal of Philosophy*, 92(3), pp. 567–83.

Miller, E. (2015). 'Humean Scientific Explanation', *Philosophical Studies*, 172, pp. 1311–32.

Peirce, C. S. (1910). 'Note on the Doctrine of Chances', in *Philosophical Writings of Peirce*. New York: Dover, 1955.

Roberts, J. (2008). *The Law-Governed Universe*. New York: Oxford University Press.

Schaffer, J. (2003). 'Principled Chances', *British Journal for the Philosophy of Science*, 54(1), pp. 27–41.

Schwarz, W. (2014). 'Proving the Principal Principle', in Wilson, A. (ed.), *Asymmetries of Chance and Time*. Oxford: Oxford University Press, pp. 81–99.

Schwarz, W. (2016). 'Best System Approaches to Chance', in Hájek, A., and Hitchcock, C. (eds.), *The Oxford Handbook of Probability and Philosophy*. Oxford: Oxford University Press, pp. 423–39.

Shumener, E. (2019). 'Laws of Nature, Explanation, and Semantic Circularity', *British Journal for the Philosophy of Science*, 70, pp. 787–815.

Sober, E. (2010). 'Evolutionary Theory and the Reality of Macro Probabilities', in Eells, E., and Fetzer, J. H. (eds.), *The Place of Probability in Science: In Honor of Ellery Eells (1953-2006)*. Dordrecht: Springer, pp. 133–61.

Strevens. M. (1999). 'Objective Probability as a Guide to the World', *Philosophical Studies*, 95(3), pp. 243–75.

Thau, M. (1994). 'Undermining and Admissibility', *Mind*, 103(412), pp. 491–503.

Tooley, M. (1977). 'The Nature of Laws', *Canadian Journal of Philosophy*, 7, pp. 667–98.

van Fraassen, B. (1989). *Laws and Symmetry*, Oxford: Oxford University Press.

Verhaegh, S. (2018). *Working from Within: The Nature and Development of Quine's Naturalism*. New York: Oxford University Press.

Ward, B. (2003). 'Sometimes the World is not Enough: The Pursuit of Explanatory Laws in a Humean World', *Pacific Philosophical Quarterly*, 84, pp. 175–97.

Ward, B. (2005). 'Projecting Chances: A Humean Vindication and Justification of the Principal Principle', *Philosophy of Science*, 72(1), pp. 241–61.

5

Generalizing the Problem of Humean Undermining

Heather Demarest and Elizabeth Miller

5.1 Introduction

For Humeans, all facts about the world supervene on a global mosaic of local qualities, the distribution of fundamental intrinsic, categorical properties arrayed in spacetime. It follows that many facts—even ones intuitively 'about' particular, localized macroscopic parts of the world—turn out to depend on surprisingly global fundamental bases.[1] For instance, the fact that this chunk of salt is soluble in water depends on facts about special science kinds that reflect, or arise from, patterns in fundamental local qualities spread across all of space and time. We investigate some counterintuitive consequences for such a Humean picture.

To get a feel for the problem, suppose there were a perfect microphysical duplicate of this chunk of salt in a world that shared our microphysical dynamical laws, yet (perhaps due to atypical initial conditions) did not happen to dissolve when submerged in water. Atypical initial conditions also allow that such a world might not contain any other samples of salt (or, more dramatically, any other macroscopic solids at all). In such a world, on the standard Humean account, there would be no simple, informative systematization that included the special science kind *salt*, nor the special science causal process of *dissolving*, and *thus* no disposition of *solubility*.[2] It seems the Humean is committed to implausible counterfactuals such as: 'If the initial conditions had been thus-and-so, then a microphysical duplicate of this chunk of salt would not have been soluble.'

The result is similar in form to better-known self-undermining counterfactuals such as, 'If this coin had landed heads each time, its chance of landing heads would have been closer to one than it actually is.' Some Humeans have suggested that the laws 'need to be held fixed' when evaluating these counterfactuals, but it

---

[1] For a more thorough explanation of this, see Miller (2018).
[2] This is particularly obvious for Humean views such as Cohen and Callender's (2009) and Markus Schrenk's (2008), according to which each domain of science has its own systematization. But, as we will explain throughout the chapter, it also arises for more 'reductive' Humean views as well. Note that there are some anti-Humean views that will face this challenge too.

Heather Demarest and Elizabeth Miller, *Generalizing the Problem of Humean Undermining* In: *Humean Laws for Human Agents*. Edited by: Michael Townsen Hicks, Siegfried Jaag, and Christian Loew, Oxford University Press.
© Oxford University Press 2023. DOI: 10.1093/oso/9780192893819.003.0006

is often unclear how exactly this ought to be implemented and motivated.[3] We show how the resources of the Mentaculus can be used to develop a precise account of counterfactual chances, and, in so doing, provide a satisfying Humean account of some self-undermining counterfactuals about chances—but only on the assumption that the Humean can make cross-world identifications of objects such as coins. We go on to argue that the Humean cannot make such identifications. We use the Mentaculus to explore what, in broad outline, a Humean solution to our more general challenge from self-undermining counterfactuals would need to look like. We conclude that it is very much an open question whether a Humean account can meet this challenge.

5.2 Humean Supervenience

Here, we briefly summarize the key bits of Humean metaphysics for our project. According to the standard, Humean picture, fundamentally, all that exists is the Humean *mosaic*. The mosaic includes:

- points of spacetime (or small regions of spacetime) and perhaps concrete occupants of these points (or regions);
- local,[4] intrinsic, categorical 'perfectly natural'[5] properties of those points or occupants;[6] and
- external spatiotemporal relations between the points or occupants.

The familiar, macroscopic properties and kinds, in addition to the familiar notions of modality (including laws of nature, causation, counterfactuals, dispositions, etc.) are not part of the fundamental mosaic; they supervene on these more fundamental features. Metaphysically possible worlds with the same Humean mosaic also have the same laws, dispositions, etc.

5.3 Local Dependence on the Global Mosaic

Chance is a familiar example of a property that is attributed locally, but that depends on the global mosaic. Consider a typical simplified example: if a coin has

---

[3] See Loew and Jaag (2020) for a discussion of this worry and for their own proposed solution.

[4] We set aside entanglement worries. For more on this, see Miller (2014) and Bhogal and Perry (2017).

[5] For the canonical exposition of perfect naturalness, see Lewis (1983). Note that perfectly natural properties, or their bearers, need not be robustly fundamental in a sense that commits Humeans to the thesis that these elements are, for example, *metaphysically prior grounds* of the global mosaic. See Barry Loewer (Chapter 6 in this volume) for an example of his 'Package Deal Account' that does not posit perfectly natural properties as metaphysically prior grounds.

[6] These perfectly natural properties may turn out to be magnitudes with further structure. See Hall (2015), Wilson (2012), Eddon (2013), and Eddon and Meacham (2015).

a 50% chance of landing heads on its next toss, this is in virtue of the fact that its microstructure and environment are part of a global pattern of tosses in which roughly half of those coins land heads and roughly half land tails.[7] Thus, the local fact about the coin toss's chance depends on the global distribution of features. In order for the toss to have the chance it does, the rest of the universe has to exhibit certain patterns. The very same coin toss, embedded in different patterns, would have a different chance of landing heads.

Dispositional properties are also locally attributed properties that depend on the global mosaic. For example, to say that a chunk of salt is soluble is to say that it has a causal profile: in certain kinds of environments, given the right circumstances, it will dissolve.[8] But, for standard Humeans, this causal profile is determined by the patterns in the global mosaic; it is only because the laws of nature systematize patterns of dissolving that the intrinsic character of *this particular chunk of salt* can be ascribed the dispositional property of being soluble.[9] This local dependence on the global can be extended further. Consider, for example, Ned Hall's (ms) observation that nothing prevents the Humean from taking the seemingly fundamental property of charge as arising from a simple, informative systematization of particle positions.[10]

The important point is that the Humean takes some things as metaphysically fundamental and takes everything else as (ultimately) derived from patterns instantiated by those fundamental things.[11] The upshot is that those further features—be they non-fundamental properties, chances, causes, dispositions, functions, etc.—while they may be instantiated locally, nevertheless depend on the global distribution. Thus, all non-fundamental properties—including special science kinds like being a predator, or being a rabbit, or being ill, or being a mother—depend on the global patterns of the Humean mosaic. For instance, if there are

---

[7] We assume the coin is fundamentally chancy. If the coin's chances are explained in terms of further microstructure and further, more fundamental, laws and chances (rather than by the global pattern of heads-lands), the argument may not go through (it will depend on the details of that further structure). Sequences of 'fundamental' coin flips are paradigmatic examples in discussions of undermining chances, but we could just as well be considering, say, sequences of stochastic decay events among fundamental particles.

[8] We bracket the issue of chance—salt is not guaranteed to dissolve. We assume that if the Humean has a satisfactory theory of chance and a satisfactory theory of dispositions, they can be unproblematically combined.

[9] Cohen and Callender (2009) and Schrenk (2008, cf. Chapter 8 in this volume) have developed Humean accounts that take each scientific domain to have its own systematization. We think this makes our challenge particularly difficult, since there are fewer resources to appeal to from within a smaller domain. Nevertheless, we argue that even the Mentaculus, whose domain (in a sense) encompasses an infinity of trajectories, is likewise unable to answer our challenge.

[10] Miller (2014) suggests applying this sort of strategy to quantum states, particularly in the context of Bohmian mechanics; Bhogal and Perry (2017) discuss the strategy's general prospects for accommodating quantum entanglement in a Humean framework.

[11] More carefully: the *truthmakers* for claims about, say, the expected outcome of this chancy event or this pellet's solubility include patterns collectively instantiated by all elements of the Humean mosaic.

not enough predators in a world, then the predator patterns in the mosaic will not be strong enough to justify their inclusion in a simple law-systematization.[12]

On the Humean picture, then, even a perfect intrinsic duplicate of an actual predator *is not* a predator—it does not belong to a class that earns the status of such a kind—in a world without robust predator–prey patterns.[13] We think this has far-reaching implications for the Humean view, particularly its approach to counterfactuals.

5.4 Counterfactuals

Counterfactuals are ubiquitous in science. They state how things *would have gone* if something *had been different*. For instance, 'If there had been an additional charged particle, then the attraction on the test particle would have been stronger.' While many philosophers who work on counterfactuals are concerned with truth conditions for ordinary utterances,[14] we set aside the linguistic concerns and focus on the causal and scientific uses of counterfactuals.[15] We also set aside counter-nomic counterfactuals that ask how things would go, say, had gravity attracted according to an inverse cube law.[16]

Counterfactuals present some well-known problems for the Humean account. We survey some of those below. We go on to show that these problems turn out to be much broader than previously noted. Recall that, according to the Humean,

[12] Certainly, there is also *some* sense of 'depend' in which *anti*-Humean laws also depend on the occurrent decoration of spacetime. After all, anti-Humeans can endorse claims of this sort: if that decoration were different in such-and-such ways, then the laws would be (would have to have been) different. For anti-Humeans, though, any (true) counterfactuals like this direct us to worlds in which the laws are different because they *have* to be, in this sense: the such-and-such decoration specified in our antecedent is *inconsistent* with (even the truth of) our actual laws. Crucially, though, Humeans don't need our antecedent's such-and-such categorical decoration of spacetime to be prohibited by our laws in order for a world with that decoration to have laws that *differ* from our own—and so, it seems, various different ('local') 'non-fundamental' facts as a result.

[13] Lewis explicitly endorses the dependence of some local properties on their embedding global distributions. In the case of pain, he writes, 'The madman is in pain in one sense, or *relative to one population*. The Martian is in pain in another sense, or *relative to another population*' (Lewis 1980, p. 221; italics added).

[14] See, for instance, the large literature on the semantics of counterfactuals in dynamic contexts, including Gillies (2007), Lewis (2018), and Starr (2014).

[15] As Loewer notes, 'English counterfactuals are expressed in a number of grammatically different ways, there are many kinds of conditionals, counterfactuals are vague, they are plausibly context relative, they have Gricean implicatures and so forth. The semantics and pragmatics of ordinary counterfactuals is a messy matter. Lewis's approach, which I will follow, is to ignore most of these difficulties' (Loewer 2007, p. 308).

[16] Thus, our project is orthogonal to the suggestion of Bhogal and Perry (2017, p. 93, n. 12) that Humeans need two kinds of modality. More generally, counter-nomic counterfactuals ask what would be the case if some of our actual physical laws were violated. But the cases we are considering here plausibly do not involve any explicit violation of our actual fundamental physical laws, but only antecedents that posit 'local' facts that have implications for what the Humean laws of the world can be, such as, 'Had only one particle been massive.'

many different things—ranging from dinner parties to solubility and from preda-
tion to currencies, etc.—depend on the laws. And since the laws, in turn, depend
on the existence of robust patterns, the existence of, say, a chance event, or a dinner party, depends on a lot more than a single coin or a collection of people.
Essentially, the Humean is unable to consider counterfactuals that specify only
'local' states of affairs in these cases.

The standard Lewisian way to evaluate a counterfactual, $A \;\Box\!\!\rightarrow C$, is to consider the closest world in which the antecedent, A, obtains.[17] If the consequent, C,
also obtains in that world, then the counterfactual is true; if not, the counterfactual is false. The closeness of the world is determined contextually, but with an
emphasis placed on avoiding 'miracles' or violations of natural law. Lewis also
thought it was important for the closest world to match the actual mosaic as
much as possible. Well-known problems for this standard account have led many
philosophers to propose modifications of various sorts—most significantly,
requiring the mosaic to match the actual world's past but not its future.[18] Here, we
focus in on this idea of violations of law. Specifically, we are interested in cases
whose antecedents direct us to worlds with mosaics not systematized by the laws
of our actual world.

We first consider the case of chances, since the problem of undermining
chances for Humeans is familiar and interestingly different from the cases we
address later. Consider a world that has only one coin which will be tossed exactly
one thousand times. Now consider the following chance counterfactual:

• **Chance:**   If the first 999 tosses of this coin had landed heads, then the 1,000th
toss would have a chance of ½ of landing heads.

Intuitively, this is true. Yet, on a natural understanding of the Humean picture, it
turns out false. According to the Humean, a coin has chance ½ of landing heads
only if it inhabits a world in which approximately half of all coins like this land
heads-side up when tossed.[19] So there is *no* metaphysically possible world at all—
let alone any near one—in which 999 of 1,000 coin tosses come out heads and yet
the last remaining toss has chance ½ of landing heads-side up. The only world we
can be taken to when we consider the antecedent, 'if the first 999 tosses had
landed heads,' is a world in which the last toss is very likely—perhaps guaranteed,
given the laws there—to land heads, rendering our consequent false.

---

[17] We bracket the caveats about equally close worlds, or ever-closer worlds. See Lewis (1973) for
details.

[18] See, for instance, Elga (2000) and Bennett (2003).

[19] Note that the specifics of this derivation may differ in the details. For a frequentist Humean, a
coin has a chance of ½ just in case the ratio of heads-lands to tosses is ½. For a more sophisticated
Humean, a coin has a chance of ½ just in case the global mosaic has contents optimally systematized
(balancing simplicity, informativeness, and fit in a canonical language) by indeterministic laws with a
value of ½. We evaluate a third kind of analysis, the Mentaculus, below.

In other contexts, some Humeans have argued that holding the laws fixed, *as laws*, or *as true*, will fix problematic undermining counterfactuals.[20] Setting aside worries about how to motivate such a move, we argue that it will not help in this case, for the simple reason that the antecedent specifies a world that is not systematizable by actual laws. Thus, we cannot consistently require that the actual laws be laws in those worlds.[21] Interestingly, these proposals focus on *nested* counterfactuals, which generate a longstanding, related issue for Humean laws. Thus, even if they are successful, much more needs to be said in order to address the problems for non-nested counterfactuals we raise here.

First, consider requiring that antecedent worlds have the actual laws of nature as laws. We'll have *no* way to analyze a large class of counterfactuals. To see this, consider what happens when we apply the idea to the chance counterfactual: 'If the first 999 tosses had landed heads, then the 1,000th toss would have a chance of ½ of landing heads.' If we require that the actual laws be laws in the worlds we counterfactually consider, we will be restricted to worlds that have an even balance of heads- and tails-lands. Thus, we cannot consider a world with *only* 1,000 tosses, 999 of which land heads. The problematic world is ruled out by fiat and we are left without an analysis.

What if we merely require that the actual laws are *true* in counterfactual worlds? We would need to consider a world in which the heads frequency is near 1, and thus the Humean chances in that world are near to 1 as well, but *also*, somehow, it is *true* that the chance of heads is ½. This is a puzzling situation since, for Humeans, chances are supposed to closely track frequencies. According to Lewis, any world in which our undermining future does come to pass is one that 'contradicts the truth about present chances' (1994, p. 482). We find it implausible that incompatible chances could be truthfully ascribed to the same token event in this way.[22]

But let us grant that there is no strict, logical inconsistency here: our laws can count some single event as having chance ½ even though it figures in a heads-only global sequence. Indeed, on some competing views of chance, the chance values do not logically constrain frequencies at all.[23] Then the counterfactual, 'if

---

[20] See Dorst (2020), Loew and Jaag (2020), and Bhogal (2020).

[21] Bhogal (2020) argues we ought to give up metaphysical consistency and consider worlds in which the laws of nature are not Humean laws of nature. Even if metaphysical inconsistency were not too steep a price for many Humeans, we argue below that such a proposal is not even straightforwardly logically consistent because it is unclear what it would mean for the same toss to have both a chance of ½ as well as a chance of 1.

[22] Note that this isn't a case of reference-class relativity, where conflicting chances may be ascribed to the same event as a result of conditionalizing on different portions of the mosaic.

[23] Here is a way to approach a proposal like this. For the Humean, the space of metaphysically possible worlds includes all combinatorially possible distributions of fundamental properties across elements in or of spacetime. So, in our example, the space will include all the combinatorially possible sequences of heads and tails across any given series of flips. Following Hall (ms), we can circumscribe the nomologically possible worlds as those *consistent* with (the *truth* of) our actual laws—regardless of

the first 999 tosses had landed heads, then the 1,000th toss would have a chance of ½ of landing heads,' would be true (good), but so would, 'if the first 999 tosses had landed heads, then the 1,000th toss would be near certain to land heads' (bad), as would variants specifying any chance value at all in the consequent (really bad). Arguably, this would render such counterfactuals useless; at the very least, it would scupper the Humean project of *reducing* chances to the occurrent decoration of spacetime. We conclude that 'holding fixed the laws' does not give Humeans a way to formulate and evaluate our undermining chancy counterfactual as true.

Humeans have directed much of their efforts to showing that this is, in fact, the right result.[24] According to these Humeans, it is a defining feature of (belief about) anything that counts as chance that it stands in a certain relationship to (belief about) rational subjective credence. Allegedly, cases of undermining threaten this relationship: in such cases, Humean laws assign a non-zero probability to some course of events even though, in light of these very same laws, we are (rationally) certain that this course of events will not actually happen. Humeans' standard fix is to block any threat of mismatch between *relevant* objective probabilities and rational credences by setting both to zero—by denying that their laws assign salient non-zero chances to all self-undermining futures like the one we consider. The so-called 'New Principle' makes this explicit: we should align our credence with the *conditional* probability that our 1,000th toss lands heads, *given* that the chance of heads is ½. But according to Humeans, for the chance of heads to be ½ is, in part, for the global mosaic to be such that (roughly) half of all tosses land heads. So our credence here should be zero, matching the likelihood that a *contradiction* obtains: all 1,000 coin flips land heads, but only half do.[25]

---

whether our actual laws qualify as optimal systematizations of, and thus laws in, all these worlds. Now, our actual laws say that coin flips have chance ½ of landing heads-side up. On the proposal we are considering, the truth of this is consistent with *all* of our metaphysically possible worlds: after all, this is a claim *just* about *chances*: it doesn't explicitly say anything at all about occurrent sequences of heads or tails, nor does it even implicitly constrain (frequencies across) our various combinations of fundamental properties. Instead, perhaps we just say that any metaphysically possible intrinsic duplicate of a coin (in our world) with chance ½ of landing heads has the same chance in any other worlds, regardless of any frequencies there. (So maybe a duplicate can count as having this same chance even in a world in which objects behave entirely differently—not landing heads or tails or even flipping at all.)

[24] For discussion, see Lewis (1994) and Hall (2004).

[25] Fernandes (Chapter 4 in this volume) argues that it is a weakness of the Humean view that it cannot allow arbitrary (but unlikely) divergence between chances and frequencies since, she claims, scientific practice presupposes that such divergence is possible. Roughly Fernandes argues that any attempt to permit some crucial divergence—enough to accommodate our chancy counterfactual, for instance—will permit *too* much: it will undermine Humeans' claim that their metaphysical reductionism about chance has a key advantage when it comes to motivating the connection between (beliefs about) objective chances and subjective credences formulated in the Principal Principle (or some variant thereof). We think Humeans are wrong to claim any such advantage to begin with—see also Hall (2004). Still, there is an important sense in which our chapter is a generalization of this sort of challenge to Humeanism. Specifically, we suggest that (i) threats of undermining arise not just with chances but also with other locally ascribed but globally based features of the world; and that

To put it another way, the default Humean move is to dismiss our case as one in which information about chances is inadmissible, since it is manufactured so that we already know that there are 1,000 flips in total. Any intuition that our counterfactual about chance is true, or even coherent and well-functioning in these circumstances, is more or less inconsequential. In more realistic, more interesting cases, relevant information about chances is (more) admissible, and Humeanism can deliver all the right verdicts—purportedly, accommodating all of the *important* intuitions.[26] The situation here is reminiscent of the philosophical discussion around anti-Humean arguments from 'lonely' worlds. Anti-Humeans claim to describe (sparse) worlds that match entirely in their total categorical decorations (frequencies) while being subject to different laws, thereby disproving Humean supervenience. Humeans must dismiss any apparent possibility here as *merely* apparent. But many are happy enough to do so: by their lights, any intuition of distinct laws in such cases is inconsequential, and any suggestion that we need to accommodate such an intuition is question-begging.[27]

5.5 Mentaculus

Interestingly, by positing a mathematical measure over worlds, a new proposal called the *Mentaculus* does have the resources to evaluate chance counterfactuals like ours. In this section, we show how. Unfortunately, the strategy does not generalize, so we will go on to show how other kinds of counterfactuals (even some chancy ones) remain in need of further resources. Importantly, simply *dismissing* the alleged data is not such an easy option for our other cases.

(ii) Humeans need to be able to accommodate the truth of at least some counterfactuals about such features. Importantly, we are also arguing that (iii) even before Humeans try to accommodate the truth of our counterfactuals, they face the non-trivial challenge of *coherently formulating* and *entertaining* many of the counterfactuals in question. We can apply a version of our worry in (iii) for the case of chance: If chances just *are* artifacts of global patterns or frequencies, then what does (or could) it even *mean* for chances and frequencies to diverge in a world—so that, say, our 1,000th flip has chance ½ of coming up heads even though we have had 999 heads outcomes? Again, we sympathize with Lewis's (1994, p. 482) claim that, on a natural understanding of the Humean account, such a scenario is incoherent.

[26] Perhaps, for example, Humeans can recover the right rational credence for a case in which we do not know how many coin flips there are in total. For Humeans, if we know that (according to our actual laws) coin flips have chance ½ of landing heads, then we automatically know a lot about the global character of our actual world—about half of the coin flips across spacetime land heads—and so about the general character of nearby worlds 'consistent' with our laws as well. But to generate any contradiction, we have to know more about the occurrent decoration of spacetime: that there are only 1,000 flips in total, or that the frequency of heads across the first 999 flips is representative of the global frequency, etc. Without that, then knowing that the first 999 flips land heads does not threaten chance ½ for our next flip. After all, Humeans can point us to a near world in which 999 tosses land heads but the 1,000th still has chance ½ of doing so—since there are, in that world, many more than 1,000 tosses in total.

[27] For a compelling Humean response to such cases, see Beebee (2000).

The *Mentaculus* is a precise, probabilistic framework that specifies the laws of nature. It has been developed persuasively and in a great deal of detail by David Albert (2000) and Barry Loewer (2007; 2008; 2009). The Mentaculus includes three postulates:

1. Low Entropy Initial Macrostate: The actual world began in a macrostate of very low entropy.[28]
2. Statistical Postulate: Each way the actual world could have begun, consistent with the low entropy condition was equally likely—or, more precisely, the region of phase space corresponding to possible low-entropy initial conditions conforms to the Lebesgue measure.
3. Deterministic, Newtonian Microevolution: the positions and velocities of the fundamental particles evolve according to Newtonian laws.[29]

The Mentaculus describes a set of 'trajectories' or 'worlds,' each of which began in the same low-entropy macrostate that the actual world did, though each in a different particular microstate. These different worlds evolve deterministically and uniquely. Some of them share our current macrostates, and some of them do not, though none of them shares our exact microstate. Since this set of worlds is well defined in phase space, there are precise measures that correspond to various macrostates (though, as we argue below, picking out the relevant macrostates in other worlds is not straightforward). For instance, of all the worlds (restricted to the low-entropy initial state) containing macrostates of rolling a die, one half of them land in a microstate of an odd number, a sixth of them land in a microstate of showing five pips, etc. This picture, then, yields ratios for any process. Simply divide the measure of worlds with the relevant outcome by the measure of worlds with the relevant input. While the Mentaculus can be interpreted in a variety of metaphysical ways,[30] when we treat this overall picture as a Humean way of describing the actual world, we can count those ratios as 'chances' due to the role they play in science and guiding our credence. Thus, we can (in theory) derive the chance of any counterfactual event.[31] David Albert (2015, pp. 7–8) describes how this process works:

[28] Albert and Loewer emphasize that nothing about the fundamental direction of time hinges on this use of 'began,' but that issue is orthogonal to our purposes here, so we use this simpler formulation.
[29] While these assumptions are in fact false (our world is quantum-mechanical and general relativistic, not Newtonian), we will follow the practice in the literature of ignoring those complications. This is justified because the arguments concerning counterfactuals and entropy are very likely to carry over to the more complicated theories.
[30] See Demarest (2016; 2019) for further discussion.
[31] A broadly statistical mechanical approach to counterfactuals has been developed in different ways by Kutach (2002), Albert (2000), Loewer (2007), and Fernandes (forthcoming). All of these accounts rely on the background of the Mentaculus. These approaches to counterfactuals can be contrasted with linguistic accounts, such as Stalnaker (1968) and Lewis's (1973) approach and with

Start (then) with the initial macrocondition of the universe. Find the probability distribution over all of the possible exact microconditions of the universe which is uniform, with respect to the standard statistical-mechanical measure, over the subset of those microconditions which is compatible with that initial macrocondition, and zero elsewhere. Evolve that distribution forward in time, by means of the exact microscopic dynamical equations of motion, so as to obtain a definite numerical assignment of probability to every formulable proposition about the physical history of the world. And call that latter assignment of probabilities the *Mentaculus*.

The crucial feature of the Mentaculus, as far as our chancy counterfactuals are concerned, is that it posits further structure—namely, measures—over the worlds under consideration. The Humean can take those measures over sets of worlds to be chances for particular events, *regardless of whether or not those chances are how the worlds within the set themselves would be systematized*. For Humeans who embrace the Mentaculus, chances remain artifacts of laws that best systematize, and so supervene on, the totality of actual categorical facts. Importantly, though, this best system builds in its distinctive further structure when it includes our initial low-entropy macrostate among the laws.[32] This is what allows frequencies within some 'non-actual worlds' to diverge from our laws' chance ascriptions. Since a more standard Humean treatment of chance lacks this further ingredient, it lacks the resources to accommodate the possible divergence between frequencies and chances required by our counterfactuals.[33]

Consider, again, the case of a world with only 999 tosses of a coin. Plausibly, the Mentaculus includes continuously many such worlds. Plausibly, of those worlds that go on to have one more coin flip, measure ½ of them will land heads, and measure ½ will land tails.[34] This is what allows us to conclude that the chancy counterfactual is true: 'If the first 999 tosses had landed heads, then the 1,000th

---

Bennett's (2003) approach, all of which require similarity judgments and which allow for 'miraculous' exceptions to the laws of nature. Tomkow and Vihvelin (ms) have developed Bennett's account in light of statistical mechanics. While these authors provide illuminating analyses of many other issues that arise for counterfactuals—ranging from the direction of time to the fixity of the past and control of the future—they have not yet satisfactorily addressed the issues we raise here.

[32] Of course, the viability of this non-standard Mentaculus approach hinges on whether Humeans can make the case that such information earns a place in the best system of laws for our world. This issue is perhaps particularly pressing given other existing challenges for Humeans, including that of accommodating the particular sort of 'informativeness' at work behind familiar division between initial conditions and dynamical hypotheses. See Hall (2015).

[33] In principle, nothing prevents other Humeans from adding a similar structure; to our knowledge, none have done so. (For any candidate along these lines, moreover, we would want to ask this question: How *substantive* are any alleged differences between this proposal and the Mentaculus itself?) Regardless, the Mentaculus serves as a nice case study.

[34] There are many 'plausibly' clauses here because of the technical promissory notes concerning the way that trajectories evolve in a phase space with six dimensions for each particle.

toss would have chance ½ of landing heads.' We rely on the *measure over* a relevant *set* of worlds, not the chance values *at* any particular world.

Humeans are not eager to introduce metaphysical entities, and they may worry that this account is *ad hoc*—a modification of the metaphysical picture merely to render certain counterfactual judgments true. So why should a Humean countenance the plurality of Mentaculus worlds? Because these worlds are nothing more than a useful way of systematizing *actual* patterns. By supplementing our microphysical dynamics with a characterization of the low-entropy macrostate and a uniform measure, we sacrifice a little simplicity but gain a *far* more informative characterization of our actual mosaic. The alternative microphysical trajectories are a straightforward consequence of this addition. Thus, they have no independent existence; they are not on an equal footing with our world. Instead, the Mentaculus takes seriously David Lewis's characterization of other worlds as tools for evaluating counterfactuals *about our* world (1986, p. 22):

> It's the character of our world that makes some A-worlds be closer to it than others. So, after all, it's the character of our world that makes the counterfactual true.... But it is only by bringing the other worlds into the story that we can say in any concise way what character it takes to make the counterfactuals true.

John Heil makes a similar point (2013, p. 172):

> When a philosopher resorts to talk of possible worlds to support a claim about what is or might be the case, it is worth asking what feature of the universe as we have it might make the claim true. The danger is that easy talk of possible worlds screens us off from serious ontology.

When we appreciate the true purpose of the Mentaculus, we need not consider the trajectories within it as independent entities that, in turn, ground their own set of laws and to which we are beholden in our counterfactual reasoning. Rather, the trajectories take their place in the Mentaculus *only because*, together, they represent facts about the actual world and its laws in a powerful, unified way.

5.6 Undermining Generalized

Chances aren't the only things whose existence depends on the global patterns. Special science kinds do as well. Consider, for example, pH. HCl is an acid. But, at least on a natural understanding of Humeanism, acidity depends upon the existence of global patterns of acids and bases systematized by our special science of chemistry. A perfect intrinsic duplicate of some actual sample of HCl would not be an acid in a world without the same patterns. Compare Lewis's characterization of pain: 'We may say that X is in pain *simpliciter* if and only if X is in the state that occupies the causal role of pain for the *appropriate* population' (1980, p. 219). He adds, 'an appropriate population should be a natural kind—a species, perhaps.' Apparently, this characterization prevents the Humean from considering an entity's pain without *also* positing the existence of a broad pattern of beings that give rise to the kind of being it is.[35] Thus, consider the following examples of counterfactuals (which should be true, but come out false):

- **Disposition:** If this chunk of salt were the only macroscopic solid, it would be soluble.
- **Special Science Kind:** If this solution of HCl contained the only complex molecules, it would be an acid.
- **Functional Category:** If a microphysical duplicate of this person (who happens to be in pain from burning his hand) were to coalesce out of thermal equilibrium, he would be in pain.[36]
- **Non-Fundamental Mass:** If there had been a single particle, it would have had mass.

Each of these antecedents directs us to some metaphysically possible world in which the Humean mosaic is much sparser than our actual one, yielding laws very different from our own. This makes trouble for Humeans, since—at the very least—it is not clear how they can recover the intuitive truth of the consequents. In a world without enough patterning to support the existence of non-fundamental science kinds, the relevant regularities do not hold. This problem has been emphasized in the context of nested counterfactuals. For instance, the above can be marshaled in support of these nested counterfactuals (which should be true, but which come out false):

- **Nested Disposition:** If there were only one macroscopic solid (an intrinsic duplicate of this salt), then if it were placed in water, it would dissolve.
- **Nested Special Science Kind:** If there were only one solution of complex molecules (an intrinsic duplicate of a solution of HCl), then if it were measured with a pH strip, it would have a pH of 3.
- **Nested Functional Categories:** If a microphysical duplicate of this person (who happens to be in pain from burning his hand) were to coalesce out of

[35] See Hawthorne (2004) and Weatherson (2007) for more detailed discussion of this point.
[36] If there are worries about whether *all* mental states, including pain, are essentially representational, we can add to the antecedent enough causal history (such as that the being coalesces out of thermal equilibrium inside of a room with a hot stove which burns his hand) for the pain state to be representational, but not enough to justify the inclusion of laws about pain in the Humean lawbook. See Davidson (1987), and the resultant literature—much too large to properly canvass here.

thermal equilibrium, then if he were asked how he felt, he would say that he
is in pain.

- **Nested Non-Fundamental Mass:** If there had been a single particle, then if
  there had been a second one, they would have been gravitationally attracted.

Much has been written on these nested counterfactuals in the context of
Humeanism. What we would like to emphasize is that the problems of nested
counterfactuals stem from a very general feature of Humeanism: our actual laws,
and their associated global patterns, are surprisingly essential to the intuitively
'local' affairs described by typical counterfactuals. Generally, any non-fundamental
feature under consideration in a Humean counterfactual will require radically
global conditions be met for plausible truth-values. If those global features are
ruled out by particular features of the antecedent, then the Humean likewise has
to relinquish the non-fundamental properties, kinds, dispositions, functions, etc.,
that relied on them. We now turn our attention to a deeper worry that threatens
any attempt to rescue Humeanism from the above problems.

5.7 How to Identify Non-Fundamental Things

Recall Albert's claim that the Mentaculus yields a definite numerical probability
for, 'every formulable proposition about the physical history of the world' (2015,
pp. 7–8).[37] We now argue that such statements are surprisingly difficult to formulate and entertain for worlds that do not share our laws. Consider the statement,
'A chunk of salt dissolves in water.' How do we go about identifying whether or
not there is a chunk of salt, whether or not there is water, or whether or not one
dissolves in the other? On a natural interpretation of Humeanism, *what it is* to be
salt is just to be part of a global pattern that is systematizable in a certain way. The
same goes for water, dissolving, etc. Note that it is open to the Humean to define
some macroscopic entities in *compositional* terms. Water is plausibly a case of
this kind: water is (metaphysically essentially) H_2O. However, the Humean has
emphasized the importance of global patterns in characterizing non-fundamental
kinds. It is *because* some compounds have the same causal profile that we group,
say, $NaCl$, $CaCl_2$, $K_2Cr_2O_7$, and $NaSHO_4$ together as 'salts.' Any disjunctive, compositional definition for a special science kind (that references only purely fundamental terms) will depend on the global patterns. Anyway, it is implausible that
such definitions could be provided for more than a few very simple cases of

---

[37] Plausibly, here Albert is interested in formulable *statements*, since a characterization in terms of
propositions is not obviously equivalent to the Lewisian requirement that the laws are statements formulated in terms of privileged predicates in the 'canonical' language.

special science kinds.[38] Every single non-fundamental property owes its existence to the patterns in which it features. Thus, for the Humean, if there is no pattern, there is no kind. Note that this is first and foremost a *metaphysical* problem about what it takes to have salt in a Humean world; it is only secondarily a problem of language or definitions.

This presents a problem for counterfactuals. Typical counterfactuals *presuppose* notions that depend on the actual laws of nature. It is not clear that Humeans, using Lewisian semantics for counterfactuals, even have the resources to evaluate these counterfactuals at all—it is not clear how we are to *entertain* their antecedents, never mind determine their truth-values. To see this, focus on the antecedent: 'If this chunk of salt had been the only macroscopic object to form.' Intuitively, the antecedent is pointing us to a world with just one macroscopic solid, this chunk of salt. Yet in that world there *are no* chunks of salt! The Humean could offer a translation: 'this chunk of salt' is to be understood as the actual, particular microstate that instantiates the actual chunk. Presumably, that microstructure, since it is specified in fundamental terms, could be duplicated in similar worlds lacking the robust patterns required for the existence of the kind, 'salt.' But arguably this is too restrictive. Certainly, we can intuitively consider antecedents that posit the existence of some chunk of salt or other: 'If the only macroscopic solid to form had been a chunk of salt.' We may prefer that antecedent if we are not concerned with any actually existing, particular microstructure.

Let us revisit the three approaches we considered in the case of chance to see how they fare in cases of non-fundamental features. First, we can hold the laws fixed *as laws*. Second, we can require that the actual laws be *true*. Third, we can add external structure to the set of worlds under consideration, as the Mentaculus does for chances. Again, requiring that the laws be held fixed *as laws* will rule out, by fiat, all of the counterfactuals we are considering. Perhaps the Humean can say something to minimize the importance of such counterfactuals. But note that the vast majority of counterfactuals we care about in ordinary contexts, and in many scientific ones, presuppose or explicitly appeal to non-fundamental properties. Furthermore, many of these counterfactuals concern local, specific states of affairs. It is at least highly counterintuitive to think that a counterfactual about a single chunk of salt's solubility requires the existence of a large number of such chunks (in order to secure the patterns underlying the single chunk's 'saltiness').

What of the proposal that we require the laws be true? It is not even clear what it is for a statement to be true if it appeals to kinds of things that *by that world's own lights* do not exist, such as, 'A chunk of salt dissolves in water.' This seems no less problematic for law statements, such as, 'Salts dissolve in water.' The Humean could, of course, declare that such laws are trivially true, but this seems to come at

---

[38] See, for instance, Fodor (1974).

the cost of making all laws trivially true, including, 'Chunks of schmalts dissolve in schmwater.' Worse, if 'negative' laws can be laws too, we'll arrive at a contradiction. 'Salts do not dissolve in water' could be a trivially true law equivalent to: it's not the case that: 'Salts dissolve in water.' More generally, the challenge here extends well beyond manufactured 'lonely' or 'sparse' scenarios that some Humeans readily discount.

- **Dispositional Kind:** If a drop in blood sugar didn't produce a spike in cortisol and adrenaline, people wouldn't get hangry.
- **Special Science Kind:** If giraffes were, on average, 6 inches shorter, then far fewer mammals would die from lightning strikes.
- **Functional Kind:** If there were no predators, then prey's rate of reproduction would be a direct function of food availability.

Maybe these counterfactuals are true, maybe not. But they are coherent, well-functioning, and plausibly evaluable. For Humeans, though, it is simply not clear *where* their antecedents direct us—or, indeed, whether there are any relevant worlds in which their antecedents obtain.

Unfortunately, the Mentaculus does not help with *this* problem. The Mentaculus provides well-defined chance values for ratios of well-defined macrostates on trajectories. But what it does *not* do is provide well-defined metaphysical principles for what *counts* as a macrostate of, say, a chunk of salt, or a predator. This is because the relevant macrostates in the actual world are special in virtue of the *actual patterns* they are embedded within. When we are asked to consider a measure over all of the microstates that could 'instantiate' some macrostate, M, we need a recipe that goes beyond the actual instantiations of the actual patterns. And, as we've argued, including (merely) possible patterns doesn't help identify macrostates in worlds without those patterns. Even so, the Mentaculus treatment in the case of chance does offer us insight into what an adequate Humean treatment of such cases would need to look like.

5.8 What a Humean Solution Might Be

The Mentaculus gives the Humean a good response to the case of undermining chances because it teases apart chances and frequencies for 'possible' events. The chance is not determined by facts about the frequencies (or systematization) within some particular 'possible' microphysical history, or 'possible world.' Rather, the chance is determined by a measure over microphysical realizers of our actual, initial, low-entropy macrostate, conditionalized on possible events. As a result, it equips Humeans to attribute chances to 'possible' events—by equating chances with ratios of worlds—even when these chances come apart from the frequencies within the worlds in our salient sets. In other words, it lets us appeal to the actual laws—the ones systematizing our actual world—and make associated chance attributions even when the antecedents of our counterfactuals point us to sets of worlds with different laws and chances of their own.

However, the measures that the Humean Mentaculus identifies with chances are only as well defined as the macrostates they concern. So, if the Humean is to succeed in providing an analysis of chances and counterfactuals, they need somehow to *apply*, or *appeal to*, the macroscopic properties, kinds, dispositions, etc. of the actual laws even when the actual laws aren't the laws *of* the worlds under consideration. Unlike what happens when we restrict all relevant counterfactual scenarios to ones that share our same actual laws *as laws*, this should let us keep the worlds we need for counterfactual analysis. We need a way to *apply* the dispositions, special science kinds, etc. to objects in 'possible' worlds, even when those worlds lack the global base that grounds or subvenes the dispositions, special science kinds, etc. While the suggestion of 'holding the laws fixed' points us to a kind of rigidity, it is one poorly suited to this problem because it requires too much of our 'possibilities.' Our arguments show that what is important is the ability to rigidify the *non-fundamental properties*—be they chances, dispositions, special science kinds, functional kinds, etc., without extending that rigidity to some systematization of an entire world. As we have seen, it won't work to require the actual laws to be laws, and it won't work to require the actual laws to be true—or at the very least, it is just not *clear enough* what it means to require them to be true, since, recall, some terms in the law statement will not refer to anything in too sparse a world. Rather, we need a way of using the robust patterns of the actual world in order to identify and characterize merely possible *objects* in the localized way distinctive of scientific reasoning and counterfactuals, without requiring extraneous corresponding global conditions.

Jeff King (2007, pp. 80–6) addresses an analogous worry, which arises for his own account of propositions; his response may provide Humeans with a helpful resource here. On his view, propositions do not exist in worlds without language— just as sentences, on the standard view of linguistic entities, do not exist in worlds without language users. King offers an account of how propositions can be true at linguistically impoverished worlds by developing ideas from Robert Adams (1981) and Kit Fine (1985) in the actualism literature. These authors are concerned with propositions such as: ⌜Socrates does not exist.⌝ Since the proposition is possibly true (Socrates is not a necessary being), we need a way to express that truth. However, on King's preferred view of structured content, propositions get their constituents from objects that exist. Intuitively, propositions about Socrates require the object *Socrates* in order to be truth-evaluable. Thus, King distinguishes between two notions: TRUE and **true at**. On his view, ⌜Socrates does not exist⌝ is not TRUE, but it *is* true at some worlds.[39] Intuitively, the counterfactuals we considered above are true, even though the worlds in which the antecedents hold are systematized by laws that would render them false (or meaningless). However, if we could appeal to something like King's idea, perhaps we could make precise a way to *use* the kinds, functions, and properties of the actual laws in order to evaluate propositions and counterfactuals (even at worlds lacking the requisite metaphysical structure).

We think the Mentaculus points toward the general form of a solution. Recall that the Mentaculus posits a measure over a set of worlds. This measure allows for precise calculations *as long as we are able to identify the relevant macrostates*. The numbers associated with certain events are not determined by the patterns in the world in which those events take place, but rather by the numbers in the structure placed over the worlds. Thus, in a similar fashion, perhaps there is a way to associate a possible object's dispositions, kinds, properties, functional roles, etc. with a structure placed *over* the worlds, rather than with patterns emerging wholly from within its own world.

One way to do this might be to stipulate some long disjunctions of actual and merely *possible* arrangements of fundamental properties that 'count'—or that *would* or *should* count—as, say, chunks of salt, or acids, or giraffes. In principle, this strategy could work for any 'Lewisian' Humean, since the characterization could be given in terms of perfectly natural properties and relations. However, we doubt that such a project could be carried out in a fully general way. Another related option might be to highlight relevant (non-fundamental, categorical) macrostates within the Mentaculus framework. In order for this approach to yield an attractive *Humean* proposal, there would need to be a nonmodal characterization (purely in terms of the actual mosaic) of which nonactual microstates count as instantiating a particular macrostate. An alternative might be to extract some distinctive functional role played by actual members of some relevant class, and then use that role—rather than intrinsic categorical bases—to highlight other members of the class across Lewisian possible worlds. But while we see (dimly) how this sort of strategy *might* work for chances or acids, we do not even know how to begin for, say, giraffes.

All these suggestions are extremely speculative, and working out the details will require great care. The most important point, though, is that there is a nontrivial challenge for Humeans here. Intuitively, we need something that lives in the space between microphysical duplicates of actual instances, which are too restrictive—there can be microphysical alterations to an actual chunk of salt that leave it as an instance of salt—and possible objects that instantiate a robust

---

[39] Carroll (2018) suggests a similar move, though the focus of his account is the idea that we can shift which laws we appeal to, depending on features of the conversational context.

enough possible pattern, which are too restrictive in another way—there can be fairly lonely chunks of salt. What exactly this might be is very much an open question; we are not sure whether Humeans can rise to the challenge.

References

Adams, R. (1981). 'Actualism and Thisness', *Synthese*, 49, pp. 3–41.

Albert, D. (2000). *Time and Chance*. Cambridge, MA: Harvard University Press.

Albert, D. (2015). *After Physics*. Cambridge, MA: Harvard University Press.

Beebee, H. (2000). 'The Non-Governing Conception of Laws of Nature', *Philosophy and Phenomenological Research*, 61, pp. 571–94.

Bennett, J. (2003). *A Philosophical Guide to Conditionals*. Oxford: Oxford University Press.

Bhogal, H. (2020). 'Nomothetic Explanation and Humeanism about Laws of Nature', in K. Bennett and D. Zimmerman (eds.), *Oxford Studies in Metaphysics,* 12. Oxford: Oxford University Press, pp. 164–202.

Bhogal, H., and Perry, Z. (2017). 'What the Humean Should Say about Entanglement', *Noûs*, 51, pp. 74–94.

Carroll, J. (2018). 'Becoming Humean', in W. Ott and L. Patton (eds.), *Laws of Nature*. Oxford: Oxford University Press, pp. 122–38.

Cohen, J., and Callender, C. (2009). 'A Better Best System Account of Lawhood', *Philosophical Studies*, 145: pp. 1–34.

Davidson, D. (1987). 'Knowing One's Own Mind', *Proceedings of the American Philosophical Association*, 60, pp. 441–58.

Demarest, H. (2016). 'The Universe Had One Chance', *Philosophy of Science*, 83, pp. 248–64.

Demarest, H. (2019). 'Mentaculus Laws and Metaphysics', *Principia: an international journal of epistemology,* 23, pp. 387–99.

Dorst, C. (2020). 'Why do the Laws Support Counterfactuals?' *Erkenntnis*, 87, pp. 545–66, https://doi.org/10.1007/s10670-019-00207-1.

Eddon, M. (2013). 'Quantitative Properties', *Philosophy Compass*, 8, pp. 633–45.

Eddon, M., and Meacham, C. (2015). 'No Work for a Theory of Universals', in B. Loewer and J. Schaffer (eds.), *A Companion to David Lewis*. Oxford: Wiley-Blackwell, pp. 116–37.

Elga, A. (2000). 'Statistical Mechanics and the Asymmetry of Counterfactual Dependence', *Philosophy of Science*, 68, pp. 313–24.

Fernandes, A. (forthcoming). 'Time, Flies, and Why We Can't Control the Past', in B. Loewer, E. Winsberg, and B. Weslake (eds.), *Time's Arrow and the Probability Structure of the World*. Cambridge, MA: Harvard University Press.

Fine, K. (1985). 'Plantinga on the Reduction of Possibilist Discourse', in J. Tomberlin and P. van Inwagen (eds.), *Alvin Plantinga*. Dordrecht: D. Reidel, pp. 145–86.

Fodor, J. (1974). 'Special Sciences (Or: The Disunity of Science as a Working Hypothesis)', *Synthese*, 28, pp. 97–115.

Gillies, A. (2007). 'Counterfactual scorekeeping', *Linguistics and Philosophy*, 30, pp. 329–60.

Hall, N. (2004). 'Two Mistakes About Credence and Chance', *Australasian Journal of Philosophy*, 82, pp. 93–111.

Hall, N. (2015). 'Humean Reductionism about Laws of Nature', in B. Loewer and J. Schaffer (eds.), *A Companion to David Lewis*. Oxford: Wiley-Blackwell, pp. 262–77.

Hall, N. (ms). 'Humean Reductionism About Laws of Nature', URL = <https://philpapers. org/rec/HALHRA>.

Hawthorne, J. (2004). 'Why Humeans Are Out of Their Minds', *Noûs*, 38, pp. 351–8.

Heil, J. (2013). 'Contingency', in T. Goldschmidt (ed.) *The Puzzle of Existence: Why Is There Something Rather Than Nothing?*. New York: Routledge, pp. 167–81.

King, J. (2007). *The Nature and Structure of Content*. Oxford: Oxford University Press.

Kutach, D. (2002). 'The Entropy Theory of Counterfactuals', *Philosophy of Science*, 69, pp. 82–104.

Lewis, D. (1973). *Counterfactuals*. Oxford: Blackwell.

Lewis, D. (1980). 'Mad pain and Martian pain' in N. Block (ed.) *Readings in the Philosophy of Psychology*, 1. Cambridge, MA: Harvard University Press, pp. 216–22.

Lewis, D. (1983). 'New Work for a Theory of Universals', *Australasian Journal of Philosophy*, 61, pp. 343–77.

Lewis, D. (1986). *On the Plurality of Worlds*. Oxford: Basil Blackwell.

Lewis, D. (1994). 'Humean Supervenience Debugged', *Mind*, 103, pp. 473–90.

Lewis, K. (2018). 'Counterfactual Discourse in Context', *Noûs*, 52, pp. 481–507.

Loew, C., and Jaag, S. (2020). 'Humean Laws and (Nested) Counterfactuals', *Philosophical Quarterly*, 70, pp. 93–113.

Loewer, B. (2007). 'Counterfactuals and the second law', in H. Price and R. Corry (eds.), *Causation, Physics, and the Constitution of Reality: Russell's Republic Revisited*. Oxford: Oxford University Press, pp. 293–326.

Loewer, B. (2008). 'Why There *Is* Anything Except Physics', in J. Howhy and J. Kallestrup (eds.), *Being Reduced: New Essays on Reduction, Explanation and Causation*. Oxford: Oxford University Press, pp. 149–63.

Loewer, B. (2009). 'Why Is There Anything Except Physics?', *Synthese*, 170, pp. 217–33.

Miller, E. (2014). 'Quantum Entanglement, Bohmian Mechanics, and Humean Supervenience' *Australasian Journal of Philosophy*, 92, pp. 567–83.

Miller, E. (2018). 'Local Qualities', in K. Bennett and D. Zimmerman (eds.), *Oxford Studies in Metaphysics*, 11. Oxford: Oxford University Press, pp. 224–40.

Schrenk, M. (2008). 'A Theory for Special Science Laws', in H. Bohse and S. Walter (eds.), *Selected Papers Contributed to the Sections of GAP.6, 6th International Congress of the Society for Analytical Philosophy, Berlin, September 2006*. Paderborn: Mentis, pp. 121–31.

Stalnaker, R. (1968). 'A Theory of Conditionals', in N. Rescher (ed.), *Studies in Logical Theory: American Philosophical Quarterly Monographs, 2*. Oxford: Blackwell, pp. 98–112.

Starr, W. (2014). 'A Uniform Theory of Conditionals', *Journal of Philosophical Logic*, 43, pp. 1019–64.

Tomkow, T., and K. Vihvelin (ms). 'The Temporal Asymmetry of Counterfactuals', https://philarchive.org/archive/TOMTTA-5.

Weatherson, B. (2007). 'Humeans Aren't Out of Their Minds', *Noûs*, 41, pp. 529–35.

Wilson, J. (2012). 'Fundamental Determinables', *Philosopher's Imprint*, 12, pp. 1–17.

6

Are Humean Laws Flukes?

Barry Loewer

It is widely, although perhaps mistakenly, believed that the contemporary heir to Hume's metaphysics is David Lewis. Lewis developed and defended a view he calls 'Humean Supervenience' (HS), which holds, as Hume is said to have held, that there are no necessary connections in nature.[1] According to Lewis the world consists of a distribution throughout the entirety of spacetime of instantiations of 'perfectly natural properties/quantities.'[2] Lewis tells us that perfectly natural properties/quantities are intrinsic to the points or point-sized individuals that instantiate them and are categorical. By this he means that a property instantiated in one spacetime region places no restriction on what properties can be instantiated in entirely distinct regions. So, any perfectly natural properties instantiated in distinct regions are co-possible. The assumption that all perfectly natural properties are categorical enables Lewis to formulate a principle of recombination according to which, given a spacetime, every mathematically possible way of combining instantiations of perfectly natural properties to fill the spacetime is a possible world, and every possible world is such a combination.[3] This is why perfectly natural properties earn the title 'the metaphysical joints of reality.'

Lewis calls the distribution of perfectly natural properties/quantities 'the Humean Mosaic' (HM) and says that it is up to physics to inventory the perfectly natural properties that appear in our world. His examples of perfectly natural properties are mass, charge, and spin.[4] Humean Supervenience (HS) further claims that the only fundamental relations are metrical and that all contingent truths at a world including truths about laws, counterfactuals, causation,

---

[1] Lewis called his metaphysical view 'Humean Supervenience', named after Hume, whom he called 'the great denier of necessary connections' (Lewis 1986b, p. ix). Galen Strawson (2015) has argued that this is a misnomer since according to him Hume didn't deny the existence of necessary connections but, rather, held the view that we can't know anything about them. Strawson suggests that Humean Supervenience is better thought of as Lewisian Supervenience. However, the name for Lewis's view has become so entrenched in the literature that I doubt that Strawson (or anyone else) has the power to change it.

[2] Space–time is comprised of a collection of points and the distance relations among them. Lewis considers the possibility that in addition to points there may be point-sized entities occupying points. But as Hall (2010) points out, his metaphysics is better off without them.

[3] For Lewis's principle of recombination see Lewis (1986a) and (1994).

[4] There are perfectly natural properties that are not and cannot be instantiated in the actual world. Lewis calls these 'alien properties.'

Barry Loewer, *Are Humean Laws Flukes?* In: *Humean Laws for Human Agents.* Edited by: Michael Townsen Hicks, Siegfried Jaag, and Christian Loew, Oxford University Press. © Oxford University Press 2023. DOI: 10.1093/oso/9780192893819.003.0007

objective probability, nomological necessity, the locations of mountains, and the states of economies, etc. supervene on the HM. In other words, possible worlds that completely agree on their HMs also agree with respect to their laws, counterfactuals, casual connections, chances, dispositions, and so on. Lewis's program for establishing HS is to propose and argue for accounts of laws, counterfactuals, chance, causation, and so on in terms of the spacetime distribution of perfectly natural properties. Laws play the central role in these accounts. His 'best systems account' (BSA) says that certain true propositions are laws in virtue of their being entailed by the scientifically best systematization of the Humean mosaic.[5] According to Lewis the scientifically best systematization is specified by axioms that are true and that optimally balance simplicity and informativeness. Laws are propositions that describe regularities and patterns entailed by the best system. According to the BSA it is in virtue of the systematizing role of laws that they are capable of performing the functions of laws, e.g. explaining, supporting counterfactuals and grounding causation, and so on.

The reason for appealing to simplicity and informativeness in characterizing the law-determining best system is that these are among the criteria that have been employed in the history of physics to evaluate proposals for law-specifying fundamental theories. Furthermore, it is evident why they are scientifically desirable features of a fundamental theory. Lewis suggests that the informativeness of a theory is measured in terms of possibilities excluded and seems to think of simplicity syntactically. While informativeness and simplicity are indeed virtues in a scientific system, both need to be given characterizations that are more in tune with scientific practice than Lewis's brief accounts of them. His account of informativeness is particularly unfortunate and immediately leads to problems.[6] Further, there are additional criteria that scientists appeal to in evaluating theories which should be added to criteria for systematizations. Later in this chapter I will sketch a version of the BSA that is independent of HS and there say a bit more about criteria for law-determining systems.[7]

The BSA includes probabilistic laws by letting the language in which candidate theories are formulated include terms for probability functions. By specifying

---

[5] See Lewis (1986b, Introduction) for his account of HS and his account of laws. Further discussion of Lewis on laws is in Loewer (1996, 2007, 2012). Lewis does not say exactly what further features a proposition entailed by the best system are needed to earn the title 'law', although he seems to think that laws must be generalizations and dynamical. I don't think this is quite right since there may be restrictions on initial conditions and propositions like symmetry principles that may be entailed by the best system and play the role of laws and so should be considered to be laws.

[6] The immediate problem is the threat of trivialization. If Fx is a predicate true of all and only actual entities and if individuals are world-bound as Lewis holds, then $\forall x Fx$ is true only at the actual world and so is maximally informative. Since it is also simple, it is the best systematization of the actual world, and since it entails all truths, it makes all general truths laws. Lewis's response is to restrict the language in which candidate systems are formulated to include only predicates and functions that refer to perfectly natural properties. For better ways of responding to the problem see Loewer (2020) and the last section of this chapter.

[7] See Loewer (2020), Jaag and Loew (2020), and Hicks (2018) for more science-sensitive accounts.

probabilities, a candidate system can gain a great deal of informativeness while still being relatively simple.[8] For example, consider a long sequence of the outcomes of measurements of x-spin on y-spin electrons, half of which are 'up' and half 'down'. Typically, a simple description of the sequence in a language lacking probability functions will not be very informative, and an informative description will be very complicated. But the proposition that the probability of a measurement of x-spin on a y-spin electron yields an 'up' result is 0.5 and that the measurements are independent may be both simple and informative. Lewis suggests measuring the informativeness of a theory that assigns probability in terms of 'fit', which he identifies with the probability of the actual history of the world given the theory.[9] The BSA may include laws that entail both dynamic and initial condition probabilities and so allows for probabilities whether the laws are indeterministic or deterministic.[10]

Opposed to Humean accounts of laws like Lewis's Humean BSA are accounts that in some way involve necessary connections. There are two main varieties of anti-Humean accounts: i) governing views, and ii) powers views. Each of these employs necessary connections, although differently.[11]

While talk of 'governing' echoes the theological birth in the seventeenth century of the concept of scientific law, few of its contemporary philosophical defenders make an overt appeal to theology to explicate it.[12] Rather, they understand laws to be features of reality over and above occurrent events that necessitate them by in

[8] Lewis says, 'Consider deductive systems that pertain not only to what happens in history, but also to what the chances are of various outcomes in various situations—for instance, the decay probabilities for atoms of various isotopes. Require these systems to be true in what they say about history...Require also that these systems aren't in the business of guessing the outcomes of what, by their own lights, are chance events; they never say that A without also saying that A never had any chance of not coming about' (Lewis 1994, p. 480). Lewis proposes evaluating the informativeness of a probabilistic theory in terms of the "fit" of the world on the theory, i.e. the likelihood of the world on the theory. This is problematic since it is plausible that the likelihood of the actual world on any plausible candidate theory is infinitesimal. For an alternative proposal see Loewer (2001, 2004).

[9] This may not be the best way of measuring the informativeness of a probabilistic theory. Probabilities inform by guiding credences and the informativeness of a credence may be assessed by its accuracy.

[10] It is a very attractive feature of Lewis's BSA account that can accommodate objective probabilities. At one point Lewis claimed that genuine chances are incompatible with determinism (Lewis 1980). This was because he was thinking of genuine chances as dynamic probabilities. However, the BSA is very naturally extended to systems with deterministic laws by construing probabilities over possible initial conditions or possible histories. See Loewer (2001).

[11] Governing accounts are associated with Armstrong (1983), Dretske (1977), and Tooley (1977). A related primitivist view is Maudlin's (2007). Powers views are associated with Shoemaker (1980), Ellis and Lierse (1994), Ellis (2001, 2002), and Bird (2005, 2007). A third view on which laws involve necessary connections is Marc Lange's counterfactual account on which lawful propositions are those that are stable under certain conditions (Lange 2009). For discussion of Lange's view see Loewer (2011).

[12] See Harrison (2019) for a discussion of the theological origin of the concept of laws of nature. Two philosophers who do make the connection between theology and governing explicit but for different reasons are John Foster (2005) and Nancy Cartwright (2005). Foster in *The Divine Lawmaker* argues that God's will is required to make sense of the governing role of laws, and Cartwright in 'No God; No Laws' appeals to the connection between laws and theology in her argument that there are no laws of nature and for a return to a more Aristotelian account of science.

some way governing them.[13] 'Governing' is meant to be a relation that makes laws responsible for the regularities they govern. Some proponents of governing views go a bit further, saying that a dynamical law governs by taking the state of a system (or the state of the entire universe) at a time (or on a Cauchy surface) and evolving it to subsequent states, thus forging necessary connections between the earlier and later states and connecting their account of laws with accounts of time.[14] The resulting necessary connections between states at different times are said to be nomologically necessary rather than metaphysically necessary since laws are contingent. However, the fact that a law implies its associated regularity is supposed to be metaphysically necessary.[15]

According to powers accounts properties are or possess powers whose instantiations produce necessary connections among events and laws are regularities that hold in virtue of the exercise of these powers. Since these regularities hold in virtue of the natures of the properties they connect, they are metaphysically necessary, although which powers are instantiated is contingent. Not all laws can be the result of the activity of powers since there are also laws that describe how powers compose to specify how systems evolve. For example, electrons have both gravitational and electromagnetic powers which combine in interactions with other electrons. These composition principles are in addition to the powers possessed by properties.[16]

Proponents of governing and powers views can agree with Lewis that as a matter of fact the lawful truths of our world can be systematized by a Lewisian best system, and that looking for a systematization is a good way to look for laws. But unlike Lewis's BSA they do not think that a proposition's place in a best systematization is what makes a proposition express a law. Instead, they hold that it is the fact that laws *govern* or *produce* necessary connections that makes them laws. Anti-Humeans maintain that it is because laws involve necessary connections that they are able to account for the world's patterns and regularities and to play the role of laws in explanation, supporting counterfactuals, and induction.

Most of the recent literature on the metaphysics of laws concerns arguments pro and con Humean and anti-Humean views. In this chapter I will discuss two of the most persistent objections against Humean accounts and especially against Lewis's BSA. The first objection is that Humean laws are too weak to play the

---

[13] This type of account of laws was developed by Armstrong (1983), Dretske (1977), and Tooley (1977).

[14] Tim Maudlin (2007) proposes an account along these lines. The account also forges a connection between accounts of laws and accounts of the nature of time. This connection is discussed in Loewer (2012).

[15] Lewis raised the question of why a governing law necessarily implies its associated regularity since they are distinct matters. He famously quipped that calling the relation between properties "contingent necessitation" no more supports this connection than calling someone "Armstrong" makes him have strong biceps (Lewis 1983).

[16] See Andreas Hüttemann (2014) for a discussion of this point.

explanatory role that laws play in science. The second is that Humean metaphysics makes it surprising that our world contains regularities that are systematizable by a Lewisian best system. In Galen Strawson (2015)'s words on Humean metaphysics, 'it would be a fluke' for the world to be systematizable or for there to be lawful regularities at all. Those who make these objections typically think that they are related, since it is the absence of necessary connections that is responsible both for the alleged explanatory deficiency of Humean laws and for the apparent flukiness of laws on Humeanism. Furthermore, both objections are claimed to create problems for squaring Humeanism with the rationality of inductive inferences, thus connecting Humean views about laws with Hume's famous problem of induction. Some necessitarians believe all these problems are resolved by adopting either a governing or powers account of laws.[17]

Nina Emery formulates the worry that Humean laws are explanatorily deficient in this way:

> It seems plausible, then, to think that the mosaic, in some sense, explains the laws. Why are the laws what they are? Surely, for the Humean the answer to this question must be: because the mosaic is the way it is. But again, one of the key roles of laws in science, is to explain both particular features of and patterns across the mosaic. So it seems that the Humean is committed to an explanatory circle: the laws explain features of the mosaic and the mosaic explains the laws.
>
> (Emery, 2021, p. 441)

Tim Maudlin succinctly puts the objection as follows:

> If the laws are nothing but generic features of the Humean Mosaic, then there is a sense in which one cannot appeal to those very laws to explain the particular features of the Mosaic itself: the laws are what they are in virtue of the Mosaic rather than vice versa.   (Maudlin 2007, p. 172)

The "circularity argument" advanced by Emery and Maudlin is that since Humean laws are regularities that are made true by their instances and the fact that they are true and lawful is made true by the HM which includes their instances, the Humean laws cannot turn around and explain these very instances. That would be circular explanation. They conclude that on Humean accounts laws don't explain their instances. But since they hold that laws do explain their instances, they conclude that Humean accounts are defective.

I have argued previously that the correct Humean reply to the circularity objection is to distinguish two kinds of explanation, 'metaphysical' and

---

[17] See Armstrong (1983).

'scientific.'[18] The Humean mosaic *metaphysically* explains laws since Humean laws supervene on the mosaic. But this doesn't preclude Humean laws from playing the role that laws play in *scientific* explanation. This distinction removes the circularity that is alleged to undermine explanation by Humean laws. On the face of it there are important differences between metaphysical and scientific explanations. Metaphysical explanations connect explanans with explanandum by metaphysical necessity and are synchronic.[19] Scientific explanations can connect explanans with explanandum contingently and are often diachronic. My response depends on how laws scientifically explain.

While anti-Humeans and Humeans may have similar views concerning metaphysical explanation, they have very different views about how laws scientifically explain. According to both types of anti-Humean accounts laws or the properties involved in laws explain by being in some sense responsible for regularities.[20] On the governing account laws are responsible for events by producing or constraining them.[21] On powers views the instantiation of one power is responsible for another instantiation. Since according to Lewis's Humean account the HM is responsible for which regularities are laws and causality, it is no surprise then that from the anti-Humean position explanation by Humean laws seems circular. Humeans have a very different account of how laws scientifically explain. While Humeans say that laws entail conditionals involving their instances, they do not produce or constrain them. According to Humeans there are two main ways that laws are involved in scientific explanations. One is explanation by unification and the other is explanation by backing causal relations and counterfactuals.

Explanation by unification works like this: Particular events are unified by lower-level laws that describe a salient pattern they exhibit and lower-level laws are unified by more general laws which in turn are unified by even more general laws. I don't have a general account of unification but examples are easy to find.[22] Kepler's laws unify the motions of the planets, Galileo's law of the pendulum unifies certain periodic motions, Newtonian laws of motion and gravitation unify the motions of celestial and terrestrial objects, e.g. projectiles and pendula, and a

---

[18] This reply to the circularity objection was originally made in Loewer (2012) and subsequently received much criticism and defense in, e.g., Miller (2015), Marshall (2015), Hicks and van Elswyk (2015), Hicks (2021), Bhogal (2020b), Lange (2013, 2018), and Emery (2019).

[19] Some philosophers (e.g. Schaffer (2009, 2017) and Fine (2012)) add that certain metaphysical explanations claim that the explanans *ground* the explanandum. In this case the mosaic grounds the laws.

[20] Some anti-Humeans think of laws as themselves *causes* of associated regularities. John Foster writes, "the only way of making sense of the notion of law…is by construing a law as the causing of the associated regularity" Foster (2001). This echoes the theological origin of the notion of laws on which God enforces lawful regularities.

[21] "The laws can *operate* to *produce* the rest of the Mosaic exactly because their existence does not ontologically depend on the Mosaic" (Maudlin 2007, p. 175). This makes the relation between a law and the events it governs seem to be something like causation.

[22] Accounts of unification can be found in Kitcher (1981) and Friedman (1974).

quantum theory of gravity, if there is one, will unify (and correct) Newtonian theory and quantum mechanics. The best system of the world is the system that best scientifically unifies the whole world.[23]

The second way laws are involved in scientific explanations is by backing causal relations that explain one event in terms of others. For example, the breaking of the window may be explained by the fact that the throwing of a rock caused the window to break. The causal relation between the throwing and the breaking is backed by dynamical laws. The analysis of causal claims is a controversial matter but everyone agrees that causation in some way involves laws. As long as an account of causation is compatible with HS, as Lewis's account is, Humean laws can scientifically explain by backing causal explanations. Neither their roles in unification or causation require laws to govern or be responsible for their instances, so the circularity argument is defanged.

On Lewis's BSA there is no further scientific explanation of the axioms of the world's best system.[24] Humeans have to accept that scientific explanation ends there. But there is a metaphysical explanation of why a particular axiom system is best. In a classical mechanical world Newton's laws are best because they are components of an axiom system that best scientifically systematizes the Humean mosaic of that world. The HM metaphysically explains the laws. The laws scientifically explain events in the HM by unifying them and backing causal explanations. I don't think that my reply will quell all worries about circularity. Some anti-Humeans have responded to my defense by rejecting the distinction between scientific and metaphysical explanation or by providing examples in which they claim that scientific explanation is transmitted across metaphysical explanation so as to restore the circularity.[25] Later in this chapter I will sketch a descendant of Lewis's BSA that avoids the circularity objection in a different way.

The second anti-Humean argument attempts to show that if Humeanism were true it would be a cosmic accident for the world to be systematized by a Lewisian system and so for there to be any laws at all. Arguments along these lines have been suggested by John Foster and Galen Strawson. Foster makes the point this way:

---

[23] See Duguid (2021) and Bhogal (2020a) for excellent discussion of the circularity argument and for how understanding scientific explanation as unification provides a response.

[24] A non-Humean can say that the fundamental regularities are explained by non-Humean laws that make them true, but Humeans reject this as a kind of *virtu dormitiva* "explanation" and in any case raises the question of what explains the existence of these non-Humean laws (or the powers that underlie them).

[25] For example, Lange (2013) has argued that Loewer (2012)'s defense is a distinction without a difference. According to him, Humeanism about laws still leads to circular explanation because scientific explanations are in some cases transmitted across metaphysical explanation. Lange's argument is rebutted by Dorst (2019a) and Hicks (2021).

What is so surprising about the situation envisaged – the situation in which things have been gravitationally regular for no reason – is that there is a certain select group of types, such that (i) these types collectively make up only a tiny portion of the range of possibilities, so that there is only a very low prior epistemic probability of things conforming to one of these types when outcomes are left to chance. (Foster 2004, p. 68)

And Strawson says:

One is presented with all these massy physical objects, out there in space-time, behaving in perfectly regular ways, and then one is told that there is, quite definitely, no reason at all for this regularity; absolutely nothing about the nature of reality which is the reason why it continues to be regular in the particular way in which it is regular, moment after moment, aeon after aeon. It is, in that clear sense, a pure fluke. It is, at every instant, and as a matter of objective fact, a pure fluke that state n of the world bears precisely the relation to the previous state of the world that one would expect, in line with the previous pattern of regularity.

(Strawson 2014, p. 30)

Foster and Strawson are arguing that if there is no non-Humean law responsible for a regularity, then it is a fluke that the regularity obtains, and if one thinks that there are no non-Humean laws, then one should assign very low prior probability to any regularity with unexamined instances. If it really were a consequence of Humeanism that allegedly lawful regularities are flukes, then it does seem to follow that candidate Humean laws with infinitely many instances should have 0 prior probability and so they would not be confirmable by their instances. The consequences for inductive inference are even more dire than this. It is not just, as Armstrong claimed, that on Humean accounts of law, induction is not rational but that it is positively irrational. Strawson claims that if a regularity is a fluke, the fact that so far observed instances conform to it provides no reason to expect the regularity to continue. In fact, if one thinks that a regularity is a fluke, then one has reason to expect the regularity not to continue.

Strawson seems to think that if there are no necessary connections among fundamental properties, they are randomly distributed in spacetime. It is correct that if fundamental property instantiations are randomly distributed, it would be a fluke for those regularities we think of as laws to hold. But it is a mistake to think that Humeanism implies that fundamental property instantiations are randomly distributed and a mistake to think that Humean laws are flukes. A fluke is a sequence of events that form a pattern that is unlikely without an explanation. An example is the proverbial gorilla randomly hitting a typewriter keyboard and typing the first act of *Hamlet*. If we think that typing each letter is probabilistically independent of typing another, then it is enormously unlikely that the text of

Hamlet results. And if we believed that the gorilla is striking keys at random and discover that he typed the first act, we would have no reason to think he would type the next act. But if there is an explanation—for example, the typist is not a gorilla but a theater professor in a gorilla suit—then we no longer consider the event a fluke and we would have reason to expect that if the 'gorilla' continued typing it is likely he would type Act 2. Strawson thinks that in a Humean world the existence of an apparently lawful regularity is like a gorilla typing the first act of Hamlet. A regularity that is a fluke doesn't explain its instances and doesn't inductively support generalizing it beyond its instances. It is not a law.

But Strawson's claim that Humean laws are flukes is defective. It confuses metaphysical independence with probabilistic independence. Humeans hold that fundamental property instantiations are metaphysically independent, but that is no reason to believe that they are probabilistically independent or even that they possess probabilities. The best system for a particular HM might entail laws that assign probabilities to instantiations of fundamental properties. But this probability distribution over the HM need not and typically will not entail that fundamental property instantiations are probabilistically independent. So, the fact that property instantiations are metaphysically independent does not entail that that there is an objective probability distribution on which they are probabilistically independent. And if the fundamental laws do imply that the instances of a regularity are not probabilistically independent, then of course that regularity is not a law of the BSA. The conclusion is that the distribution of properties in the HM is not a fluke and neither are the regularities entailed by its best system.

So far, I have argued that Humean laws can scientifically explain their instances and that Strawson's argument that they are 'flukes' fails due to its confusing metaphysical independence with probabilistic independence. Recently Dustin Lazarovici proposed an interesting argument related to Strawson's that also aims to show an incompatibility between Humeanism and the existence of BSA laws. According to Lazarovici it would be surprising for a Humean mosaic to be systematizable and so to have BSA laws. He claims that this is because *typical* Humean mosaics fail to have simple informative systems:

> It is typical for Humean worlds to have no Humean laws. Almost all Humean worlds are too complex to allow for any systematization. The challenge to the Humean theory is thus not to account for why we find these particular laws in our universe but why we find any laws at all. Conversely, if we live in a world that is regular enough to be described by laws of nature, the best explanation is the existence of something in the fundamental ontology that makes it so. (Lazarovici 2020)

Lazarovici suggests that while it might be a mistake to claim that if Humeanism is true then it is unlikely (objectively or subjectively) that the actual Humean mosaic is systematizable, it is nevertheless the case that systematizable mosaics are atypical and that is a problem for Humeanism. "Atypical" is a technical term adapted from work in the foundations of statistical mechanics where it refers to a behavior or property that is very infrequent.[26] An example is that violations of the second law of thermodynamics in isolated systems with sufficiently many degrees of freedom are atypical. Lazarovici argues that systematizable mosaics are atypical in the class of all mosaics since there are uncountably many mosaics (every distribution of fundamental quantities at spacetime point is a mosaic) but that there are at most only countably many scientific systems. He seems to think that this means that there are only countably many systematizable mosaics. If this were so, then he claims that that it is a problem for Lewis's BSA. Lazarovici invokes a principle which says that if a theory entails that a condition or phenomenon is atypical, then it incurs an explanatory deficit and that is a reason to reject it if there is a competing theory that explains the condition or phenomenon. I think that this principle is plausible when applied to scientific theories but less clear that it holds for metaphysical theories. In any case, he concludes that since, on the BSA, mosaics that are scientifically systematizable are atypical, the BSA should be rejected—at least if there is an alternative metaphysics on which the world's being systematizable is not atypical. He thinks that there are anti-Humean views for which this is the case.

Lazarovici's argument, like Strawson's, expresses the thought that the existence of lawful regularities in Humean worlds would be a fluke. Ingenious as his argument is, I think it is no more successful. One problem with it is that it doesn't follow from the fact that there are only countably many systems that there at most countably many scientifically systematizable worlds. For example, there are uncountably many worlds (think of all possible initial conditions) systematized by classical mechanics.[27] One can reply, as defenders of typicality do in statistical mechanics, that when comparing uncountable sets, one has to apply an appropriate measure. So, for example, on the Liouville measure the set of worlds that satisfy thermodynamics has much greater measure than the set of anti-thermodynamic worlds even though they are both uncountably infinite. There are measures relative to which the set of unsystematizable HMs is much greater than the set of systematizable HMs, but there are also measures on which the reverse holds. In the case of statistical mechanics there are reasons to hold that the Liouville measure is the appropriate one for evaluating typicality because of its relationship to the fundamental dynamics, but no reason like this is available to select a measure on HMs.

---

[26] The notion of typicality has recently been invoked in statistical mechanics and other theories in physics in order to provide an account of explanation in such theories. For example, entropic behavior is typical among energetically isolated systems. A property or behavior is typical in a reference class if almost all members of the class have the property or behavior. It is not a probability but rather a 'counting' notion. See Wilhelm (2022) for further discussion of typicality.

[27] Systems that include probabilistic laws can also systematize uncountably many worlds. Lazarovici addresses the issue of probabilistic theories, but so far as I can see he doesn't show that a probabilistic theory can't systematize uncountably many mosaics.

Furthermore, the situation with respect to atypicality of the world having a best systematization seems at least as bad for non-Humean as for Humean accounts of laws. On governing views there is a possible world corresponding to every distribution of properties and every collection of governing laws as long as the properties are distributed so that the laws are not violated. But just as systematizable HMs are said to be atypical in the class of HMs, systematizable worlds that may have governing laws are atypical in the class of all worlds that may have governing laws. This class includes all the Humean worlds that lack governing laws and in addition includes worlds with constraining laws that are enormously complicated and gerrymandered. For example, there is a member of the class that consists of an HM identical to the actual world's except this world also contains the constraining law that all emeralds are grue. The situation at first seems a bit better on a powers metaphysics. If fundamental properties are powers, then it is typical for a world to exhibit regularities. But it doesn't follow that typical worlds are systematizable since powers interacting with one another can produce arbitrarily complicated patterns of events. Of course, it is possible to avoid this consequence if it is required that the world only contains powers that combine to produce systematizable worlds, but a Humean could likewise just posit that the actual world is scientifically systematizable. It seems that non-Humean metaphysics has no advantage over Humean metaphysics when it comes to explaining why the world is systematizable.

But if our world is systematizable as fundamental physics apparently assumes it to be, it does appear to be a profound mystery that this is so. I don't think that there can be any metaphysical guarantee that dispels this mystery. However, there is an account of laws and properties that, I think, makes the systematizability of the world a little less mysterious and has a number of other advantages over both Humean and non-Humean accounts. The account I have in mind is 'the Package Deal Account of Laws and Properties' (PDA) that I have developed in a number of prior papers.[28] The PDA is a descendent of Lewis's Humean BSA, though, as we will see, it differs in a number of important respects. Here I will sketch it and explain how it handles the two objections to Humean accounts that were discussed earlier and then how it helps alleviate the mystery that the world is systematizable.

In Lewis's metaphysics perfectly natural properties/quantities and their distribution are metaphysically fundamental. Candidates for a law-determining system are formulated in a language whose atomic predicates refer to what he calls 'perfectly natural properties.' Which properties are perfectly natural is a matter of metaphysics prior to physics, although it is the job of physics to discover them. Lewis posits metaphysically prior perfectly natural properties in order to avoid

[28] Loewer (2007) and (2020).

trivialization of his BSA. In contrast, the PDA does not appeal to the existence of metaphysically prior perfectly natural properties. Its basic idea is that the world can be described in terms of many different languages that contain different predicates that claim to refer to fundamental properties/quantities. Given a candidate for a fundamental language and the totality of truths in that language, candidates for best system are compared with respect to simplicity, informativeness, and other criteria for a fundamental theory that are derived from the aims and practice of physics. The law-determining best system of the world is the package of fundamental predicates and the system formulated in terms of the language including them that optimally satisfies these criteria. Laws are generalizations entailed by this best system.[29]

Because the PDA doesn't start with a preferred language in which to formulate candidates for best system, some other account of what the best system is aiming to systematize is needed. The basic idea is that a fundamental theory is aiming to systematize macroscopic truths and regularities, especially regularities formulated in the languages of special sciences like thermodynamics and chemistry as well as truths it counts as fundamental.[30] To accomplish this a candidate for an optimal language must contain predicates that refer to properties whose distribution throughout spacetime can serve as a supervenience base for macroscopic truths and special science truths, e.g. truths of thermodynamics, chemistry, biology, etc. The distribution of properties referred to by the fundamental predicates of the optimal language play the role that perfectly natural properties play in Lewis's account. The difference between the PDA and Lewis's account is that where Lewis begins with a collection of metaphysically prior perfectly natural properties and asks for the best systematization of the distribution of their instantiations, the PDA compares alternative packages of proposed fundamental properties and systematizations of their distributions and selects the best package. The best package is the one whose distribution of fundamental properties both serves as a super-venience base for other truths and is systematizable by a theory that best satisfies the criteria for a fundamental physical theory. As on Lewis's account, these criteria include informativeness and simplicity, but as physicists actually evaluate them. There are additional criteria involved in selecting the best language-system pair. One is that it counts in favor of a proposed package the degree with which it systematizes special sciences. Another is that laws are invariant under a variety of spatial and temporal transformations. For example, the same laws apply in distinct regions of spacetime. A third is that a good fundamental theory provides the basis for relatively simple accounts of how the macroscopic emerges from or is

---

[29] If there are ties among candidate systems, then what counts as a law is system-relative.

[30] An account of special science laws similar to the PDA has been proposed by Cohen and Callender (2009). Their view is that in a special science predicates and systematization of truths employing those predicates are selected together with the aim of optimizing the satisfaction of criteria determined by the special science.

grounded in the fundamental and provides explanations of special science regularities. We should be able to understand, at least roughly if not in detail, how arrangements of fundamental properties/quantities behaving in conformity with the fundamental laws give rise to macroscopic phenomena and laws. Fourth, while a good fundamental theory may depart from our ordinary folk concepts and beliefs about the world, such departures should be justified by satisfying the other criteria. For example, special relativity departs from our ordinary folk beliefs by claiming that there are pairs of events such that there is no fact of the matter as to whether they are simultaneous. This departure is justified by special relativity satisfying other criteria for a fundamental theory, especially empirical adequacy. There are plausibly further criteria that a fundamental theory aspires to that might be learned from examining the history of proposals for fundamental theories. Of course, the extent or whether these criteria can be satisfied for the actual world is open.

The properties in the optimal package earn their title of 'fundamental' by being a supervenience base that 'grounds' instantiations of non-fundamental properties.[31] For example, an ontology of classical particles whose fundamental properties are mass and inter-particle distances and whose distribution is systematized by Newton's laws and a law of contact looks at first (if one doesn't look too closely) promising as such a theory. Macroscopic objects and their motions are identified with configurations of particles and their motions. Of course, a lot more needs to be said to develop and defend this proposal to make it plausible that metaphysically prior fundamental properties can be dispensed with and that the PDA yields credible candidates for fundamental properties and laws. Steps in that direction are taken in Loewer (2020).

One important difference between the PDA and Lewis's BSA is that the PDA doesn't include a requirement that fundamental properties are categorical. It may turn out that the optimal package includes predicates that refer to properties whose instantiations in one region exclude the instantiations of other properties in disjoint regions. Necessary connections among properties are allowed but not required by the PDA. This enables the account to accommodate properties and quantities that refer in theories in contemporary physics that seem to be individuated in terms of their connections with other properties. For example, the standard model of elementary particles and forces plausibly individuates types of particles in terms of their relations to each other.[32] Thus, the PDA is not committed either to Humeanism or to Lewis's recombination principle. The PDA laws don't involve necessary connections even though they permit them. Although

---

[31] Exactly what it is for the instantiation of one property to ground another has recently been the subject of much discussion. See Schaffer (2016) and Sider (2020).

[32] See McKenzie (2014) for examples of fundamental theories whose properties are apparently not categorical.

this violates the letter of Lewis's Humeanism, it does so in a way that should not offend Humeanism. It shares with the Humean BSA that it is in virtue of its role in systematizing, not in virtue of necessary connections, that a proposition expresses a law. We might say that the view is half-Humean.

It is worth examining, at least briefly, how the two objections to Humean accounts of laws apply to the PDA. The first, recall, was that Humean laws are incapable of sustaining explanations of their instances since those instances play a role in explaining the laws. According to proponents of this objection this renders proposed explanations by Humean laws defective because circular. My response was to distinguish scientific from metaphysical explanation and scientific explanation and claim that laws scientifically explain their instances while instances metaphysically explain the laws. There is no circularity since different kinds of explanation are at issue. Those who don't accept this distinction or the way it was employed may still find it interesting that the objection is a non-starter according to the PDA. On the PDA even though the laws supervene on the totality of fundamental truths including its instances, this totality does not metaphysically explain the law. Rather, the laws and the fundamental truths are determined together as a package. Since the mosaic as characterized by the optimal language doesn't metaphysically explain the laws there is no circularity in the laws scientifically explaining aspects of the mosaic.

The second objection to Humeanism was that since most Humean mosaics are unsystematizable, it would be an accident or fluke if the actual mosaic is systematizable. But because the PDA is not committed to metaphysically prior categorical perfectly natural properties and a principle of recombination, this argument doesn't get a foothold. First, some of the properties deemed fundamental at a world by the PDA may not be categorical and so cannot be combined in arbitrary fashion in other worlds. Moreover, even if the fundamental properties at one world are categorical, it doesn't follow that these very properties will be counted as fundamental at other worlds. So, the fact that the perfectly natural properties at w may not be systematizable at another world w* doesn't show that there aren't other properties whose instantiations are systematizable at w*. Because the ontology/properties and the system/laws are determined together as a package, there is room to make adjustments in the fundamental language/properties to aid systemizing. This makes the existence of a system of the world a bit less mysterious than the rival views, both Humean and non-Humean. I admit, though, it falls short of dispelling the mystery entirely. It doesn't follow that on the PDA the world is guaranteed to be systematizable, even if it does mitigate to an extent the feeling that we should be surprised that our world has laws.[33]

[33] Thanks to David Albert, Harjit Bhogal, Eddy Chen, Chris Dorst, Mike Hicks, Jenann Ismael, Siegfried Jaag, Dustin Lazarovici, Christian Loew, Elizabeth Miller, and Isaac Wilhelm.

Bibliography

Armstrong, D. M. (1983). *What is a Law of Nature?* Cambridge: Cambridge University Press.

Bhogal, H. (MS). "Does Anything Explain the Regularity of the World? https://harjitbhogal.com/Does%20anything%20explain%20the%20regularity%20of%20the%20world%20Website.pdf.

Bhogal, H. (2020a). 'Humeanism about laws of nature', *Philosophy Compass*, 15(8): pp. 1–10.

Bhogal, H. (2020b). 'Nomothetic Explanation and Humeanism about Laws of Nature', *Oxford Studies in Metaphysics*, 12, ch. 6.

Bird, A. (2005). 'The Dispositionalist Conception of Laws', *Foundations of Science*, 10(4), pp. 353–70.

Bird, A. (2007). *Nature's Metaphysics*. Oxford: Oxford University Press.

Cartwright, N. (2005). 'No God; No Laws', *Dio, la Natura e la Legge*. 'God and the Laws of Nature', *Angelicum-Mondo X*, pp. 183–90.

Cohen, J., and Callender, C. (2009). 'A Better Best System Account of Lawhood', *Philosophical Studies*, 145(1): pp. 1–34.

Dorst, C. (2019a). 'Humean Laws, Explanatory Circularity, and the Aim of Scientific Explanation', *Philosophical Studies*, 176: pp. 2657–79.

Dorst, C. (2019b). 'Towards a Best Predictive System Account of Laws of Nature', *British Journal for the Philosophy of Science*, 70(3): pp. 877–900.

Dretske, F. I. (1977). 'Laws of nature', *Philosophy of Science*, 44(2): pp. 248–68.

Duguid, C. (2021). "Lawful Humean explanations are not circular" *Synthese*, 199, pp. 6039–59, https://doi.org/10.1007/s11229-021-03058-y.

Ellis, B. (2001). *Scientific Essentialism*. Cambridge: Cambridge University Press.

Ellis, B. (2002). *The Philosophy of Nature: A Guide to the New Essentialism*. Montreal: McGill-Queen's University Press.

Ellis, B., and Lierse, C. (1994). 'Dispositional Essentialism', *Australasian Journal of Philosophy*, 72, pp. 27–45.

Emery, N. (2019). 'Laws and their instances', *Philosophical Studies*, 176(6), pp. 1535–61.

Emery, N. (2021). 'Laws of Nature', in Michael J. Raven (ed.), *The Routledge Handbook of Metaphysical Grounding*. New York: Routledge, ch. 31.

Fine, K. (2012). 'Guide to Ground', in Fabrice Correia and Benjamin Schnieder (eds.), *Metaphysical Grounding*. Cambridge: Cambridge University Press, pp. 37–80.

Foster, J. (2001). 'Regularities, Laws of Nature, and the Existence of God', *Proceedings of the Aristotelian Society*, 101(1), pp. 145–61.

Foster, J. (2004). *The Divine Lawmaker: Lectures on Induction, Laws of Nature, and the Existence of God*. Oxford: Clarendon Press.

Friedman, M. (1974). 'Explanation and Scientific Understanding', *Journal of Philosophy*, 71(1), pp. 5–19.

Hall, E. (2010). 'David Lewis' Metaphysics', *Stanford Encyclopedia of Philosophy*, https://plato.stanford.edu/entries/lewis-metaphysics/.

Harrison, P. (2019). 'Laws of God or Laws of Nature? Natural Order in the Early Modern Period', in Harrison, Peter, and Roberts, Jon H. (eds.), *Science Without God? Rethinking the History of Scientific Naturalism*. Oxford: Oxford University Press, pp. 58–76.

Hicks, M. T. (2018). 'Dynamic Humeanism', *British Journal for the Philosophy of Science*, 69(4), pp. 983–1007.

Hicks, M. T. (2021). 'Breaking the Explanatory Circle', *Philosophical Studies*, 178 (2), pp. 533–57.

Hicks, M. T., and van Elswyk, P. (2015). 'Humean laws and circular explanation', *Philosophical Studies*, 172(2), pp. 433–43.

Hüttemann, A. (2014). 'Ceteris Paribus Laws in Physics', *Erkenntinis*, (2014) 79, pp. 1715–28.

Jaag, S., and Loew, C. (2020). 'Making best systems *best for us*', *Synthese*, 197(6), pp. 2525–50.

Kitcher, P. (1981). 'Explanatory Unification', *Philosophy of Science*, 48(4), pp. 507–31.

Lange, M. (2009). *Laws and Lawmakers: Science, Metaphysics, and the Laws of Nature*, Oxford: Oxford University Press.

Lange, M. (2013). 'Grounding, Scientific Explanation, and Humean Laws', *Philosophical Studies*, 164(1), pp. 255–61.

Lange, M. (2018). 'Transitivity, Self-Explanation, and the Explanatory Circularity Argument against Humean Accounts of Natural Law', *Synthese*, 195 (3), pp. 1337–53.

Lazarovici, D. (2020). 'Typicality and the Metaphysics of Laws', https://core.ac.uk/download/pdf/334436574.pdf.

Lewis, D. K. (1980). 'A Subjectivist's Guide to Objective Chance', in Richard C. Jeffrey (ed.), *Studies in Inductive Logic and Probability, Volume II*. Berkeley: University of California Press, pp. 263–93.

Lewis, D. K. (1983). 'New Work for a Theory of Universals', *Australasian Journal of Philosophy*, 61(4), pp. 343–77.

Lewis, D. K. (1986a). *On the Plurality of Worlds*. Oxford: Basil Blackwell.

Lewis, D. K. (1986b). *Philosophical Papers, vol. 2*. Oxford: Oxford University Press.

Lewis, D. K. (1994). 'Humean Supervenience Debugged', *Mind*, 103(412), pp. 473–90.

Loewer, B. (1996). 'Humean Supervenience', *Philosophical Topics*, 24(1), pp. 101–27.

Loewer, B. (2001). 'Determinism and Chance', *Studies in History and Philosophy of Science Part B: Studies in History and Philosophy of Modern Physics*, 32(4), pp. 609–20.

Loewer, B. (2004). 'David Lewis's Humean Theory of Objective Chance', *Philosophy of Science*, 71(5), pp. 1115–25.

Loewer, B. (2007). "Laws and Natural Properties', *Philosophical Topics*, 35(1/2), pp. 313–28.

Loewer, B. (2011). 'Symposium on Marc Lange's *Counterfactuals all the Way Down*', *Metascience*, 20 (1), pp. 27–52.

Loewer, B. (2012). 'Two Accounts of Laws and Time', *Philosophical Studies*, 160(1), pp. 115–37.

Loewer, B. (2020). 'The Package Deal Account', *Synthese*, 199, pp. 1065–89, https://doi.org/10.1007/s11229-020-02765-2.

Marshall, D. G. (2015). 'Humean laws and explanation', *Philosophical Studies*, 172(12), pp. 3145–65.

McKenzie, K. (2014). 'In No Categorical Terms: A Sketch for an Alternative Route to Humeanism about Fundamental Laws', in M. C. Galavotti, S. Hartmann, M. Weber, W. Gonzalez, D. Dieks, and T. Uebel (eds.), *New Directions in the Philosophy of Science*, pp. 45–61, Berlin: Springer.

Maudlin, T. (2007). *The Metaphysics Within Physics*. Oxford: Oxford University Press.

Miller, E. (2015). 'Humean Scientific Explanation', *Philosophical Studies*, 172(5), pp. 1311–32.

Schaffer, J. (2009). 'On What Grounds What', in David Manley, David J. Chalmers, and Ryan Wasserman (eds.), *Metametaphysics: New Essays on the Foundations of Ontology*, pp. 347–83. Oxford: Oxford University Press.

Schaffer, J. (2016). 'Grounding in the image of causation', *Philosophical Studies*, 173(1), pp. 49–100.

Schaffer, J. (2017). 'The Ground Between the Gaps', *Philosophers' Imprint*, 17(11), pp. 1–26.

Shoemaker, S. (1980). 'Causality and Properties', in Peter van Inwagen (ed.), *Time and Cause*, pp. 109–35. Dordrecht: D. Reidel.

Sider, T. (2020). 'Ground Grounded', *Philosophical Studies*, 177(3), pp. 747–67.

Strawson, G. (2014). *The Secret Connexion: Causation, Realism, and David Hume.* Oxford: Oxford University Press.

Strawson, G. (2015). 'Humeanism', *Journal of the American Philosophical Association*, 1(1), pp. 96–102.

Tooley, M. (1977). 'The Nature of Laws', *Canadian Journal of Philosophy*, 7(4), pp. 667–98.

Wilhelm, I. (2022). 'Typical: A Theory of Typicality and Typicality Explanation', *British Journal for the Philosophy of Science*, 73(2), pp. 561–81.

7

The Package Deal Account of Naturalness

Harjit Bhogal

7.1 Introduction

In contemporary metaphysics the distinction between *natural* properties, like *mass* and *charge*, and *unnatural* properties, like *grue*, is very familiar (though, of course, not uncontroversial). The core idea, stemming from David Lewis's 'New Work for a Theory of Universals' (1983a), is that this distinction is tightly connected to a variety of other important notions. Lewis identifies this distinction, and motivates the need to accept it, by arguing that we need to distinguish natural from unnatural properties in order to give adequate accounts of intrinsicality, laws, induction, causation, explanation, reference, and other notions. Work following this has identified even more roles for naturalness to play (e.g. Sider 2011; Dorr and Hawthorne 2013). Further, Lewis, and much of the literature following, claimed that the distinction between natural and unnatural properties is primitive.

Again, since this is familiar ground, here is not the place to go into detail. What's important for our purposes is the way in which the distinction between natural and unnatural properties reflects the distinction between properties that are legitimate to use in scientific theorizing and those that are not. (Let's, for now, restrict to fundamental science. That is, to fundamental physics, since the literature on natural properties has developed in a way that closely connects natural properties to those of fundamental physics.) It seems inappropriate for our basic physical theories to be built using extremely unnatural, gerrymandered properties, and this is because of the way that such unnatural properties shouldn't be used in inductions, in scientific explanations, in laws of nature, and so on.

Here is something else that is familiar to metaphysicians, particularly to those reading this volume—Lewis also developed a *Humean* approach to laws of nature. Humeanism in metaphysics is generally understood as the denial of necessary connections between distinct existences. A ball being scarlet can necessitate it being red, in a way that is consistent with Humeanism, because scarletness and redness are not distinct existences. But there cannot be such necessary connections between entities that are distinct. Of course, laws of nature are a challenge for such a view. If it's a law that $F = ma$, then there seems to be a necessary

Harjit Bhogal, *The Package Deal Account of Naturalness* In: *Humean Laws for Human Agents*. Edited by: Michael Townsen Hicks, Siegfried Jaag, and Christian Loew, Oxford University Press. © Oxford University Press 2023.
DOI: 10.1093/oso/9780192893819.003.0008

connection between the distinct existences *force*, *mass*, and *acceleration*. Lewis's view is one which can accept the existence of such laws but deny that they imply any necessary connections.

The way this works is that we identify a fundamental base of the world that does not contain necessary connections and then we give an account of how the laws reduce to this base.[1] Lewis calls this necessary-connection-free base the *Humean mosaic*. The Humean mosaic consists in the intrinsic physical state of each spacetime point (or each point-like object) and the spatio-temporal relations between those points. That is, the Humean mosaic is all the local property instantiations pushed up against each other in spacetime.

The laws are determined by this mosaic via the Best Systems Account (BSA). The basic idea of the BSA is that the laws are propositions that are relatively simple, but also informative about the mosaic. More precisely, consider sets of axioms. Some sets of axioms are informative about the mosaic—their deductive closure tells us a lot about the mosaic. Some sets of axioms are simple, in the sense of being syntactically simple when written down. The laws are the set of axioms that best balance simplicity and informativeness.[2] (Strictly, on Lewis's approach the laws are a subset of those axioms—the laws are the axioms which are universal generalizations. We will come back to this point later.)

Humean approaches of this kind are, as can be seen by this volume, very popular. Since Lewis, there has been a vast amount of work developing such approaches. Much of this work has been driven by the ideas of Barry Loewer—via his own published papers, but also via his huge influence on a generation of philosophers working in this area.

One particularly interesting suggestion of Loewer's is that we can combine our account of laws with our account of natural properties. The BSA claims that a best system procedure can output the laws—the laws are axioms that best systematize the mosaic. Loewer suggests that natural properties can be outputted by a similar systemization procedure. In fact, the laws and natural properties arise from the same procedure. Roughly speaking, the natural properties are the properties that are referred to in the axioms that best systematize the mosaic. Consequently, Loewer calls this the *Package Deal Account* (PDA).

Just as the BSA was a huge advance in the metaphysics of laws—whether or not you like the BSA, it did suggest a fully reductive account of laws which seemed to, broadly speaking, get the right results for what counts as a law or not—the PDA, if successful, would be a huge advance in the metaphysics of naturalness. As we noted, the standard account of naturalness is a primitivist one—there is a basic

---

[1] Strictly, Lewis doesn't use the notion of reduction; rather, his claim is that the laws supervene on the mosaic. Though modern Humeans typically do talk about the laws reducing to, or being grounded in, or holding in virtue of, the Humean mosaic.

[2] Another dimension of goodness of a system—'fit'—becomes relevant if we are considering probabilistic laws. We will simplify things by only focusing on the non-probabilistic case in this chapter.

distinction between natural and unnatural properties. A fully reductive account of naturalness would be a very substantial philosophical achievement.

However, the PDA is somewhat underexplored in the literature. Loewer's discussions of it in his (1996) and (2007) are only brief—although he has now developed the account in a bit more detail in his (2021). And there isn't much other written work on the account. (Eddon and Meacham 2015 are an exception—they survey a series of variants of the BSA which don't commit to a primitive notion of naturalness, including the PDA.) Further, Loewer's developments of the PDA differ in some substantial ways from each other—his thinking seems to have developed between 1996 and today.

Even though the account is underexplored, it's still very influential. Its influence spread not mainly via writing but rather, in the oral tradition, via informal discussion in the relevant communities of philosophers. Judging by the conversations that I have had, there is a large number of modern Humeans who think, or at least hope, that something like the PDA is the correct approach.

Given all this—the way the account is potentially important but underexplored, and the way that the ideas are influential but it isn't particularly clear how they are best developed—the PDA needs some exploration. And so, in this chapter I'm going to start this exploration. In particular, I'll map the connections between naturalness and laws on the traditional BSA and I'll consider how those connections have to be adapted in order to develop different versions of the PDA. And I'll discuss how some versions of the PDA, including, perhaps, those favored by Loewer, are an instance of a larger approach to Humeanism—one that has been very visible in the recent literature—that focuses on the role of 'ideal observers' or 'ideal scientists'.

Since this chapter is an exploration, my conclusion won't be very conclusive. But I'll suggest that these 'ideal scientists' approaches to the PDA are not successful, and while some versions of the PDA might be feasible, they don't get you quite as much as you might have started out hoping for.

7.2 The Role of Naturalness in the BSA

But let's start by going back to the traditional BSA of laws. The way I'll describe the approach won't be very strictly Lewisian—the aim here isn't to accurately represent the nuances of Lewis's views, but rather to get on the table a pretty standard version of the BSA so that we can see the ways in which the PDA must deviate.

Again, the basic idea of Humeanism is that the world consists in the mosaic of events distributed across spacetime. And the BSA says that the laws are propositions that best balance simplicity and informativeness about this mosaic. In developing this account, we can identify three distinct roles for naturalness to play.

(1) Naturalness plays a metaphysical role in characterizing the Humean mosaic. As we noted, the mosaic consists in the intrinsic physical state of each spacetime point (or each point-like object) and the spatio-temporal relations between those points. For Lewis, though, intrinsicality is very closely connected to naturalness—so closely that the Humean mosaic is, in effect, the distribution of natural properties instantiated at each spacetime point, and the spatio-temporal relations between these points. On the standard BSA the laws are reduced to this distribution of natural properties.

More generally, many people accept a connection between naturalness and fundamentality—the natural properties are fundamental; they are the basic building blocks of the world (see Tahko 2018, section 1, and Bennett 2017, section 5.7, for discussion).

So, the first role of natural properties in this picture is to help characterize the fundamental base.

(2) There is a role for naturalness to play in identifying what the relevant data are that theories need to be informative about. On the traditional BSA, a set of axioms scores better on informativeness when the axioms are informative about the mosaic—that is, about the distribution of natural properties across spacetime.

One way to put this is that there is a set of data sentences—its informativeness about those data sentences that counts toward the bestness of systems—and on the traditional BSA those data sentences are restricted to sentences that say that certain natural properties are instantiated at spacetime points. So, the data sentences are things like, 'At spacetime *point a* properties x, y, z…are instantiated.' (This, as we will see, is somewhat different from the way Lewis presented things, but it's useful to have the idea that naturalness can play a role in fixing the language of the data sentences in mind.)

There's an interesting question here about why the relevant data sentences are just about the mosaic. The basic intuition of the BSA is that the laws are those axioms that best balance simplicity and informativeness. So why would you restrict the data to the facts about the mosaic? Well, if you think that the facts about the mosaic determine all the other facts about the world, then there is at least one natural sense of informativeness where being informative about the world just consists in being informative about the mosaic. Of course, though, there are other reasonable senses of what it is to be informative about the world where a set of axioms being informative about the fundamental base does not imply that they are informative about the whole world. But, for now, we can see that the metaphysical role of naturalness in characterizing the fundamental base of the world motivates a certain conception of what the data are.

(3) The most discussed role of naturalness in the BSA is in fixing the language that the axioms can be formulated in. Part of what makes axioms better is that they are simple. And the traditional approach to measuring simplicity of axioms is syntactic—ideally there should be few axioms and they should be short. But to

measure simplicity in this way we need to fix on a particular language—the length of an axiom depends upon the language it's expressed in, and there is no reason to think that an axiom that is syntactically simple in English will be similarly syntactically simple in Urdu. So, we need to identify some privileged language.

The first step, then, is to move to a formal logical language rather than a natural language, say the language of first-order logic. But, obviously, we need to augment the logical language with non-logical vocabulary—if the language has no non-logical vocabulary then we won't be able to formulate any informative statements about the mosaic.

But here is the problem: Consider a predicate, F, that is instantiated by all and only the things that exist in the actual world (including, for example, spacetime regions and mereological fusions). Then a system with only one axiom, '$\forall x F(x)$', would be extremely simple. And also it would be, at least in one sense of informative, maximally informative because the truth of $\forall x F(x)$ rules out all non-actual worlds. But, clearly $\forall x F(x)$ should not count as a law. This is the *Predicate F Problem*. (Whether such systems really are informative is a question that we will consider more later.)

Lewis's solution is to restrict the language that the systems can be formulated in to a language where 'the primitive vocabulary...refer[s] only to the perfectly natural properties' (Lewis 1983b, p. 42). This rules out $\forall x F(x)$ because the predicate F does not refer to a perfectly natural property.

So, this is another place naturalness comes into the traditional BSA approach—it helps us fix the language that axioms are formulated in. And restricting the language in this way—to first-order logic and predicates that refer to perfectly natural properties—seems somewhat fitting given the picture of the metaphysics as just natural properties being instantiated at spacetime points.

So, again, there are three 'slots' in this approach for naturalness to play a role, (1) in characterizing the fundamental base of the world, (2) in fixing the relevant data sentences—both what the data are about and the language the sentences are formulated in, and (3) in fixing the language of the axioms.

Some Humeans, though, are unhappy with the central role that naturalness plays in this story about laws. In particular, many (but not all) Humeans are drawn to the view because of general empiricist or anti-metaphysical tendencies— for example, some Humeans want to avoid postulating necessary connections because they worry that such connections would be empirically indetectable (e.g. Earman and Roberts 2005), or perhaps because such necessary connections would be strange or 'spooky' (what Maudlin 2007, p. 71, calls 'prejudicial' motivations for Humeanism). And this suspicion towards postulating additional structure to the world carries over to natural properties. Many Humeans want to do without this appeal to a primitive distinction between natural and unnatural properties (see, for example, the views surveyed in Eddon and Meacham 2015).

In particular, Loewer (2021), following Demarest (2017), and van Fraassen (1989, p. 53), argues that the role of naturalness in the traditional BSA leads to the *mismatch problem*. Imagine that scientists have formulated something that they take to be a 'theory of everything'. The theory is widely accepted in the scientific community as the ultimate theory of the fundamental nature of the world. And the laws of such a theory very simply and informatively systematize the mosaic. According to the traditional BSA, though, it is possible that the laws of this theory of everything are not the actual laws of nature, because they are not formulated using perfectly natural properties. That is to say, the traditional BSA allows the possibility of there being a mismatch between the properties that are used in our best scientific practice and the natural properties. And the possibility of this mismatch might seem uncomfortable.

One way to put the concern is that the mismatch problem shows that the traditional BSA fails to show adequate respect for science. For example, Loewer (2021, p. 1077) says that 'it seems presumptuous for a metaphysician to say to a physicist who believes she has found a theory that optimally satisfies all the scientific criteria but not the metaphysical one that she may not have discovered the laws since the theory is not formulated in the language of perfectly natural properties.' Another way to put the concern is that the potential mismatch between science and the natural properties casts doubt upon the empirical detectability of natural properties.

Either way, though, the mismatch problem points towards some of the same doubts that Humeans have with the existence of necessary connections. Defenders of the BSA often claim that their view respects science more than anti-Humean alternatives which postulate strange entities and that there are issues with the detectability of necessary connections. The mismatch problem seems to be in the same spirit.

Because of these issues (and others that I'm not going to go into here) there is a desire among some Humeans to build a version of the BSA that doesn't appeal to a primitive distinction between natural and unnatural properties (again Eddon and Meacham 2015 provide an excellent survey of such views).

But still, there does seem to be a distinction between natural and unnatural properties. There are some properties that are legitimate to use in scientific theorizing—that play the right roles in laws, causation, induction, explanation, etc.—and some properties that are not. So, instead of just rejecting this, it would be ideal if this adapted version of the BSA could give an account of the distinction between natural and unnatural properties as well as an account of laws that doesn't take such a distinction as primitive. As we noted, that is the aim of the PDA.

Of course, though, in order for the PDA to succeed it needs to excise naturalness from the traditional BSA. We will now look at each of the roles of naturalness we just identified in turn to see how, and if, the PDA can avoid the need for naturalness to play those roles.

7.3 Developing the PDA: Role (3) and the Predicate *F* Problem

How can we develop the PDA? The idea, at its simplest, is that we adapt the BSA so that the language of the axioms is not restricted to a set of predetermined natural properties. Instead, we define the natural properties as the ones that are part of the best set of axioms—the ones that best balance simplicity and informativeness—and, further, those best axioms are the laws. That is to say, the PDA denies that natural properties play the role (3) that we identified earlier.

But now the Predicate *F* Problem rearises: If there are no restrictions on the language that the axioms can be formulated in, then it seems like a system containing the single axiom $\forall x F(x)$ would count as the best system. Consequently, $\forall x F(x)$ would be the single law of nature, and *F* would be the single natural property—clearly not the desired result.

Loewer, however, doesn't think that we need to restrict the language of the axioms in order to rule out predicates like *F*. Rather, he thinks, $\forall x F(x)$ does not count as an axiom of the best system because it is uninformative, even though its truth is inconsistent with all non-actual worlds: 'While [$\forall x F(x)$] might be maximally informative given Lewis's characterization of information as excluding alternatives this merely shows that Lewis's proposal for evaluating informativeness is not relevant to the way scientists evaluate informativeness. The information in a theory needs to be extractable in a way that connects with the problems and matters that are of scientific interest' (Loewer 2007, pp. 324–5). Similar ideas are expressed in Loewer (2021).

It's not exactly clear what this notion of 'extractability' consists in though. In his (1996), Loewer is a bit more specific. He claims we should 'measure the informativeness of a system not in terms of its content (i.e. set of possible worlds excluded) but in terms of the number and variety of its theorems' (Loewer 1996, p. 186). The discussion that follows makes it clear that 'theorem' is meant in the logical sense—systems are informative, then, in virtue of there being logical derivations starting from the axioms of the system to facts about the world.

The thought is that this conception of informativeness would rule out a system containing only $\forall x F(x)$ from being the best system. From just that axiom and logic there are very few facts about the world that we can derive. From a richer set of axioms, say the laws of Newtonian mechanics and certain background conditions, we could derive a lot more.

So, switching to this notion of informativeness seems like a sensible option for someone trying to build a PDA. But there are difficulties with this approach—perhaps these difficulties are part of the reason that Loewer was less specific about what makes for informativeness in his (2007).

Here is one concern with this approach. Say we have a fact that is informative about a particular predicate *P*. For us to be able to logically derive that fact from a set of axioms those axioms must already contain the predicate *P*. This is a problem since the PDA says that the natural properties are the ones that figure in the axioms of the best system procedure. But now it seems like just about every property that is involved in the data sentences needs to be part of the axioms of a system if that system is to be informative about the data. Consequently, the best system procedure doesn't do much to pick out the natural properties—the natural properties that result for this view will be given by the properties that the data sentences are about.

How can we deal with this problem? Here's one possible strategy in the spirit of some comments in Loewer (2021). We can suggest that the best system contains both the laws and some additional principles that do not have the status of laws—*bridge principles* that connect up the laws with other facts. Further, we can say that the natural properties are the properties that are involved in the laws—a property being involved in a bridge principle does not make it natural. This strategy opens up the possibility of there being derivations from the best system to a wide range of data sentences without it being the case that the natural properties are just determined by whatever properties the data sentences are about.[3]

A strategy along these lines sounds promising. In part because, as we mentioned in passing earlier, standard versions of the BSA already distinguish between the axioms of the best system that count as laws, and the axioms that do not count as laws. Typically, this is done to allow the best system to contain background or initial conditions, along with the laws.

Imagine our set of axioms was just the laws of Newtonian mechanics—such axioms aren't really very informative unless they are augmented with background conditions about, for example, what objects there are and their initial physical state. So, the standard BSA typically says that the axioms of the best system can sometimes include such background conditions. But, clearly, these background conditions don't count as laws.

How might we make this distinction between the axioms of the best system that count as laws and the axioms that do not? As we mentioned in Section 7.1, Lewis's approach to this was syntactic—the laws are the axioms of the best system that have the form of *universal generalizations*. The axioms that had some other logical form count as background conditions. But regardless of how we make this distinction, it seems like we need a distinction between the axioms of the best systems that are laws and the ones that are not.[4]

---

[3] Siegfried Jaag has suggested to me that a defender of the PDA might try a different strategy—claiming that the systemization procedure fixes the data language at the same time as it fixes the laws and natural properties. How exactly this can be done in a way that avoids trivialization is extremely unclear though. If not only the laws and the natural properties but also the data itself are determined internally to the PDA, then it is hard to see exactly what the problem is with systems like the one that contains only $\forall x F(x)$. After all, such a system is very informative about the facts about F.

[4] Interestingly, Loewer and Albert—in developing their influential *Mentaculus* approach—revise the BSA so as to not contain such a distinction. This is in order to allow the *Past Hypothesis*—the claim that the universe started out in a very low entropy initial condition to count as a law (see Loewer 2012; Albert 2000).

The strategy that we are considering—one that Loewer (2021) sometimes seems to accept—is to defend similar distinction in the context of the PDA. That is, claiming that in order for a system to be informative it needs to include axioms that are basic laws, but also it needs to include axioms that are bridge principles that connect the vocabulary of the basic laws to the vocabulary of the data sentences.

Such bridge principles, for Loewer, are things like principles that identify 'quartz rock at a particular location in terms of there being a certain arrangement of SiO_4 molecules' and more generally 'principles that underlie connections between fundamental and macroscopic and other non-fundamental sentences' (Loewer 2021, p. 1082). The thought, then, is that the natural properties are the properties that are part of the axioms that constitute basic laws and not those that constitute bridge principles. Thus, we can logically derive the data sentences from the axioms while still having the set of natural properties be much narrower than the properties that are involved in the data sentences.

(For Loewer, these bridge principles connect the fundamental vocabulary to facts about macroscopic states because, as we will discuss in the next section, he thinks that the data sentences are about macroscopic states. If your conception of the data sentences is different, that will affect your view of the nature of the bridge principles.)

Lots of questions arise about this strategy though. In particular, how does this version of the PDA distinguish between the axioms that are the basic laws and those that are the bridge principles? Lewis's claim that the laws are the axioms that are universal generalizations doesn't seem to help in this context.[5]

And further, even if we have a way of distinguishing the laws and the bridge principles, there are hard questions about the role of such bridge principles in the systemization procedure. For example, do axioms that constitute the bridge principles count for simplicity to the same degree as the axioms that construe the laws? If they do, then there is a concern that there will be far more bridge principles than basic laws. If bridge principles include things like principles that identify a 'quartz rock at a particular location in terms of there being a certain arrangement of SiO_4 molecules', then there will have to be similar bridge principles for huge numbers of other objects. And these bridge principles will be very complicated. Consider our current candidates for basic laws, i.e. the equations of quantum field theory. However, the principles connecting quantum field theory, say, with facts about rocks will be incredibly complicated. So, the simplicity of a system will be almost wholly determined by the simplicity of these bridge

---

[5] Neither, I think, does Hall's (2012) way of distinguishing the initial conditions part of a system and the nomic part of a system. His view is that the initial conditions and the laws play a different role with respect to simplicity and informativeness—but getting into the details of that would take us too far astray.

principles. But this seems to be the wrong result—the simplicity of the basic laws should matter a lot.

Perhaps, then, the bridge principles should not count in measuring the simplicity of a set of axioms. But then it seems that the Predicate F Problem comes back. Imagine a set of axioms where the single basic law is $\forall x F(x)$ but it's combined with lots of other axioms—complicated bridge principles connecting the predicate F to all the other facts that we care about. If the complexity of these bridge principles doesn't matter for judging the simplicity of a set of axioms, then the result is that this set of axioms really is the best system, and the property F is a natural property.

So, for this strategy to succeed, the bridge principles have to count for simplicity but not to the degree that the basic laws do. Exactly how to develop this, and how to distinguish bridge principles from the basic laws, is somewhat unclear.

But these problems came from appealing to a conception of informativeness where a set of axioms being informative involved being able to logically derive data sentences from the axioms. Loewer (1996) certainly seems to commit to this. But in his (2007) and (2021), Loewer is less committal. As we noted earlier, he uses a notion of 'extractability'—the axiom $\forall x F(x)$ is not informative because the information contained in that axiom isn't 'extractable'. But, as we noted, it's not clear what this notion of extractability comes to. Loewer doesn't give us a precise account to work with.[6]

I don't think this is an oversight or a failure on Loewer's part though. Rather, I think the reason that he doesn't give us much information about the nature of extractability or the way that $\forall x F(x)$ is uninformative is because he wants to rest upon the answers that scientific practice implicitly gives to these questions.

For example, in Loewer (2021, p. 1082), directly after discussing the way in which the axiom $\forall x F(x)$ is uninformative, he says that:

> The criteria for evaluating candidate systems are determined with an eye toward their resulting in systems that provide scientifically significant information in forms that are useable to scientists for prediction and explanation. These criteria

---

[6] In personal correspondence Loewer has suggested that his current view takes informativeness to be something like a priori deducibility (though he's undecided about exactly how this deducibility works). The overall idea is that the PDA fixes on a set of axioms, and those axioms are informative when we can a priori deduce facts about the data from those axioms. So, perhaps we can, given the axioms, a priori deduce facts about properties that are not mentioned in the axioms—thus avoiding the problem under discussion. This is an extremely interesting suggestion and there is a huge amount to say about which facts are a priori deducible from what base—see, for example, the vast and incredibly complex discussion in Chalmers (2012). Perhaps the PDA is best developed by construing informativeness as a priori deducibility, but that would require at least a whole other chapter to explore. So, for the rest of this chapter I'm going to stick to the suggestions that arise in Loewer's published work.

have been developed and refined during the history of physics since the first proposals for law specifying theories. Especially important is the extent to which a candidate system supports predictions and explanations of fundamental events and regularities, events and regularities of the special sciences and more generally of phenomena that come to be seen as important to the scientific community.

The thought, I take it, is that the precise criteria for judging the bestness of systems, and the answers to questions like the ones we have been raising about bridge principles, have been developed over the years by scientific practice (or, perhaps just the practice of physics). So, there is a sense in which we don't need a philosopher to come up with these criteria; we just need to point to those that are implicit in science. It's clear that scientists would not take $\forall x F(x)$ to be informative—so, that means that there must be some conception of informativeness, implicit in science, that rules out such an axiom. This thought runs through lots of Loewer's work on the PDA and the work of many other Humeans. We will discuss these ideas in much more detail later.

7.4 Developing the PDA: Role (2) and the Language of the Data

So, Loewer wants to deal with the Predicate F Problem by switching to a notion of informativeness on which $\forall x F(x)$ does not count as informative. As we noted, there are challenges in making sense of this notion of informativeness. But, for now, let's put this aside and assume that such a notion is available.

Even if we can, by appealing to such a sense of informativeness, avoid the result that trivial systems—like that where the only axiom is $\forall x F(x)$—count as best, there is still a concern about how the PDA could get the right results. Is it really the case that only natural properties will be involved in the axioms of the PDA? (Of course, if the PDA is true, then it is a definitional truth that only natural properties will be involved in the regularities of the PDA. The issue is about whether the properties that the PDA says are natural are good candidates for being so.)

In particular, the major concern is that the facts about the world include facts about both intuitively natural and intuitively unnatural properties. It is a fact about the world that this grass is green, but it's also a fact about the world that this grass is grue. In fact, there will be far more facts about intuitively unnatural properties than intuitively natural ones since the natural properties are a small subset of all the properties. According to the PDA, a system is informative in virtue of entailing facts about the world. But it's hard to see why the intuitively natural properties will be effective at entailing facts about predominately unnatural properties.

The obvious way to deal with this problem is by restricting the data sentences in some way. Someone developing the BSA might restrict the data sentences to only involve natural properties. But clearly this option isn't available to a defender of the PDA. Loewer, in his (1996) and his (2007), suggests two strategies for restricting the data sentences. (Loewer actually never describes exactly why such strategies are required—they are both just integrated into his description of the PDA. But I take it that the reason they are there is, at least in part, to deal with the issue we just described.)

The first strategy involves restricting the phenomena that we are interested in, and thus restricting the data sentences. The second involves restricting the language that the data sentences can be formulated in.

He explicitly appeals to both of these strategies in Loewer (2007). But let's start by considering the first strategy, restricting the phenomena that the data sentences describe.

7.4.1 Restricting to Phenomena of Interest

Again, in his (2007) Loewer suggests that the PDA should restrict the data to deal with 'truths of scientific interest'. The notion of scientific interest is a little vague; in his (1996) he has a slightly more specific proposal for how to restrict the data, in the same spirit:

> I assume that it is the job of physics to account for the positions and motions of paradigm physical objects (planets, projectiles, particles, etc.). This being so, the proposal is that we measure the informativeness of an axiom system so that a premium is put on its informativeness concerning the positions and motions of paradigm physical objects.   (Loewer 1996, p. 186; cf. Loewer 2021, p. 108)

Strictly, this doesn't tell us to restrict the data so that the only informativeness we care about is informativeness about paradigm physical objects; rather, it tells us to weight informativeness about paradigm physical objects highly, but the basic strategy is the same.

However, restricting the data in this way doesn't seem to do much to solve our problem. Even when we focus on true sentences about paradigm physical objects, there are huge numbers of true sentences about those paradigm physical objects that are stated in non-natural terms. In fact, there will be far more true sentences ascribing unnatural properties to paradigm physical objects than those ascribing natural ones. For example, for every sentence expressing that a certain paradigm object has a certain acceleration, there will be many others expressing that it has 'grueified' versions of acceleration. Restricting to a certain topic will not exclude unnatural properties from being part of the data.

7.4.2 Restricting the Language of the Data

More promising is the second strategy—restricting the language that the data sentences are formulated in. Loewer suggests this strategy in his (2007) (in combination with the strategy of the previous section), though not in his (1996). Again, the suggestion is as follows:

> Let SL be a present language of science, say scientific English (English supplemented by the languages of mathematics, fundamental physics, and the various special sciences). A candidate for a final theory is evaluated with respect to, among the other virtues, the extent to which it is informative and explanatory about truths of scientific interest as formulated in SL or any language SL+ that may succeed SL in the rational development of the science.   (Loewer, 2007, p. 325)

This strategy looks like it will do better at ensuring that the PDA only outputs laws that involve natural properties. The restriction to scientific English rules out facts about gerrymandered, very unnatural properties as being part of the data, and so it's far more plausible that the axioms that best balance simplicity and informativeness with respect to this data will be the natural properties.

But the obvious concern with this is that restricting the language of the data seems to undermine the point of the PDA. Notice that the restriction of the data to scientific English privileges a certain set of properties as special—the properties that the predicates of scientific English refer to. But the aim of the PDA all along was to pick out a set of properties as special. So, it's not clear why it is legitimate to start by privileging the properties of scientific English.

In fact, if we start by restricting the data to the language of scientific English, then we might start to wonder what the point of the whole systemization procedure is—isn't the restriction to scientific English doing all the work at picking out the right properties?

This, though, is taking the point a bit too far. Notice that there are properties that are referred to by scientific English that are not perfectly natural. For example, it's common in scientific English to talk about measurement, but *being a measurement* isn't a natural property. So even if we restrict the data to be formulated in scientific English, the best system procedure does have some work to do.

There is a reasonable version of the view, then, where we start with some privileged language in which the data are formulated—say, the language of scientific English—and then from this starting point the PDA identifies what the natural properties are. That is, the PDA tells us about the properties active in our most basic scientific theories. Of course, this view perhaps doesn't get us as far as we would have hoped—we need to start with some restriction that already rules out lots of gerrymandered, very unnatural properties; we don't generate the natural properties out of nothing. But still, the systemization procedure does do serious work.

The most obvious issue to be worked out with this approach is what exactly this initial language of the data is, and how to justify privileging this language. Loewer says that data should be formulated in scientific English or some development of scientific English, but, understood literally, this seems very implausible. Why English? Why not Cantonese? Or Telugu? Rather, we should move to some formal language. But then, of course, we still need to restrict the relevant formal language to exclude extremely unnatural properties like the predicate F, or else we have got nowhere. And exactly how we restrict the language and what the justification for such a restriction is are very hard questions—they mirror the questions that motivate the PDA approach in the first place.[7]

7.5 Ideal Scientists

It might seem like defenders of the PDA have not made much progress then—they once again face the question of how and why to restrict the systemization procedure to some privileged language. But instead of restricting the language of the axioms, like the traditional BSA, the PDA restricts the language of the data.

However, I think this isn't right. The discussion here in fact points to an interesting feature of Loewer's approach (and of the approach of many modern Humeans).

As we started discussing at the end of Section 7.3 there is a certain way in which the details we have been considering in the last two sections—of what exactly the language of the data is, of how exactly to define informativeness, of what the role of bridge principles are, and so on—don't really get to the heart of the matter, at least as Loewer develops it in his (2007) and (2021). Loewer's guiding thought, I think, is a kind of ideal observer view of naturalness—the natural properties are whatever an ideal scientist would use in their most basic physical theories.

An ideal scientist, here, isn't a scientist who just believes all and only the true theories. Rather, it's someone who ideally implements the methods that are implicit in actual science. So, take the methods for deciding what the fundamental laws are that are used in actual scientific practice, and imagine that those

---

[7] In personal correspondence Loewer suggests another option—that instead of fixing on one particular privileged language of the data, rather the data can be in any language, but the criteria which govern the bestness of systems should be invariant across translations. The thought is that actual science takes place in multiple languages, but this doesn't affect scientists' judgments about which system is best. Unless we give a substantial account of those criteria and how they can be invariant across translation, then this strategy involves appealing to the criteria that are implicit in science to solve the problem. I'll discuss this kind of strategy in the next section.

methods and their implementation are idealized in certain ways—we idealize away from our ignorance of the non-nomic facts, our limitations on computational power, and so on. On this approach, the natural properties just are the properties that a scientist ideally implementing these methods would use in her most basic physical theories.

One way to give this view some intuitive support is to note that science seems to be a very good guide to what the natural properties are. Good candidates for natural properties are things like *mass*, *charge*, and *spin*—properties that play central roles in our basic science. So, an initial idea is just to define the natural properties as whatever scientists say that they are. But this is too simple because there are many ways that scientists could be in error. They could be missing important observations that would be relevant to their theories; they could make logical or mathematical errors; they could just fail to think of relevant possibilities; and so on. So, imagine a counterfactual scientist who didn't make such errors and had all the relevant information—such a scientist would, presumably, be recognized by current scientists as being in an improved epistemic situation. The thought is that the natural properties are whatever this counterfactual scientist says they are.

This kind of ideal observer view is described clearly in Hall (2012) and seems to be the guiding idea for other modern Humeans (e.g. Jaag and Loew 2020; Dorst 2018).[8]

In fact, the traditional BSA is very commonly described in a similar way—as taking the epistemology of laws and raising the principles implicit in this epistemology to an account of the metaphysics of laws. Simplicity and informativeness are clearly relevant considerations for scientists picking out the laws—the traditional BSA makes them constitutive of the laws. For example, Carroll (1994, p. 45) says that defenders of the BSA '[let] their metaphysics be shaped by the epistemology of lawhood'. And Hall (2012) says that 'our implicit scientific standards for judging lawhood are in fact constitutive of lawhood'.

And further, as Lewis himself notes, the epistemology of laws and the epistemology of naturalness are very closely connected. In fact, Loewer adopts the phrase 'package deal' from Lewis's description of the epistemology of laws and naturalness:

> Thus my account explains...why the scientific investigation of laws and of natural properties is a package deal; why physicists posit natural properties such as the quark colours in order to posit the laws in which those properties figure, so that laws and natural properties get discovered together.   (Lewis 1983a, p. 368)

---

[8] Though there is interesting debate about exactly how idealized this ideal observer is. Hall (2012), for example, calls his ideal observer a 'limited oracular perfect physicist' and a theme in some of the work just cited is that we should understand the ideal observer to be limited and to be more 'like us' in important ways.

Although Lewis's account of the metaphysics of laws starts with a set of privileged natural properties, his account of the epistemology of laws is very different. The natural properties get discovered along with the laws as part of the ordinary scientific methods for building theories.

So, if the epistemology of laws and naturalness is a package deal, and the BSA involves taking the epistemology of laws and raising it to the level of metaphysics, then it can seem obvious that the Humean could, and should, do the same with naturalness, giving a package deal account of the metaphysics of laws and naturalness.

That this ideal observer approach is Loewer's approach is, I think, pretty clear in his (2007) and (2021). For example, he emphasizes the 'rational development of science' in his (2007), and in (2021, p. 1083) he stresses that the criteria for choosing what counts as the best system 'can be gleaned from an examination of practice in physics and the special sciences. As sciences develop these criteria may evolve and new ones develop.'

Given that this is Loewer's strategy, then a detailed and comprehensive account of the criteria for picking out the best system might not seem required. Sure, it would be nice to have an account of exactly the methods via which scientists identify the basic theory, and hence the natural properties, but even if we don't, we can still make a substantial metaphysical claim about the natural properties— that the natural properties are those that would result from an ideal implementation of scientific methods, whatever they happen to be.

In fact, Loewer (2021, p. 1083) lists nine (!) criteria that are relevant to picking out the best system, including things like: 'Many sub-systems can be treated as almost isolated so that in typical circumstances the laws apply to them neglecting their environments', and '[the system] enables predictions, explanations and understanding of a wide variety of phenomena via systematic perspicuous principles that connect fundamental to non-fundamental descriptions.' Further, Loewer (2021, p. 1082) notes that this isn't 'anything like a complete list of the criteria operative in fundamental physics for evaluating candidate systems or an account of how to balance them'. And 'All [of the criteria] are in need of clarification and elaboration' (Loewer 2021, p. 1083). It's pretty clear that Loewer doesn't intend to commit to anything particularly specific about the criteria for judging systems.

This is very different from Lewis's version of the BSA, which is clearly committed to strength and simplicity—and to particular precisifications of these notions—as being the criteria by which we judge bestness of systems.

There are two pretty substantial problems with this kind of ideal observer strategy though. Firstly, the strategy leaves it unclear why we should care about the laws and the natural properties. Sure, there is this particular practice of scientific investigation into the basic laws and natural properties. And there is an ideal implementation of the methods implicit in this practice. But why should we care about what those methods output? After all, there is the practice of astrology, and, presumably, an ideal implementation of those methods, but I take it we have no particular reason to care about that. So, what is the difference in the case of science?

Perhaps the most obvious suggestion is that the scientific methods are actually effective at getting at the facts about the laws and natural properties while astrology is not. But, notice that this move is not available to the defender of the ideal observer view because for them there is no realm of laws and natural properties that are independent of the methods of science. So, it's true that the scientific methods are effective at getting at the laws and natural properties while astrology is not, but that's only by stipulation. A defender of astrology could equally define laws and natural properties as the output of the ideal implementation of those methods, and so claim that astrology is effective at getting at the laws and natural properties while science is not.

So, there needs to be some other justification for why we care about the output of the ideal implementation of scientific methods for discovering the laws and natural properties. And, of course, there are options available. Perhaps we care about the ideal implementation of scientific methods because the laws and natural properties outputted by those methods are pragmatically useful to us— they help us achieve our epistemic and non-epistemic goals.

But, in order to give this alternative story for why we should care about the ideal implementation of scientific methods, we need to know what those scientific methods actually are. In order to argue that those methods output things that are pragmatically useful to us we must look in detail at the actual methods—we cannot remain noncommittal about those methods in the way Loewer does.

Why is this? Why can't we remain noncommittal? Can't we just argue that we know that actual science is extremely pragmatically useful, even if we can't specify exactly what methods it uses? And hence, we have reason to think that the ideal implementation of those methods will output things that are pragmatically useful.

When considering this line of thought, it's important to remember that what is currently under investigation are the basic scientific laws and the natural properties. That is what the PDA is supposed to output—the laws and properties of our most fundamental science. So, the question under consideration is why we should care about the ideal implementation of scientific methods for discovering these basic laws and natural properties.

Restricting to our basic science, then, the line of thought is that our actual basic science is extremely pragmatically useful, so we have reason to think that the ideal implementation of the methods for discovering these basic laws and natural properties is pragmatically useful.

However, it's just extremely unclear whether the actual laws and natural properties of our basic science are particularly pragmatically useful to us. We just cannot use our basic scientific theories to make predictions about, or manipulate, most of the things that we care about. It's not quantum field theory that we use to work out how to make an airplane fly or how to perform heart surgery.

Of course, science as a whole is extremely pragmatically useful to us. It outputs lots of extremely useful heuristics. And higher-level sciences, like thermodynamics, chemistry, biology, and so on, have great pragmatic value. But it's very unclear whether our basic science is especially pragmatically useful. Consequently, it's rather doubtful that the ideal implementation of our methods for outputting the basic laws and natural properties will be pragmatically useful. (Perhaps someone might argue that our current basic science is not especially pragmatically useful, but our ideal basic science will be. Maybe this is true, maybe not—I don't see much evidence one way or another about what would happen in this situation that is very far from the actual world.)

Consequently, I don't think an appeal to the utility of current science can help us much in establishing why we should care about the ideal implementation of the scientific methods for identifying the laws and natural properties of our basic sciences.

In order to say why we should care about the ideal implementation of these scientific methods, we need more details about exactly what they are and how they work. If we have such details, then, perhaps, we can give a pragmatic story about why we should care about them.

But further, once we have done this—once we have a pragmatic story for why we care about the output of ideal scientific methods—there is an additional problem: the ideal observer view seems to lose its force. Imagine a situation where the methods that output things which are pragmatically useful to us and the ones that an ideal scientist would implement are different. If the reason that we care about the methods of science is their pragmatic utility, then in such a situation we shouldn't care about what the ideal scientist says. Once we have given this pragmatic story, it is what seems to do all the work, and the appeal to the ideal observer drops out. Similar reasoning applies to any alternative story we might give for why we should care about the output of the ideal scientist.

The moral here is that we need to be committal—we need to give a substantial account of what the criteria for bestness of a system are and why we should care about what those criteria output. If we instead just rely on whatever an ideal scientist would say, then it's hard to see why we should care.[9]

---

[9] Notice that the way an ideal scientist was characterized in this discussion was as someone who ideally implements certain scientific methods. This fits with the general strategy discussed in this section of taking the methodology of actual science and raising it to a metaphysical account of laws and natural properties. We could characterize what an ideal scientist is very differently—for example, perhaps an ideal scientist is someone who generates theories that are maximally pragmatically useful to us (cf. the ideal advisor in Callender, Chapter 1 in this volume). Clearly if this is what an ideal scientist is, there is no problem in explaining why we should care about what they output. But if we characterize ideal scientists in this way, it's not at all clear what the connection is between such an imagined

The second concern with the ideal observer strategy is perhaps more direct. The idea that we can give a reductive account of the natural properties as part of the output of the ideal implementation of scientific methods assumes that we can characterize those methods without already assuming facts about naturalness. I think this is rather doubtful though.

For example, one of the criteria Loewer (2021, p. 1083) lists as being important in the scientific practice of deciding on a best theory is that the theory 'enables predictions, explanations and understanding of a wide variety of phenomena via systematic perspicuous principles that connect fundamental to non-fundamental descriptions'. This seems right—our fundamental theory should have explanatory power over a wide range of phenomena, both fundamental and non-fundamental. But the concern is that, in order to recognize whether a theory provides explanations and understanding, we already need to have some grip on the distinction between natural and unnatural properties.

In particular, many putative explanations are bad explanations and do not yield understanding precisely because the properties involved are unnatural. For example, take some particular event e that occurs this year. Imagine that the laws are deterministic so that there is some set of possible initial conditions of the universe $\{i_1, \ldots, i_n\}$, such that e occurs if and only if one of these initial conditions held. Then consider some property P that holds of the universe if and only if the initial conditions of the universe are one of $i_1, \ldots, i_n$. Given this, the fact that the universe has property P seems like it might explain the event e. After all, the event e holds if and only if the universe has property P. And the universe having property P leads to the event e occurring. Nevertheless, it's clear that this is a very bad explanation and that's because the property P is unnatural—it is just not the type of property that is appropriate for explaining.

Cases like this suggest that we can't characterize explanatory power independently of judgments of naturalness. And so, if the ideal scientific methods for deciding on a best theory involve judgments of explanatory power, then they cannot be characterized independently of judgments of naturalness.

More generally, as we noted at the start of the chapter, in the literature on naturalness it has been argued that the concept of naturalness is involved in a huge range of other concepts like similarity, intrinsicality, induction, causation, explanation, reference, and so on. When we see the wide-ranging roles that naturalness seems to play, it becomes rather implausible that we could characterize ideal scientific methods without mentioning things that presuppose naturalness.

Of course, it is open to the defender of this approach to the PDA to deny that naturalness does, in fact, play this role with respect to explanation. But then they need to give an alternative story about the badness of explanations involving the

scientist and the actual practice of science. Loewer's consistent appeals to the actual practice of science as being relevant to how the PDA works would be somewhat puzzling on this approach.

property *P*. And similarly, they could deny that naturalness plays a role with respect to similarity, induction, causation and so on, giving a story about each of those concepts that doesn't presuppose naturalness. This seems like a very hard task. It's seriously hard to see how to, for example, give a characterization of scientific practices of explanation and induction without already presupposing exactly the thing you are trying to give an account of in the PDA—that is without presupposing a distinction between the properties that it's legitimate for science to theorize in terms of and those that are not legitimate for science to theorize in terms of.

For these two reasons, the ideal observer strategy is, I think, not successful. More plausible is a strategy which is more committal—which really specifies the criteria that make theories best. Though as we noted in Sections 7.4.1 and 7.4.2, there are very big challenges in properly developing this strategy. But, I suspect, this is the path the defender of the PDA should take.

7.6 Role (1) and the Metaphysical Underpinnings of the PDA

Even if such challenges are met, there is another way in which the PDA faces challenges that do not face the traditional BSA. In particular, Lewis's version of the BSA was built upon the Humean mosaic—the mosaic, as we discussed earlier, is the distribution of perfectly natural properties that are instantiated at spacetime points. The mosaic provides the fundamental basis on which the BSA is built. But what is the metaphysical picture underlying the PDA?

On the PDA, laws and natural properties are clearly metaphysically non-fundamental, but what things are metaphysically fundamental? Perhaps the best answer for the defender of the PDA is agnosticism—the details of the underlying metaphysics don't matter so long as there is enough structure to act as truthmakers for the data sentences. This answer is suggested by some parts of Loewer (2021).

This approach does mean, however, that a lot of the traditional motivations for the BSA don't apply to the PDA. The BSA was, for Lewis, part of a project of defending a Humean view about the metaphysical basis of the world—that there are no necessary connections between distinct existences; the world is just one thing after another. The BSA gave an account of laws that followed from that view of the metaphysics.

But a PDA that is agnostic about the underlying metaphysics is not clearly a Humean view. The fundamental metaphysics might contain necessary connections, or it might not. Perhaps, in order to retain its Humean character, the PDA should not be completely agnostic about the underlying metaphysics—perhaps defenders should say that the world, at the fundamental level, does lack necessary connections. But this raises another concern: If the underlying metaphysics of the world is Humean—if, for example, it consists in just certain properties being instantiated at spacetime points with no necessary connections—then doesn't this metaphysics just provide us with a set of privileged properties? Shouldn't we just say, as Lewis does, that the natural properties are the ones that make up this underlying metaphysics?

The defender of the PDA view has to say that the answer to this question is no. Even if the underlying metaphysics privileges a set of properties, those are not the *natural* properties—they are not the properties that are part of accounts of intrinsicality, laws, induction, causation, explanation, reference, and so on. Rather, it is the properties that are the output of the PDA that have these connections. This view is, of course, defensible—in fact, it's in the spirit of Dasgupta's (2018) argument (that Loewer 2021 mentions) that even if there were some primitive natural properties they should not guide our scientific theorizing—but it's another substantial commitment for the defender of the PDA that the best candidate for the natural properties is the output of the PDA and not the properties that make up the underlying metaphysics.

7.7 Conclusion

In the traditional Humean picture, the notion of naturalness plays multiple roles—as part of the mechanics of the BSA and as giving an account of the underlying metaphysics. The PDA, in aiming to integrate a reductive account of naturalness with a reductive account of laws, has to develop an approach where naturalness doesn't play these roles. Consequently, the PDA is not just a small adaption of the traditional Humean BSA; it's a radically different, and novel, approach.

This approach faces many difficulties that the traditional BSA does not. However, I do think there are versions of the PDA that might be defensible. In particular, a version that is (i) agnostic about the underlying metaphysics and (ii) heavily restricts the language of the data sentences might be feasible. As we have noted, though, such an account still faces significant challenges. But perhaps more importantly, even if successful, this account is just far more restricted than we might have originally hoped. Such an account starts by privileging some particular language, and thus some particular properties—part of the work that we might have hoped the PDA would do has to be done before the account gets going.

In light of these problems, you might naturally want to rest more heavily on the methods implicit in science as outputting the natural properties. But, I argued, this ideal observer view is not particularly promising.

One last suggestion: Maybe the PDA is a good tool for a slightly different task than the one we have been considering so far. Loewer uses the PDA to give an account of the natural properties, where the natural properties are construed as having a very close relation to the properties of fundamental physics. This is in line with most of the literature on naturalness. But we can make a similar distinction between the natural and unnatural properties in the special sciences too. And it's not easy to see how this special science naturalness could arise from an account of fundamental-level naturalness. Perhaps the PDA could be used as an account of this higher-level, special science naturalness and not as an account of fundamental-level naturalness. So, for example, the laws and natural properties of biology could, perhaps, arise from an application of the PDA where the data are formulated in terms of the perfectly natural properties and restricted to the paradigm phenomena of biology. I suspect that building a PDA of this kind, upon a base where we assume some fundamental notion of naturalness, allows the view to be developed more smoothly and compellingly. But that's a discussion for another time.

References

Albert, D. Z. (2000). *Time and Chance*. Cambridge (MA): Harvard University Press.

Bennett, K. (2017). *Making Things Up*. Oxford: Oxford University Press.

Carroll, J. (1994). *Laws of Nature*. Cambridge: Cambridge University Press.

Chalmers, D. (2012). *Constructing the World*. Oxford: Oxford University Press.

Dasgupta, S. (2018). 'Realism and the Absence of Value', *Philosophical Review*, 127(3), pp. 279–322.

Demarest, H. (2017). 'Powerful Properties, Powerless Laws', in Jacobs, J. (ed.), *Putting Powers to Work: Causal Powers in Contemporary Metaphysics*. Oxford: Oxford University Press, pp. 38–53.

Dorr, C., and Hawthorne, J. (2013). 'Naturalness', in Bennett, K. and Zimmerman, D. (eds.), *Oxford Studies in Metaphysics, Vol. 8*. Oxford: Oxford University Press, pp. 3–77.

Dorst, C. (2018). 'Toward a Best Predictive System Account of Laws of Nature', *British Journal for the Philosophy of Science*, 70(3), pp. 877–900.

Earman, J., and Roberts, J. T. (2005). 'Contact with the Nomic: A Challenge for Deniers of Humean Supervenience about lLws of Nature, Part II: The Epistemological Argument for Humean Supervenience', *Philosophy and Phenomenological*, 71(1), pp. 1–22.

Eddon, M., and Meacham, C. J. G. (2015). 'No Work For a Theory of Universals', in Schaffer, J., and Loewer, B. (eds.), *A Companion to David Lewis*. Oxford: Blackwell, pp. 116–37.

Hall, N. (2012). 'Humean Reductionism about Laws of Nature' (unpublished manuscript), https://philarchive.org/archive/HALHRAv1.

Jaag, S., and Loew, C. (2020). 'Making Best Systems Best for Us', *Synthese*, 197, pp. 2525–50.

Lewis, D. K. (1983a). 'New Work for a Theory of Universals', *Australasian Journal of Philosophy*, 61, pp. 343–77.

Lewis, D. K. (1983b). *Philosophical Papers, Vol. I*. Oxford: Oxford University Press.

Loewer, B. (1996). 'Humean Supervenience', *Philosophical Topics,* 24(1), pp. 101–27.

Loewer, B. (2007). 'Laws and Natural Properties', *Philosophical Topics*, 35(1/2), pp. 313–28.

Loewer, B. (2012). 'The Emergence of Time's Arrows and Special Science Laws from Physics', *Interface Focus*, 2(1), pp. 13–19.

Loewer, B. (2021). 'The Package Deal Account of Laws and Properties (PDA)', *Synthese*, 199, pp. 1065–89.

Maudlin, T. (2007). *The Metaphysics Within Physics*. Oxford: Oxford University Press.

Sider, T. (2011). *Writing the Book of the World*. Oxford: Oxford University Press.

Tahko, T. (2018). 'Fundamentality', in Zalta, E. (ed.), *The Stanford Encyclopedia of Philosophy* (Fall 2018 Edition), https://plato.stanford.edu/archives/fall2018/entries/fundamentality.

van Fraassen, B. (1989). *Laws and Symmetry*. Oxford: Oxford University Press.

8

Properties for and of Better Best Systems

Markus Schrenk

8.1 Introduction

Many advocates of the Better Best System Account (BBSA) and variations thereof suggest that Lewis-style best system competitions can successfully be executed for any arbitrary but fixed set of predicates/properties. This affords the possibility to launch system analyses separately for each of the special sciences.

However, predicates/properties of these sciences are unlike the perfectly natural properties Lewis envisaged in that they are non-fundamental, maybe non-natural, and not intrinsic. Moreover, they might be dispositional, i.e. already equipped with a nomological profile. The latter fact conflicts with the idea that it is only via the BBSA that nomological facts emerge. Furthermore, the special sciences' predicates might be vague or even without extension. All these are challenges pertaining to these predicates/properties in themselves.

Further challenges arise for the BBSA when it comes to eligible sets of such properties/predicates: What are the boundaries between the different sets of properties that demarcate the sciences? Also, the BBSA is in danger of depicting the whole of science as a patchwork of unrelated, maybe even contradictory systems. Is there a unity or a hierarchy to be found after all? The latter issues concern the interrelations across separate best systems and their properties. Relating to scientific progress, there are internal issues as well: as a science develops, it hosts different sets of properties. System analyses for different property sets, however, might well be incommensurable. How can the BBSA account for this?

This chapter aims to give answers to most of these challenges.

8.2 The Starting Points for Better Best Systems

The doctrine of *Humean Supervenience* is that

> all there is to the world is a vast mosaic of local matters of particular fact, just one little thing and then another,...For short: we have an arrangement of qualities. And that is all there is. There is no difference without a difference in the arrangement of qualities. All else supervenes on that.   (Lewis 1986b, pp. ix–x)

Markus Schrenk, *Properties for and of Better Best Systems* In: *Humean Laws for Human Agents*. Edited by: Michael Townsen Hicks, Siegfried Jaag, and Christian Loew, Oxford University Press. © Oxford University Press 2023. DOI: 10.1093/oso/9780192893819.003.0009

The qualities/properties are meant to be perfectly natural and fundamental.[1] They are intrinsically instantiated by space–time points or point-sized occupants thereof.[2] They are also quiddistic: the world has no necessary or other modal connections at its fundament, there is 'just one little thing and then another'. This is the Humean core within the doctrine of *Humean Supervenience* which marks the metaphysical background thesis for Lewis's whole philosophy.[3]

If *everything else* supervenes on the mosaic of these perfectly natural properties, how, then, do *laws* supervene? Suppose you knew everything about the distribution of perfectly natural properties in space–time and you organized your knowledge in various competing deductive systems that mention only predicates which refer to perfectly natural properties. Then, in short, a true contingent generalization is a law of nature if and only if it appears as an axiom or a theorem in the one deductive system that, amongst all the competing systems, achieves the best balance of simplicity, strength, and fit. To have strength is to bear a great deal of informational content about the world; to be simple is to state everything in a concise way; and to fit is, for the probabilistic laws, to accord, as much as possible, with the actual world history.[4]

The introduction of perfectly natural properties to his *Best Systems Account* (BSA)—and, in fact, thereby to the overall doctrine of *Humean Supervenience*—came late. Formerly, Lewis

had been persuaded by Goodman and others that all properties were equal: it was hopeless to try to distinguish 'natural' properties from gruesomely gerrymandered, disjunctive properties.   (Lewis 1999, pp. 1–2; cf. Lewis 1983a)

---

[1] As an aside on fundamentality that Lewis (2009, p. 204) identified with perfect naturalness: often, the term 'fundamental' means more than 'being independent of something even more basic'; it also often means giving rise to (or being the building block of) something less basic. Theoretically, we could distinguish these two directions of fundamentality (call them 'fundamental$_\downarrow$' and 'fundamental$^\uparrow$'; cf. Schrenk 2007, p. 129, n. 3; and more recently Leuenberger 2020) and they might well come apart: Lewis's idlers (cf. Lewis 2009, p. 205) might be fundamental$_\downarrow$ in being independent of anything else but not fundamental$^\uparrow$ in that they are not constitutive for any (non-fundamental$_\downarrow$) property. The other direction is more straightforward: non-fundamental$_\downarrow$ chemical properties might be fundamental$^\uparrow$ to (non-fundamental$_\downarrow$) biological properties. This indicates, however, that an identification of perfect naturalness with fundamentality should at most be limited to being fundamental$_\downarrow$.

[2] Cf. Lewis (1986a, p. 14). For brevity, I henceforth assume supersubstantivalism about space–time and consider properties as being instantiated at space–time points or, later, as *being* sets of space–time points.

[3] On the nature of Lewis's properties, i.e. their naturalness/fundamentality, intrinsicality, and sparseness, see Lewis (1983b, pp. 13–14), Lewis (1986a, pp. 50–69), and Lewis (2009, pp. 204–5). For Lewis's most encompassing statement on quiddism (categoricalism), see Lewis (2009, pp. 208–12). Note that the only exceptional fundamental extrinsic/relational properties Lewis allows, namely spatio-temporal relations, play no direct role for my purposes.

[4] Lewis develops his theory of laws in Lewis (1973, pp. 73–4), Lewis (1999, pp. 41–3), and Lewis (1994, pp. 231–2).

Yet Lewis was persuaded to accept natural properties/predicates which, amongst other things, fix a technical bug within his BSA[5] that occurs if one allows 'gruesomely gerrymandered, disjunctive', i.e. abundant, properties. Barry Loewer brings it to the point:

> There is a problem concerning the languages in which the best systems are formulated. Simplicity, being partly syntactical, is sensitive to the language in which a theory is formulated, and so different choices of simple predicates can lead to different verdicts concerning simplicity. A language that contains 'grue' and 'bleen' as simple predicates but not 'green' will count 'All emeralds are green' as more complex than will a language that contains 'green' as a simple predicate.
> (Loewer 1996, p. 109; for a model exemplifying this trouble
> cf. Schrenk 2016, pp. 139–41)

That is, 'a single system has different degrees of simplicity relative to different linguistic formulations' (Lewis 1999, p. 42). We are even in danger of getting an artificial, unwanted winner amongst the competing systems:

> The system axiomatized by "(x)Fx" (where "Fx" is a predicate true of all and only individuals that exist at the actual world) apparently maximizes simplicity and informativeness and so counts as the best system of the world. The consequence is that all true generalizations are laws.
> (Loewer 2007, p. 324; also cf. Lewis 1999, p. 42)

At this point it seemed to Lewis that the most natural move is to give up property/predicate egalitarianism. To formulate it in nominalistic fashion: not just any set of entities is a natural property:

> Eventually I was persuaded, largely by D. M. Armstrong, that the distinction I had rejected [between natural and non-natural properties] was so commonsensical and so serviceable—indeed, was so often indispensable—that it was foolish to try to get on without it.
> (Lewis 1999, pp. 1–2; cf. Lewis 1983a; cf. Armstrong 1978)

Thus Lewis came to accept that there is an elite set of sparse natural properties within the abundant properties (where naturalness is not further analysable). Once this assumption has been made, it is only a canonical step to 'let the primitive vocabulary that appears in the axioms refer only to perfectly natural properties'

---

[5] Amongst other considerations: in Lewis (1983a) he highlights further philosophical areas where natural properties are serviceable or indispensable, next to laws and causation, for example, the content of language and thought.

(Lewis 1999, p. 42). The reasonable background assumption being that they will include neither *grue* nor the *F-property*.

At this point, proponents of the *Better Best System Account* (BBSA)[6] put forward that, in order to solve the technical bug, one does not have to assume the existence of Lewis-style 'perfectly natural properties'. Restricting the abundance of properties *in some way* and also thereby excluding grue, *F*, etc. is sufficient to yield appropriate best systems.

Many BBSA proponents seize this opportunity and suggest to launch, independently, best system analyses for different choices of fixed sets of predicates/ properties. Not of just any random sets, of course (although one might do so), but a set for the chemical properties and, separately, for the biological properties, and, separately, the economical properties (etc. for all the special sciences), and thereby gain the laws of chemistry, the laws of biology, etc. A neat side effect.

Some current proponents of a BSA also wish to change other (or additional) aspects of Lewis's original BSA: on top of exchanging the envisaged set of natural properties for other sets, some change (or add to) the parameters of simplicity, strength, and fit. Loewer, for example, suggests comprehensiveness, unity, and symmetry (cf. Loewer 2012, p. 120).[7]

Thus many features of Lewis's account—at the least, kinds of properties and competition parameters[8]—seem to be individually, modularly exchangeable.[9] In fact, his BSA can, in the light of these suggested revisions, be taken as a function with a larger number of free variables than Lewis had envisaged:

---

[6] Ideas in the spirit of the BBSA have been put forward, amongst others, by Taylor (1993, p. 97), Roberts (1998), Albert (2000), Halpin (2003), Hoefer (2007), Loewer (2007; 2012; 2021), Schrenk (2007; 2008; 2014; 2017), Cohen and Callender (2009; 2010), Frisch (2011; 2014), Weslake (2014), and Blanchard (Chapter 10 in this volume). Of course, there is a considerable divergence between these authors.

[7] A summary of parameters is in Loewer (2021, p. 1083). See also Hoefer (2007, pp. 571–2), Hicks (2018), Dorst (2018), Jaag and Loew (2020) for other amendments of parameters. The latter philosophers encourage us to add to simplicity and strength further (anthropocentric) parameters in order to get laws that are 'best for us' human cognizers. Here's a suggestion: we could then also radicalize the opposite direction: if we keep the old metaphysical assumption of the existence of perfectly natural properties, we might just as well drop simplicity and (maybe also) strength and label, unfiltered, all the regularities within the distribution of natural properties 'laws'. If the original BSA is anyway not 'for us', the world might well have weird laws like Reichenbach's golden cube law.

[8] See Eddon and Meacham (2015, pp. 118–20) for an exhaustive list of possible variations that have been given in the literature.

[9] This can be taken as an indicator of the ingenuity of Lewis's account. Speaking of which: One might want to adopt the best systems idea in other fields of inquiry as well. Maybe the *laws of metaphysics* can be found in the best system of competing ones (cf. Hall, Chapter 12 in this volume). Or take ethics: let '*best system rule utilitarianism*' be the theory that says that an action is right if it conforms to a rule that belongs to that system of rules which, if obeyed by all people, would overall/ holistically maximize happiness, i.e. which would lead to a greater good than any other competing system of rules does. (For a different comparison to ethics see Callender, Chapter 1 in this volume.) Lewis himself also used best system/best balance rules elsewhere, for example for reference fixing of theoretical terms within a theory (cf. Lewis 1984b, p. 228).

Take $f(p, d, c)$ to be the best-system function that takes as input: (i) a set p of freely chosen albeit subsequently fixed properties/predicates; (ii) the distribution d of these properties in world w, i.e. d is itself a function of p and w, where, for simplicity, we assume that w is just our world; and (iii) a set of freely chosen but fixed criteria c, including their weighing.[10]

$f(p, d, c)$'s output is the one deductive system—including as axioms and theorems the laws of the (scientific) realm demarcated by the respective property choice p—that best axiomatizes the distribution d according to criteria c and their weighing. We might also immediately take the laws as $f(p, d, c)$'s value.[11] In other words, if the d for the chosen p is not too disorderly in w and there is, according to c, a best axiomatization, then $f(p, d, c) = L_p$, with L_p being the laws of the science of which p are the properties; $L_{Phy}, L_{Chem}, L_{Bio}$ would be the laws of physics, chemistry, biology. This function $f(p, d, c)$ is sometimes imagined to be realized by an omniscient being.[12]

Lewis, of course, took p and c to be constants: p = perfectly natural properties/ predicates, and c = the balance of (his) simplicity and strength and fit. Proponents of the BBSA wiggle at least with p, proponents of pragmatic best system analyses modify at least with c,[13] some do both.

However, some such variations cannot be made as innocently as here suggested. The exchange of perfectly natural properties/predicates for any kind of properties/predicates has consequences elsewhere. In what follows, this chapter describes these challenges and adds novel answers to those that have already been given. The focus is on challenges related to changes of variable p, the properties/ predicates. Questions arising from revisions and variations of c, the competition criteria, are left aside.

(Note that I will use the abbreviation 'BBSA' for the whole idea, i.e. the better best system analysis. I will speak of 'a BBSA' or 'BBSAs' when I mean individual, separate competitions with separate predicate/properties sets and the set of law statements/laws they yield. Note, too, that I mainly focus on variations of the BSA that do indeed allow for many such separate and parallel competitions for many

---

[10] I could have written $f(d(p, w), c)$ instead, where $d(p, w)$ delivers the distribution (mosaic) of properties p in world w. Since, for our purposes, only our world matters, I leave w out of the equation and I have decided to give the properties p a more prominent place: *they* are the protagonists in this chapter. (For a similar formulation of the function $f(p, d, c)$ see Loewer 2021, p. 1081.)

[11] Some, like David Albert (2000), Barry Loewer (2012), and Ned Hall (unpublished, pp. 13–14), think that not only regularities should get the status of laws (as Lewis thought) but also, for example, particular facts about the initial conditions of our world, for example, particular facts about the Big Bang. Thus, we could have another variable for the BBSA function f to consider: are only regularities or also particular facts eligible for law status? For brevity, I omit this variable, yet note that initial conditions of all sorts are probably particularly important for special science laws to obtain at all: no inhabitable planet, no biological laws!

[12] Cf. Lewis (1973, p. 74), Beebee (2000, p. 574), Schrenk (2008, p. 126), Albert (2015, p. 23), Hall (2015, p. 265), etc.

[13] As alluded to in Cohen and Callender (2009, pp. 5ff.), even the competition criteria might be a function of the science under concern, i.e. of the chosen predicates: $c(p)$.

(special) sciences. However, I will, of course, also mention Albert's (2000) and Loewer's (2007; 2012; 2021; Chapter 6 in this volume) physics-centred accounts.)

8.3 Objectivity

Some subjectivity or anthropocentricity seems to be involved if it is up to us to choose property sets p. Does this lead to the lunacy Lewis (1994, p. 479) warned us of? It doesn't. We can secure the objectivity of systems so that a relativism to the degree of the social construction of the laws is avoided (or other worrisome anti-realistic features that would, for most, be undesirable regarding a theory of lawhood):

(i) Nature, although she does not dictate the vocabulary, still dictates which discernible patterns/distributions of properties d that can be seen through the lens of each vocabulary. That is, given a predicate set p, the regularities that exist in the world are factually and objectively given by nature.[14]

Ilkka Niiniluoto, although not a player in the BBSA business, puts this nicely. The truth or falsity of a sentence written in vocabulary p

> depends on the structure of the world relative to the descriptive vocabulary.... True sentences are not in a relation of correspondence to a non-structured reality, but rather to a structure consisting of objects with some properties and relations.... we may say that "states of affairs" are relative to languages which are used to describe them. But it does not follow that facts depend on our knowledge or that truth is in any sense "epistemic:" as soon as a language L is given, with predicates designating some properties, it is up to the world—not to us—to "decide" what sentences of L are factually true.   (Niiniluoto 1987, p. 141)[15]

He adds that the world 'has "factuality" in the sense that it is able to resist our will' (Niiniluoto 1987, pp. 141–2).

---

[14] However, laws will get a more anthropocentric flavour if we let the parameters c maximize the cognitive benefit for creatures like us (cf. Jaag and Loew 2020). Note that this was an issue already for Lewis (cf. 1994, p. 479).

[15] Compare Niiniluoto's statement to Kuipers's (2000) *constructive realism*, Giere's (2006) *perspectival realism*, Tambolo's (2014) *sophisticated realism* (attributed to Feyerabend), and, last but not least, Cohen and Callender's *explosive realism*: 'On this view, the world permits possibly infinitely many distinct carvings up into kinds, each equally good from the perspective of nature itself, but differentially congenial and significant to us given the kinds of creatures we are, perceptual apparatus we have, and (potentially variable) matters we care about' (Cohen and Callender 2009, p. 22). Here is a worry: How can the world answer to distinct carvings up if it does not have its own, mind-independent structure, if it were *in itself* just an amorphous blob? Structure is acknowledged for the receiving end of the (epistemic) relation: the kinds of creatures we are have a certain perceptual apparatus. It seems that for the possibility to interact, the giving end of the relation, the world, has to have some structure, too. (Above, Niiniluoto suggests that reality is structured.)

(ii) We can assume (as Lewis does in 1973, p. 73) that, for any vocabulary set p and distribution d of the respective properties, the winners of the respective BBSA already objectively exist as abstract (Platonic) objects.[16] That is, our 'choosing' a vocabulary/predicate set p to be systematized equals picking the already exiting set of laws $f(p, d, c) = L_p$.

Note that, although all abundant properties are in principle legitimate, some vocabularies will possibly be so inapt to systematize the world—because, so to speak, there are no regularities to be seen through their lenses—that no winning system, far ahead of its competitors, will (abstractly) exist. Note also that if you wish to systematize a set of properties (and their distribution) which contains *grue* or the above, F, $f(p, d, c)$ will deliver the best system for that set. For sure, it might be a silly set of predicates/properties to focus on and the best system you get in return (if any) is probably the useless one you deserve: 'Properties like F and the ensuing threatened trivialization of MRL [the Mill–Ramsey–Lewis view of laws; here BSA] are ruled out for lack of interest rather than any intrinsic deficiency' (Cohen and Callender 2009, p. 23).

8.4 Which Predicates, Which Properties?

In Lewis's BSA, the predicates refer to all and only the intrinsic, quiddistic, point-size instantiated, fundamental, perfectly natural properties, i.e. we have a straightforward, canonical language-to-world fit: natural properties and their instantiations are objectively (world-)given and the only permitted predicates trace, language-to-world, precisely those properties. This (stipulated) tight fit is the reason why, within the description of the Lewisian BSA, we can smoothly move from talk about properties to predicates and vice versa.

And this possibility of a smooth transition is taken for granted in most formulations of the best system idea: at least simplicity (and its balance with strength) is most often phrased in terms of language/predicates (Loewer 1996, p. 109), while

---

[16] The number of these abstract (Platonic) objects is large indeed: If properties are conceived of as sets of things (as Lewis does and I tend to do in what follows), then already the number of abundant *fundamental* (one place) properties is huge: the abundant fundamental properties are precisely the members of the power-set of all space–time points. Yet we do not stop here: next, every possible set of properties is eligible to enter a best system analysis. That is, every member of the *power-set of the power-set of space–time points* is eligible. For each, so our assumption from above, there is, as abstract (Platonic) object, an outcome $f(p, d, c) = L_p$. Thank God, there is space in Platonic heaven! Note that, for simplicity, I focused on abundant *fundamental* properties, i.e. those that instantiate at space–time points. The BBSA's *non-fundamental* abundant properties would (on the background of Humean Supervenience) have to be identified with sets of space–time regions/mereological sums of space–time points/4D-worms. Also, where original Lewisian laws—all Fs are Gs, say—ultimately just express subset relations between (natural) sets of space–time points—here: $S_F \subseteq S_G$, i.e. F's extension (a set of space–time points) is a subset of G's extension (another set of space–time points)—BBSA laws express relations or intersections of 4D-worms with other 4D-worms.

for considerations of strength the focus on the distribution of properties in the world is needed. Indeed, we had the p in $f(p, d, c) = L_p$ variably mean both: sets of properties or sets of predicates, whichever was appropriate.

Yet, when turning from Lewis's BSA and his natural properties to the BBSA, the transition from the BBSA's arbitrary (non-natural) predicates to the respective properties is not straightforward. Indeed, it is hard to find a semantics for the predicates of the special sciences that both satisfies the anti-reductionism of the BBSA and allows the kind of free recombination required for a Humean BSA-style view of laws.

Before I go into details, it is useful to assume nominalistically (with Lewis and probably most BBSA proponents) that properties are (just) sets of objects (actual and merely possible) and that predicates have such sets as their extensions. Talk of instantiation of a property by an object o then boils down to o's set-membership, and reference or denotation for predicates is replaced by predicate extension.

Where, in the case of Lewis, perfectly natural properties and, thus, the respective predicate's extensions are elite sets of space–time points, the extensions of the BBSA's special science predicates are sets of objects (within the doctrine of Humean Supervenience, objects are probably mereological sums of space–time points/regions or 4D-worms).

Coming back to the question of the relation between predicates and properties in BBSAs, we may, first, safely assume that, for any (special science) predicate you like, there is an abundant property that is its extension (for an estimate of how many abundant properties there are see note 16; for the possibility of empty predicates see below). Second, you can give a name (a predicate) to any such abundant property. A one-to-one match is secured (for any predicate there is a property and vice versa), and this suffices for the possibility of a smooth transition in $f(p, d, c)$ from predicates to properties and back.

We could leave things here and be content with the established one-to-one match. Yet it is instructive to look a little closer at semantic issues for the special science predicates BBSA proponents envisage: Not any semantic theory will suit the BBSA's agenda and the one that does needs to handle some *prima facie* objections.

(i) *Semantic externalism* (SE). An influential way to think of the semantics of kind-terms and scientific predicates is in terms of *semantic externalism*. Roughly, semantic externalism goes like this: In baptism situations, point to a prototypical member of the kind/property. Anything which is just like *that*—i.e. which shares some (intrinsic) features with *that*—is also a member of the kind/property and, thus, it, too, shall belong to the respective predicate's (kind-term's) extension. Now, at least for scientific properties/kinds, SE seems to presuppose that the decisive (intrinsic) features to be shared are nature-given (and not up to us). Yet this would jeopardize BBSA's freedom of predicate choice. For then we'd better feed only predicates into $f(p, d, c)$ that have those nature-given extensions. If we don't choose predicates that trace the nature-given intrinsic features, it is at best unclear whether $f(p, d, c)$ delivers the laws *of nature* or rather some anthropocentric *unnatural* regularities.

A further worry would be that some semantic externalisms tend to assume that the nature-given intrinsic features are of a dispositional nature, i.e. they have dispositional essences. These essences would equip their bearers with nomological profiles and a BBSA would be obsolete.[17] (This problem will recur in (iii) below.)

Moreover, assuming nature-given extensions on the level of the special sciences would amount to what Lewis assumes for fundamental properties, and if we admit the former, why not the latter? Doing both would provoke all sorts of follow-up questions like the one that asks about the relation of the theorems one could derive from the fundamental system and the laws the BBSAs would come up with on the non-fundamental level.[18]

(ii) *Reference magnetism*. In general, any semantic claim that entails that nature herself has a given property structure threatens BBSA's mission. So, too, might *reference magnetism*—a further in principle possible semantic theory for the scientific predicates' extensions under concern. Here, roughly, such predicates are said to pick out classes of things that, as closely as possible, 'respect the objective joints in nature' (Lewis 1984, pp. 227–9). In other words, special science predicates are supposed to hit at least 'fairly natural properties' (Lewis 1999, p. 42). Needless to say, and for the same reasons given against semantic externalism, reference magnetism is hardly a suitable semantic theory for the BBSA.

(iii) *Description theories of reference*. Having renounced magnetism and externalism, we are still left with the question which semantics is viable for the BBSA, i.e. how, for the BBSA, the relation between special science predicates and their properties/extensions is established.

Famously, semantic externalism turned an old orthodoxy on its head and proclaimed: extensions first, then, second, a language that traces these extensions. The most straightforward alternative to semantic externalism is to assume again the prior doctrine, a *description theory of reference*: it's the predicates' intensions

---

[17] For an option how non-Humeans could profit from best system analyses, one that makes us aware of the possibility of systematic disturbances and interferers (known as finks or masks in the dispositions literature), see Vetter (2015, pp. 289–90); for further arguments that dispositionalism and a (variant of the) best system account go well together after all, see Demarest (2017). Also see Demarest and Miller (Chapter 5 in this volume). Note that on Barry Loewer's recent version of the package deal account, fundamental properties need not be categorical (cf. Loewer 2021, p. 1085; Chapter 6 in this volume). Such dispositions/powers would, however, not compromise Humeanism because, within the package deal, laws and fundamental properties are generated together, i.e. properties would not have nomological roles prior to the emergence of the laws (see Bhogal, Chapter 7 in this volume).

[18] Siegfried Jaag suggested to me that this position—natural properties on multiple levels plus the according (B)BSAs—could be an interesting third way between Lewis's dismissiveness regarding special science laws and the BBSA's anything goes. The hierarchy and coordination problem might be delegated to nature: we could trust that she will not contradict herself within her laws on multiple levels.

(i.e. causal/nomological roles that belong to their meaning or semantic content) that fix their extensions.

Special science predicates do seem to have such intensions:

Molecule x is *an acid* (belongs to the extension of 'acid') *iff* x is capable of donating a hydrogen ion, i.e. when x is added to water, the concentration of H^+ ions in the water increases;

event y is *an earthquake* (belongs to the extension of 'earthquake') *iff* y is a tremor resulting from an energy release in the earth's lithosphere creating seismic waves;

z is *a tiger* (belongs to the extension of 'tiger') *iff* z is a carnivorous mammal and a solitary but social predator with fur and stripes and claws and teeth; etc.[19]

The problematic issue with this reference-fixing mechanism is not, as according to externalism or magnetism, that the world is (tacitly) assumed to be equipped with natural properties, but that *being capable of something, resulting from* and *creating something, being carnivorous* and *a predator*, etc. are all causal/dispositional/nomological roles that objects have to fulfil in order to be members of the predicates' extensions.[20] Yet, if these objects are already chosen for the nomological roles they fulfil, then the BBSA seems, in some sense, to be obsolete: the BBSA was meant to deliver the nomological facts, yet, somehow, they are already present prior to the BBSA.[21]

Note that while the function $f(p, d, c)$ uses the intensional roles attached to predicates p to discern their extensions, in the according mosaic d only the predicates' extensions, i.e. the 'naked' *properties*, are systematized. That is, intensional roles are mere reference-fixers. The discerned properties/extensions in the mosaic are, *per se*, void of such roles. (This is unlike the possible dispositional essentialism which might be associated with (i), semantic externalism.)

Yet the reference-fixing roles of the predicates' intensions do at the very least introduce a bias into the mosaic of objects the BBSA is supposed to systematize, however small it might be. For if the predicates' extensions are already selected such that their members fulfil (pre-nomological) roles, then this will leave traces

---

[19] For the intensions of the predicates 'acid', 'earthquake', and 'tiger' I have, on 15 May 2020, consulted and copied from Wikipedia's respective entries. Some intensions might be analytical: that tigers are mammals maybe. For more on the alleged analyticity of attached roles, see Michael Strevens (2012, pp. 167–90). There, Strevens argues that natural kind concepts 'should consist entirely of empirical hypotheses concerning the kind, and thus should neither contain nor imply analytic truths' (Strevens, 2012, p. 178).

[20] Jesse Mulder (2018, pp. 677–9, 683–5) makes a similar point introducing the notion of modally 'thick' conceptions of objects.

[21] Or not immediately. See the considerations at the end of this section.

on the mosaic under concern. (An *innocently given* mosaic is a myth.) In the extreme case, the BBSA (re)delivers exactly what has been in the predicates' intensions. To use one of the examples given above: the BBSA might deliver 'all acids increase the concentration of H^+ ions when added to water'.

How bad would this be? There are ways for BBSA proponents to respond. For a start, they could point out that the scientific predicates' intensions are to be treated *merely* as epistemic agents' nomological *hypotheses*, inscribed into the predicates. Yet, these 'hypotheses' do not count as the 'real' laws; only those nomological roles the BBSAs will deliver do. The first are epistemic conjectures, the second the metaphysical truths. In other words, if, in the extreme case, the BBSA (re)delivers exactly what has been in the scientific predicates' intensions, this should count as scientific success: your (predicate-inscribed) law hypotheses would have been confirmed! Also, note that the predicates chosen have, after all, been established through scientific research and, to quote van Fraassen:

> any scientific theory [and its predicate set] is born into a life of fierce competi-
> tion, a jungle red in tooth and claw. Only the successful theories [predicate set]
> survive—the ones which in fact latched onto the actual regularities in nature.
>
> (van Fraassen 1980, p. 40; my addenda)

There is a remaining worry: it would, after all, be bad if success were predeter-mined, i.e. if the BBSA were, for any set of scientific predicates, always to (re) deliver *all and only* the intensions of these predicates. Yet this is very unlikely to be the case.

First, note that the predicates' intensions are most probably not exhaustive. The best system might well list some additional axioms or theorems which are not yet captured by the predicates' prior intensions. There might, for example, be some global matters, exceeding the roles singular objects fulfil due to their properties: theorems concerning global matters like conservation principles, the Pauli exclusion principle, the second law of thermodynamics, the Lotka–Volterra equations etc. are probably not to be found in any predicate's intension.[22]

Second, as scientific progress shows, some of the predicates' intensions (nomo-logical role hypotheses) might well be wrong: scientists most probably will err. That such discrepancies between prior conjectured roles and anterior ideal BBSA outcomes are likely diminishes the danger of predetermined success.

However, for this discrepancy to be possible, further thoughts about extension fixing are in order: if intensions ought to fix extensions but some or all of the intensions are *wrong-headed*, how is any extension obtained? Some roles attached

---

[22] For further nomological facts that escape nomological roles inscribed into predicates/properties see Vetter (2009), Schurz (2011), or Hildebrand (2020).

to predicate meaning might pick out some objects, some roles others, some none. Which objects are then eligible for reference? Here, a Canberra Plan strategy is probably helpful for BBSA proponents: take all and only the best deservers, fulfilling most roles or most roles reasonably well (cf. Lewis 1984, p. 223). Of course, 'the notion of a near-realization is hard to analyse' but, said Lewis believably, it is 'easy to understand' (Lewis 1972; 1999, p. 253).

A BBSA on sets of such imperfect predicates (that are nonetheless successful in securing their extension in a Canberra-style way) should then be informative because not only might new laws emerge, they might also correct erroneously assumed nomological role hypotheses. 'Newtonian mass', for example, might, with some of its roles, pick out objects (extensions) for which a BBSA would deliver the roles today associated with 'Einsteinian relativistic mass'.[23]

Some predicates might even have an empty extension—'phlogiston', for example. No object fits the predicate's intension. Thus, the best system obtained from a BBSA on a set of predicates containing 'phlogiston' (amongst other predicates) will—for its total lack of strength and its potential to diminish simplicity— not have any generalization containing 'phlogiston' in its set of axioms or theorems. This is a welcome result.

An issue unrelated to problematic prior existing nomological roles is *vagueness*. Some of the predicates the special sciences use might be vague and, thus, they demarcate no sharply circumscribed extension: When does life start? Is a mushroom a plant? Is a bacterium an animal? (Although it is unrelated to nomological roles, I mention the problem of vagueness here because it poses a problem for extension-fixing that finds a solution which is similar to the one discussed for imperfect role profiles.)

BBSA proponents can use the following recipe to handle vague predicates (cf. Lewis 1979, 352). Suppose that for some vague predicate p there is some leeway for a variety of possible extensions/delineations, probably with some shared core set. In other words, there are various possible precise extensions $e_1, e_2, e_3, \ldots$ for p. Name them $p_1, p_2, p_3, \ldots$ and let all these (now precise, not vague) predicates $p_1, p_2, p_3, \ldots$ figure in your BBSA. The system that will turn out best might mention some, most, or all of those predicates in its axioms. If it mentions only one of the p_i you should think of your initial (vague) p being replaced by that p_i.[24]

To sum up, if BBSA proponents engage in semantics at all and aim to answer the question of how the relevant predicates' reference/extension is fixed, they might be in for a surprise: every semantics here considered—(i) *semantic externalism*, (ii) *reference magnetism*, and (iii) *description theories of reference*—has

---

[23] Admittedly, should the Canberra Plan's *best deserver theory* presuppose magnetism and/or a naturally given ground within which to find a deserver, then it is no solution for BBSA purposes.

[24] What if too many of our chosen properties have no extension or are too vague? Maybe no system would come out as uniquely best. Unlike Lewis, who could blame the trouble on unkind nature, we would probably have to blame ourselves for a bad predicate choice.

some friction with the BBSA's agenda. The first two semantic options seem to be incompatible with the BBSA; the latter can be made to cohere, although the ways suggested here might well be a beginning only.

Here's a pressing remaining worry. At one point the question will be asked, what kind of entities ultimately fulfil roles within description theories of reference? An intermediate answer like 'those described by physics' will not do, for the question can be repeated for a physical BBSA just as well: What entities, what kind of stuff in the world, fulfils the roles of a (physical) BBSA's properties? Whatever the substratum is, if BBSA proponents deny that it has any extra-mental structure (cf. note 15), then it is hard to see how roles can fix reference: if what is out there is gunky and has no structure (no hooks and eyes) that makes our descriptions true (to latch onto), then even the descriptive theory becomes a questionable semantics for BBSAs (see Hall unpublished, p. 28, n. 14, for a similar worry).

Note that there is a deep, general philosophical worry here and one specific to the BBSA. The first is how we know of and relate to a world without any structure at all (be that with language or our cognitive apparatus in general). The second is a specific dilemma for the BBSA: If—because of the BBSA's specific reference problem above—you grant the world structure after all (its own joints), then you might just as well return to Lewis's original best system analysis including natural properties. If you don't: see the first deeper worry! There are answers to the BBSA specific problem. Loewer, for example, takes the first horn and argues that even if there is a natural structure (even if there *must be* such a structure), 'it is not at all obvious that these arguments would show that it is the job of physics to locate these [perfectly natural] properties or that scientific laws in any way involve them' (Loewer 2007, pp. 325–6). We cannot rehearse Loewer's arguments here, yet it seems clear that they are necessary to save the BBSA's endeavour because taking the second horn leads us to a deeper philosophical puzzle.

8.5 Demarcation

Above, the metaphysics of properties and semantics for predicates were at issue. The next difficulty does not concern predicates/properties as such but sets of predicates/properties. To be more precise, it concerns the question of which and how many kinds of properties we should include in which sets. So far we pretended that inserting, *separately*, sets of properties—P_{Phy}, P_{Chem}, P_{Bio}, etc., one for physics, one for chemistry, one for biology, etc.—into $f(p, d, c)$ is a straightforward matter. This, unfortunately, is not the case, since many sciences involve multilevel explanations: it would, for example, not suffice to only choose what seem to be biological properties when our target is the biological laws. The reason is that most special sciences need, for their generalizations, and thus for their prospective laws, not only their very own properties but also properties from lower levels. Take, for example, the biological rule that humans cannot survive much longer than a couple of days without water, i.e. H_2O + certain dissolved isotonic salts.[25]

Thus, if we were to stick into $f(p, d, c)$ only what seem to be pure biological properties ('being human') without the chemical ('being H_2O') and the physical ('being Na^+' and 'being Cl^-'), we could not possibly get back the above mentioned candidate nor other (more realistic) examples. That is, we would get back a system of laws (if any) that would not have much in common with what we would expect judging from the viewpoint of our current biological science. So, for a bio-BBSA, we'd better hand over also some chemical and physical properties so that there is at least the chance to get many of the laws we expect from biology.

This seems to be a fairly innocent adjustment, but it is not, and a follow-up problem needs to be solved. Still taking biology as the example: if we include too many properties also from the sciences 'from below', we are again in danger of provoking biological bankruptcy. This is now because some regularities that contain only physical or chemical properties, maybe that *NaCl* dissolves in water, might sneak into the axioms or theorems of the respective best system. If so, that regularity would, until further notice, count as a biological law.

In the worst case, the properties from below (chemistry/physics) might deliver material so rich that the higher-order regularities we are after (biology) get ignored for reasons of simplicity and strength: the physical and/or chemical regularities might, alone, capture so much information (strength) in such a simple way that biological regularities are not to be found in the axioms of the best system. For they, the biological ones, have, conversely, little strength (they probably cover only a very small part of the universe: earth) and their inclusion would cost simplicity.[26] This problem might be aggravated if a theory uses higher- and lower-level predicates where respective higher-level properties supervene on the lower-level ones, e.g. biological properties on chemical ones.

To sum up: we need to include into the property set for, say, a biological BBSA some physical and chemical properties; otherwise we can, in principle, not get the expected biological laws that cross the disciplines' boundaries. Yet, including physical and chemical properties might result in a best system that totally ignores biological regularities and, thus, does not yield any of the laws we would expect but, rather, physical or chemical laws.

---

[25] The literature, especially in areas like neurobiology or biochemistry, abounds with realistic examples. Take, for example, regularities that describe chemical transmissions at synapses where electrical signals (a *physical* feature) in one neuron (a *biological* entity) are transformed into *chemical* signals within the synapse which are converted back into electrical signals in a second neuron (cf. Shepherd 1988). Above, I stick for simplicity to the toy model of water and salt. Note that interdisciplinary sciences like biochemistry, physical chemistry, or biochemistry aggravate our problem.

[26] All this has to be taken with a grain of salt, of course: we know neither the exact way $f(p, d, c)$ calculates simplicity and strengths nor do we know d or how properties depend on each other.

A possible way to deal with this problem is to admit the richer vocabulary but at the same time add a further mechanism to $f(p, d, c)$ in order to weed out the intuitively non-biological laws.

A first step could be to let $f(p, d, c)$ assign, by fiat, zero strength to any regularity that mentions no properties of the target science. Then, systems cannot directly accumulate strength via non-biological regularities. However, the above biological rule that human animals cannot survive much longer than a couple of days without water does count towards strength as it contains at least one property of biology.[27]

This, however, would not suffice. A second filter would have to be implemented into $f(p, d, c)$: Since simplicity and strength are (also) features of the overall system (not just singular statements), it might well happen that, although a (best) system can gain no strength from non-biological regularities individually, it still lists them as axioms because they have mediating functions for the other axioms and theorems. That is, adding them might still increase the quality of the overall system. Thus we would have to demand, additionally to the first filter, that only the axioms/theorems of a system that do contain biological predicates count as biological law.[28] (Why does the second filter not suffice on its own? Can't we drop the first? We can't. Remember that without the first filter systems with only physical or chemical regularities might be winners. Apply the second filter to such systems and you are left empty-handed.)[29]

Compared to the original BSA, the BBSA needs two additional rules/filters to yield the intended laws: regularities that do not mention predicates/properties of our target science get zero strength by default *and*, should such regularities nonetheless be part of the axioms or theorems of the best system, they should be filtered out by decree.

Here is a complaint: Both the problem I discuss and the suggestion I offer seem to be tied to the idea that there is an important and deep division between the special sciences, one we should not assume. The division between our sciences is most probably a contingent, anthropocentric, epistemic, and historically grown phenomenon.

I agree, and I do not wish to suggest that there is any a priori (or naturally) given division. However, what I do claim is that if, as a BBSA proponent, you want to get laws that trace such anthropocentric, historically grown special sciences as biology and chemistry, then something like the two-steps mechanism I advise is essential. Maybe you prefer to wait until the (ideal) endpoint of these sciences is

---

[27] I shall ignore possible logical tricks here that might have to be ruled out: a regularity that says 'if physical property p_1 and (biological property b or not biological property b) then physical property p_2' should not count either, of course. Only regularities that mention biological properties *substantially*, however that can be spelled out.

[28] Too ad hoc? Remember that, similarly, Lewis decreed that analytic truths shall be no laws.

[29] By the way, once we add these two filters to c, not only is d in our function $f(p, d, c)$ itself a function of p, but c takes the chosen properties p as variables, too. Taking note 10 on board, we get $f(d(p, w), c(p)) = L_p$.—Whether these filter suggestions survive counterexamples remains to be seen.

reached (see Section 8.7 below), but still, if they survive as distinct sciences, you need such mechanisms.

Of course, if you are not interested in *these* sciences at all and are curious to see what happens if you let $f(p, d, c)$ do its job (almost) unrestrictedly on $p = p_{Phy} \cup p_{Chem} \cup p_{Bio}$, then you need no demarcation or filters. (I admit, it would be interesting to see the result!)

Above, we asked the question whether the two filters are effective if we have a target theory that uses higher- and lower-level predicates/properties where the higher-level ones supervene on the lower-level ones. What happens anyway in such a case? Still no system with only lower-level generalizations can win because these generalizations get zero strength (filter 1). That is good. There are, however, problematic systems: look, for example, at a system S that contains all the lower-level 'laws'. ('Laws' in quotation marks because they are not laws unless they belong to a winning system. What I assume is that they would be the axioms/theorems if we organized a separate competition for this lower level alone). Also, let S contain the supervenience-tracing bridge principles for all higher-level properties/predicates. Finally, let S contain a description of the initial state of the universe (or the earth at some point in history). How good, in terms of competition, is system S? Because of the two established filters, it earns no points for the lower-level 'laws'. Yet, it earns credit for the bridge principles (for a caveat see note 30 below) and for what you can derive with them together with lower-level laws and the initial state description. Now, (i): could such a system be the winner of the whole competition? And (ii): would that be an unwanted result? I turn to (ii) first: One thing that is unproblematic is that the filters won't deter the bridge principles. These principles would turn out to be laws because they contain at least one target-science-predicate. This is not troublesome. Consider, for example, that genes are certain amino acids or that haemoglobin is a certain macromolecule etc. Such 'bridge principles' seem perfectly acceptable as biological laws.

What is problematic is that bridge principles could turn out to be *the only biological/higher-level science laws* and that pure biological candidates are left out. Take the laws of evolution as an example: that heritable characteristics of biological populations change over successive generations according to certain rules. For sure, if, indeed, all biological properties supervene on physical ones, then this 'law' of evolution is encoded in physics together with the bridge laws, plus the initial state (or so it might, but let us assume that this encoding claim is correct; cf. note 36 on the debate surrounding imperialists and anarchists). Yet this 'law' is neither explicitly expressed in the axioms nor does it follow as a theorem from the axioms alone. In other words, if the (total) supervenience claim is right and if such a bridge-laws-system wins, then, despite the filters, we might get the unwanted or at least unexpected result that 'laws' we would assume to be central to biology are really not.

Turning to question (i), it is, however, not entirely clear *that* the bridge-laws-system S is indeed the competition's winner. A competing system, S^*, that contains S's axioms plus, on top, many purely biological axioms surely loses against S in so far as it makes things less simple without adding strength (still assuming, regarding strength, that the pure biological regularities follow from physics, bridge principles, and the initial state). But now take a system S^{**} that has some but not all bridge principles (therefore being simpler than S while losing strength) and that has, instead, many pure biological regularities (including all those that, now that some bridge principles are missing, are no longer encoded in the remaining bridge principles plus physics plus initial state). Because of the addition of the latter axioms, S^{**} is less simple but it gains strength. In comparison to S, did S^{**} gain and lose strength and simplicity in such a way that it turns out to be better or worse? I know of no a priori argument how to settle this question.

Thus, it seems that the worst that could happen is that, despite the filters, system S is the winner of the higher science competition (cf. (ii)). That this could be the case is, however, unclear (cf. (i)). In the worst case, supervenience-tracing bridge principles would become the higher-level science's laws: not too bad a result if the higher-level science indeed supervenes on the lower.[30]

8.6 Unity, Hierarchy, and Cross-Discipline Inter-System Frictions

Although many advocates of BBSAs want the special sciences to be autonomous, they might not want a too radical disunity, i.e. an unrelated patchwork of sciences: There are interesting inter-scientific explanations and a total disunity of the sciences would leave that mysterious.

(i) *Nested sets of properties.* As we have just seen, some unity is already achieved in that we include properties from one science in a BBSA of another science. In fact, this is not only unity-forming but also hierarchy-building, for note the following asymmetry: fundamental physics never uses chemical properties in its generalizations, chemistry never biological ones, etc. It's only always the other way round. Thus, the physical property set (or some subset thereof) is an essential subset to the set of properties relevant for the chemical best system analysis and

---

[30] Siegfried Jaag (in personal communication) made me aware of two twists that could turn out to be favourable for the higher-level sciences: if strength is measured in terms of number of excluded possible worlds and bridge principles are deemed metaphysically necessary, then they get zero strength and the above system S is unable to compete. Also, a pragmatic interpretation of strength (or alternatives thereof) could yield this result: being extremely complex, bridge principles would not be helpful for creatures like us and thus get little weight (or whatever the pragmatists' alternative measure is).

also of the set of properties relevant for the biological best system analysis. The property set for chemistry (or some subset thereof) is an essential subset to the set of properties relevant for the biological best system analysis, and so forth. This nesting of property sets thus establishes both a unity *and* a hierarchy of the sciences that starts with physics and a set of only its very own properties.[31]

(ii) *Direction of explanation for exceptions.* The following considerations are extrinsic to the core BBSA business but they might be used to establish an ordering of the (separate) sciences: Special science laws hold *ceteris paribus* only, i.e. they can and do have exceptions.[32] Exceptions, however, need explanations when they occur (cf. Pietroski and Rey 1995; and also Fenton-Glynn 2016). Now, typically, the explanation for why a special science law did not hold in a specific case is not only sought in a *different* scientific realm: the gene mutation that caused this raven to be white is a matter of, say, X-rays hitting the DNA molecule and thereby knocking out the plumage colour coding. There are also striking asymmetries: a chemical reaction might fail because some macro-molecules break due to physical forces and fields; biological regularities might fail because some necessary chemical reaction was missing, etc. Yet, if, at CERN, some measurement contradicts an established law, no one will look for a biological explanation.

This asymmetry of explanation can be utilized as a further ordering mechanism: we can claim that the sciences that typically explain exceptions to the laws of another science are to be regarded as more fundamental (or, at most, same-level). Note that, if this is a possible move, a top-down (rather than a bottom-up) approach of the hierarchy of sciences has been given: the direction in which we look for explanation is downwards.[33]

(iii) *Ontological dependence.* Of course, both (ii), the possibility of the just praised top-down explanatory hierarchization, as well as (i), the nesting of property sets, might have its reason in the bottom-up ontological hierarchy suggested next.

Developed by many proponents in an anti-reductionist spirit, the BBSA remains, as such, silent about possible 'grounding' relations between its free-standing special sciences and their laws. Keeping this anti-reductionist attitude

---

[31] Note that all this is contingent. In worlds where substantial minds practise telekinesis this hierarchy might well be lost. For our world, however, the world answers fairly favourably to such hierarchical BBSAs.

[32] This issue, the one of *cp*-laws, exceeds the scope of the present chapter, but solutions specifically targeted at the BBSA have been offered elsewhere (Schrenk 2007, 2.1.3 and 3.3.4; 2008; 2014; 2017).

[33] At least psychology is a possible exception to the idea that interference goes only upward: psychosomatically induced illnesses seem to be a case of downward influence. Thus, such influences could count as explanations for exceptions on the normal functioning physiological body. Of course, we could try to counter with some reductionist move here, saying that psychology is, in the end, pure brain physiology, but such a reductionist move is not (or not so easily) available for a BBSA proponent. Note also that, curiously, physics is the science either where all exceptions are immediately falsifications (so that the search is open for new physical laws) or where exceptions are explained again with reference to physics only.

and the autonomy of the sciences there still can turn out to be some ontological dependence. Here is how that can be: Which generalizations classify as laws is decided autonomously within each separate system analysis (this is the autonomy BBSAs promise). However, focusing on truth (rather than on nomicity), the *mere truth* of the regularity statements of one special science (most probably) depends on the laws and properties of more fundamental special sciences. Here's, roughly, how: let $\forall x \, (Fx \supset Gx)$ be a true regularity of one special science S_1 (with nomological status conferred to it by its own BBSA competition). Now, entities with properties F and G typically have parts $c_1, \ldots, c_n$ with their own properties (and relations) P_i of a more fundamental special science S_2: plants are composed of molecules, which are composed of atoms, etc.: 'If we have reason to believe anything in science, it's that macroscopic entities are constituted by microscopic ones and their relations' (Cohen and Callender 2010, p. 432). Now, the set of properties $\{F, G\}$ of the macroscopic entities might supervene on the set of all the properties P_i of the microscopic parts in the following way. It is necessary that for all x_1 and x_2 with parts $c_1, \ldots, c_n$ and, respectively, $c_1', \ldots, c_n'$: if there is a total match for all P_i between $c_1, \ldots, c_n$ and $c_1', \ldots, c_n'$, then Fx_1 iff Fx_2 and Gx_1 iff Gx_2 and, resulting from the laws amongst the P_i: $\forall x \, (Fx \supset Gx)$.[34]

It is, of course, a contingent matter whether these supervenience relations hold true. Yet, if they do, the special science regularities' *nomicity* is still autonomously conferred via the respective system analysis, while the regularities' *mere truth* ontologically depends on the more fundamental laws/regularities.[35] This dependence unifies and hierarchically structures the (still autonomous) special sciences.[36]

(At the end of this section that aimed at unity a deep worry for any BBSA proponent who grants autonomy to multiple sciences and their laws has to be mentioned: Different sets of laws (for the different sciences) support different counterfactuals and ground different (objective) chances. Yet, these different chances and counterfactuals might impose conflicting constraints on rational belief. That such incompatibilities are no vague possibility but can in fact occur

---

[34] Here's a metaphor that might ease understanding: the laws for the properties of the parts 'push' those parts around in such a way that the Fs and Gs of the macro-objects (which are merely piggybacking on the parts and their properties) are 'pushed around' accordingly and thereby happen to make true that all Fs are Gs.

[35] Mike Hicks has (in personal communication) proposed an interesting option for BBSA proponents who believe that such supervenience claims are true. Suppose the chemical supervenes on the physical. Then the mosaic of physical property instances or physical objects is just as much chemistry's supervenience base as Lewis's original mosaic of instances of perfectly natural properties is (for him) the basis of very much the whole rest of the world. If this is so, we can set up the whole (original) Lewisian machinery shifted just one level up. (Including the one he envisages for theoretical terms (cf. Lewis 1984, pp. 226–9), which would solve the semantic issues we discussed in Section 8.4.)

[36] I defend and give details of this picture in Schrenk (2017). How it is to be positioned within the debate between the 'imperialists' and the 'anarchists' I cannot say. See Weslake (2014, pp. 249–51), who coined these terms and defends, together with Cohen and Callender (2009, p. 14; 2010, pp. 435–9) and Frisch (2011, sections 5–7; 2014, sections 3–5), anarchism; and see Albert (2000) or Loewer (2007; 2012) for imperialistic accounts.

when different vocabulary sets are admitted for different BBSAs has been shown (for chances) by Meacham (2014). A reply to save BBSAs from such inconsistency comes from Hoefer (2014). Still, this worry has not been sufficiently dissolved.[37] It jeopardizes any established unity by the threat of incompatibility.)

8.7 Diachronic Trouble: Scientific Progress

Einstein can explain his temporal precursors' findings: Newtonian mass is merely an approximation to Einsteinian relativistic mass, Euclidean space(–time) is merely a local approximation to a non-Euclidean manifold. To have an example from biology too: Watson and Crick can explain findings of their temporal precursor Mendel. And yet, for proponents of the BBSA, a certain type of scientific progress brings challenges with it.

The difficulty lies not in the progress which occurs when science discovers 'merely' new laws, when old hypotheses are rejected and new ones are established. I said '*merely* new laws' because the challenges occur when progress is made due to changes *also* (or first and foremost) in predicates/properties sets: 'Higgs boson', 'buckminsterfullerene', 'double helix', etc. are all terms not known to nineteenth-century physics, chemistry, biology. For BBSAs this is problematic: the outcome of $f(p_o, d_o, c)$ with old predicate/property set p_o (and the according mosaic d_o) might well be incommensurable with the outcome of new $f(p_n, d_n, c)$ with new predicates p_n. The competition criterion of simplicity is a language-relative issue and, thus, BBSA proponents cannot account for that kind of scientific progress. They find themselves in the Kuhn–Feyerabend predicament of paradigm shifts (here: shifts of property sets) that have as a consequence that the old and the new systems talk past each other (cf. Kuhn 1962 and Feyerabend 1962, pp. 74, 90).[38]

I have two suggestions to make this less worrisome. The first is to hijack and transfer Barry Loewer's optimism regarding physics to the other sciences: In elaborating his package deal account (see Loewer 2007, p. 325; 2021; Chapter 6 in this volume; cf. Bhogal, Chapter 7 in this volume), he claims that we have good reasons to believe that physics will make further progress and postulate new theoretical terms. This should lead to a *physics+* with its vocabulary. Loewer then

---

[37] This worry is also an issue in the aforementioned debate between anarchists and imperialists (see note 36 above).

[38] This same-discipline friction can also occur synchronically, not only diachronically as here taken as a sample case: 'Different communities of scientists, using the language that is best for them, might arrive at different laws' (Callender, Chapter 1 in this volume); and, indeed, 'The [BBSSA proponents] must be a kind of Carnapian or Kuhnian with respect to theory change, explaining the change from one theory to another as always the result of explanatory/pragmatic needs and not rational compulsion' (Cohen and Callender 2009, p. 31). In personal communication Craig Callender suggested that it should not be incumbent upon a theory of laws to solve Kuhn's problem. He points out that, rather, the BBSA gives us an opportunity to sharply state Kuhn's problem.

extrapolates this development: the vocabulary our physics is ideally heading towards is the one that shall ultimately figure in $f(p, d, c)$. This is the vocabulary (and no other) that yields, via $f(p, d, c)$, the (true) laws of nature (physics) as axioms or theorems (cf. Loewer 2007, p. 325).

Now, say about chemistry, biology, etc. what Loewer says about physics: Extrapolate our chemistry's, biology's, etc. development; these sciences head towards future sciences and the ideal endpoints contain their ultimate vocabulary. The (true) laws of the special sciences are the axioms or theorems of BBSAs on the basis of these final vocabularies. If this is tenable, the Kuhnian predicament has lost some of its bite.[39]

The second suggestion is to unify the old and the new property set, p_o and p_n, of two different stages of a science. Then, so the hope would be, $f(p_o \cup p_n, d, c)$ should yield, as the winner, a system whose laws quantify only over the p_n because regularities containing those, as opposed to regularities quantifying over p_o, should be superior in strength. This hope can be justified via the considerations regarding vague predicates and predicates with imperfect prior nomological roles we offered in Section 8.4, (iii). In the best case, the outcome of $f(p_o \cup p_n, d, c)$ is identical to the outcome of $f(p_n, d_n, c)$. Via this detour, the old and the new system would be compatible and the new system would be superior.

8.8 Conclusion

The most substantial change that is involved in moving from Lewis's BSA to the BBSA is replacing one elite set of perfectly natural properties/predicates by arbitrary (special science) sets of properties/predicates. This change is not as unproblematic as might have been assumed. This chapter's goal was to make us aware of the possible challenges and also how to tackle them.

On the issue of objectivity, we have highlighted that worrisome anti-realistic features to the degree of the social construction of the laws are avoided because it is still nature that dictates which discernible patterns/distributions of properties d can be seen through the lens of each freely chosen vocabulary. We have also assumed that, for any vocabulary set p and distribution d of the respective properties, the winners of the respective BBSA, $f(p, d, c) = L_p$, already exist abstractly.

---

[39] With some justification, an eager scientific/metaphysical realist might want to point out that, for Loewer as well as for proponents of BBSAs that utilize his idea (and for pragmatist solutions as well), it is somewhat miraculous why there is a trajectory in science at all (to a *best* best system) if this is not because nature herself has a given structure (natural properties and a natural mosaic) which scientific progress successfully traces. Yet this challenge from scientific or metaphysical realism (similar to no-miracles arguments) must wait for another time. (Lewis's original system is at a clear advantage here!) Compare this, to the final worry in Section 8.4 and to what has been said in note 15.

BBSA proponents hardly ever engage in semantics for the predicates of their concern (at least not when dealing with the BBSA). Doing so revealed that semantic externalism and reference magnetism proved to be untenable for the BBSA's purposes. The old (internalist) orthodoxy—that intensions fix extensions— has some friction with the BBSA's agenda: predicates come already equipped with the nomological roles the BBSA is supposed to deliver. This can, however, be made to cohere with the BBSA. Still, it has been revealed that, unlike Lewis's mosaic, the BBSAs' mosaics are an already biased (*theory-laden*, if you wish) *given*.

Other than the original BSA, the BBSA needs to implement two additional filters within its competition mechanisms because more properties need to be included in the envisaged special sciences' property sets than expected, especially properties of sciences other than the one under concern. Such filters prove to be necessary because, say, biological BBSAs might otherwise deliver physical laws. The filters we suggested are: (i) regularities that do not mention predicates/ properties of our target science get zero strength by default, *and* (ii) should such regularities nonetheless find their ways into the axioms or theorems, they should be filtered out by decree.

For those proponents of the BBSA who value the unity and/or hierarchy of the sciences we suggested establishing them via (i) *nested sets of properties,* (ii) the *direction of explanation for exceptions,* or (iii) *an ontological dependence like supervenience.* The latter suggestion could be a middle way between *imperialist* and *anarchist* positions regarding BBSAs and physics.

Finally, for Kuhnian troubles regarding scientific progress we suggested hijacking an idea Loewer proposed for physics: chemistry, biology, etc. head, too, towards future sciences, and the ideal endpoints should contain the ultimate vocabulary. BBSAs of these final vocabularies deliver *the* laws of these sciences.[40]

References

Albert, D. Z. (2000). *Time and Chance.* Cambridge, MA: Harvard University Press.

Albert, D. Z. (2015). *After Physics.* Cambridge, MA: Harvard University Press.

Armstrong, D. M. (1978). *Universals and Scientific Realism, Vols. I and II.* Cambridge: Cambridge University Press.

Beebee, H. (2000). 'The Non-Governing Conception of Laws of Nature', *Philosophy and Phenomenological Research*, 61(3), pp. 571–94.

[40] I am grateful to Siegfried Jaag for the many helpful discussions of Humean Supervenience, BSAs, and BBSAs, and general Lewisian themes, to Craig Callender and Mike Hicks for valuable comments on earlier versions of this chapter, and to Güney Alp Sapmaz for preparing the bibliography and the script. The chapter was written within the context of the DFG Research Group *Inductive Metaphysics* (FOR 2495). I would like to thank the DFG for their financial support. Finally, I wish to thank the editors for their work on this volume.

Cohen, J., and Callender, C. (2009). 'A Better Best System Account of Lawhood', *Philosophical Studies,* 145(1), pp. 1–34.

Cohen, J., and Callender, C. (2010). 'Special Sciences, Conspiracy and the Better Best System Account of Lawhood', *Erkenntnis,* 73, pp. 427–47.

Demarest, H. (2017). 'Powerful Properties, Powerless Laws', in Jacobs, J. (ed.), *Putting Powers to Work: Causal Powers in Contemporary Metaphysics.* Oxford: Oxford University Press, pp. 38–53.

Dorst, C. (2018). 'Toward a Best Predictive System Account of Laws of Nature', *British Journal for the Philosophy of Science,* 70(3), pp. 877–900.

Eddon, M., and Meacham, C. J. G. (2015). 'No Work For a Theory of Universals', in Loewer, B., and Schaffer, J. (eds.), *A Companion to David Lewis.* Chichester: Wiley-Blackwell, pp. 116–37.

Fenton-Glynn, L. (2016). 'Ceteris paribus laws and minutis rectis laws', *Philosophy and Phenomenological Research,* 93(2), pp. 274–305.

Feyerabend, P. (1962). 'Explanation, Reduction, and Empiricism', in Feigl, H., and Maxwell, G. (eds.), *Minnesota Studies in the Philosophy of Science, Vol. II.* Minneapolis: University of Minnesota Press, pp. 28–97.

Frisch, M. (2011). 'From Arbuthnot to Boltzmann: The Past Hypothesis, the Best System, and the Special Sciences', *Philosophy of Science,* 78(5), pp. 1001–11.

Frisch, M. (2014). 'Why Physics Can't Explain Everything', in Wilson, A. (ed.), *Chance and Temporal Asymmetry.* Oxford: Oxford University Press, pp. 221–41.

Giere, R. N. (2006). *Scientific Perspectivism.* Chicago: University of Chicago Press.

Hall, N. (2015). 'Humean Reductionism about Laws of Nature', in Loewer, B., and Schaffer, J. (eds.), *The Blackwell Companion to David Lewis.* Oxford: Blackwell, pp. 262–77.

Hall, N. (unpublished). 'Humean Reductionism about Laws of Nature' (extended version of Hall 2015), https://philarchive.org/archive/HALHRAv1.

Halpin, J. F. (2003). 'Scientific law: A perspectival account', *Erkenntnis,* 58, pp. 137–68.

Hicks, M. T. (2018). 'Dynamic Humeanism', *British Journal for the Philosophy of Science,* 69(4), pp. 983–1007.

Hildebrand, T. (2020). 'Individuation and Explanation: a Problem for Dispositionalism', *Philosophical Studies,* 177, pp. 3863–83.

Hoefer, C. (2007). 'The Third Way on Objective Probability: A Skeptic's Guide to Objective Chance', *Mind,* 116(463), pp. 549–96.

Hoefer, C. (2014). 'Consistency and Admissibility: Reply to Meacham', in Wilson, A. (ed.), *Chance and Temporal Asymmetry.* Oxford: Oxford University Press, pp. 69–80.

Jaag, S., and Loew, C. (2020). 'Making best systems best for us', *Synthese,* 197, pp. 2525–50.

Kuhn, T. S. (1962). *The Structure of Scientific Revolutions.* 2nd edn (1970). Chicago: University of Chicago Press.

Kuipers, T. A. F. (2000). *From Instrumentalism to Constructive Realism*. Dordrecht: Kluwer.

Leuenberger, S. (2020). 'The Fundamental: Ungrounded or All-Grounding?', *Philosophical Studies*, 177, pp. 2647–69.

Lewis, D. K. (1972). 'Psychophysical and Theoretical Identifications', *Australasian Journal of Philosophy*, 50(4), 249–58. Reprinted in Lewis (1999), pp. 248–61.

Lewis, D. K. (1973). *Counterfactuals*. Oxford: Blackwell.

Lewis, D. K. (1979). 'Scorekeeping in a Language Game', *Journal of Philosophical Logic*, 8, 339–59.

Lewis, D. K. (1983a). 'New Work for a Theory of Universals', *Australasian Journal of Philosophy*, 61(4), pp. 343–77.

Lewis, D. K. (1983b). 'Extrinsic Properties', *Philosophical Studies*, 44(2), pp. 197–200.

Lewis, D. K. (1984). 'Putnam's Paradox', *Australasian Journal of Philosophy*, 62, pp. 221–36.

Lewis, D. K. (1986a). *On the Plurality of Worlds*. Oxford: Blackwell.

Lewis, D. K. (1986b). *Philosophical papers, Vol. II*. Oxford: Oxford University Press.

Lewis, D. K. (1994). 'Humean Supervenience Debugged', *Mind*, 103(412), pp. 473–90.

Lewis, D. K. (1999). *Papers in Metaphysics and Epistemology*. Cambridge: Cambridge University Press.

Lewis, D. K. (2009). 'Ramseyan Humility', in Braddon-Mitchell, D., and Nola, R. (eds.), *Conceptual Analysis and Philosophical Naturalism*. Cambridge, MA: MIT Press, pp. 203–22.

Loewer, B. (1996). 'Humean Supervenience', *Philosophical Topics,* 24(1), pp. 101–27.

Loewer, B. (2007). 'Laws and Natural Properties', *Philosophical Topics*, 35(1/2), pp. 313–28.

Loewer, B. (2012). 'Two Accounts of Laws and Time', *Philosophical Studies,* 160(1), pp. 115–37.

Loewer, B. (2021). 'The Package Deal Account of Laws and Properties (PDA)', *Synthese*, 199, pp. 1065–89.

Meacham, J. G. (2014). 'Autonomous Chances and the Conflicts Problem', in Wilson, A. (ed.), *Chance and Temporal Asymmetry*. Oxford: Oxford University Press, pp. 46–68.

Mulder, Jesse (2018). 'The Limits of Humeanism', *European Journal for Philosophy of Science*, 8(3), pp. 671–87.

Niiniluoto, I. (1987). *Truthlikeness*. Dordrecht: Kluwer.

Pietroski, P., and Rey, G. (1995). 'When Other Things Aren't Equal: Saving Ceteris Paribus Laws from Vacuity', *British Journal for the Philosophy of Science*, 46(1), pp. 81–110.

Roberts, J. (1998). ' "Laws of Nature" as an Indexical Term: A Reinterpretation of Lewis's Best-System Analysis', *Philosophy of Science*, 66(9), pp. 502–11.

Schrenk, M. (2007). *The Metaphysics of Ceteris Paribus Laws*. Frankfurt: Ontos.

Schrenk, M. (2008). 'A Lewisian Theory for Special Science Laws', in Bohse, H., and Walter, S. (eds.), *Selected Contributions to GAP.6, Sixth International Conference of the Society for Analytical Philosophy, Berlin, 11.–14. September 2006*. Paderborn: Mentis.

Schrenk, M. (2014). 'Better Best Systems and the Issue of CP-Law', *Erkenntnis*, 79, pp. 1787–99.

Schrenk, M. (2016). *Metaphysics of Science. A Systematic and Historical Introduction*. London: Routledge.

Schrenk, M. (2017). 'The Emergence of Better Best System Laws', *Journal for General Philosophy of Science*, 48(4), pp. 469–83.

Schurz, G. (2011). 'Review of Alexander Bird: Nature's Metaphysics', *Erkenntnis*, 74(1), pp. 137–42.

Shepherd, G. M. (1988). *Neurobiology*, 2nd edn. New York: Oxford University Press.

Strevens, M. (2012). 'Theoretical Terms without Analytic Truths', *Philosophical Studies*, 160(1), pp. 167–90.

Tambolo, L. (2014). 'Pliability and Resistance: Feyerabendian Insights into Sophisticated Realism', *European Journal for Philosophy of Science*, 4(2), pp. 197–213.

Taylor, B. (1993). 'On Natural Properties in Metaphysics', *Mind*, 102(405), pp. 81–100. Reprinted in *The Philosopher's Annual*, XVI, pp. 185–204 (1993).

van Fraassen, B. (1980). *The Scientific Image*. Oxford: Oxford University Press.

Vetter, B. (2009). 'Review of Nature's Metaphysics: Laws and Properties', *Logical Analysis and History of Philosophy*, 12, pp. 320–8.

Vetter, B. (2015). *Potentiality. From Dispositions to Modality*. Oxford: Oxford University Press.

Weslake, B. (2014). 'Statistical Mechanical Imperialism', in Wilson, A. (ed.), *Chance and Temporal Asymmetry*. Oxford: Oxford University Press, pp. 241–57.

9

Predictive Infelicities and the Instability of Predictive Optimality

Chris Dorst

9.1 Introduction

Recent neo-Humean theories of laws of nature have placed an increasing emphasis on the characteristic epistemic roles played by laws in scientific practice. In particular, these theories seek to understand laws in terms of their predictive utility to creatures in our epistemic situation. In contrast to other approaches, this view has the distinct advantage that it is able to account for a number of pervasive features possessed by putative actual laws of nature, such as their dynamical form, widespread applicability, and various spatiotemporal symmetries. However, this view also faces a unique challenge: given that it attempts to characterize the laws in terms of their predictive utility, any respects in which putative actual laws are suboptimally predictively useful are inherently problematic. Such 'predictive infelicities' can easily be found among our best physical theories.

From the perspective of neo-Humeanism, these predictive infelicities are not necessarily damning, but they do require explanation. Accordingly, in this chapter I propose four strategies that the neo-Humean can appeal to in order to explain away these infelicities. These strategies illustrate how predictive infelicities might be expected to arise even if laws are rightly thought of as optimal predictive principles.

However, it will turn out that one of these strategies raises some problematic issues of its own. Roughly, the core neo-Humean idea is that features of the laws are responsive to features of our epistemic situation. But of course our epistemic situation can change—indeed, it does so all the time. Most of these changes are modest, in that the laws remain just as suitable for prediction before and after. But it is also conceivable that our epistemic situation would change quite radically, such that the standards of predictive utility differ on either side of the change. When this happens, the neo-Humean faces a worrying question: What happens to the laws? As we will see, this question is particularly worrisome for pragmatic views that, like neo-Humeanism, tie lawhood to contingent facts about our current situation.

Chris Dorst, *Predictive Infelicities and the Instability of Predictive Optimality* In: *Humean Laws for Human Agents.*
Edited by: Michael Townsen Hicks, Siegfried Jaag, and Christian Loew, Oxford University Press.
© Oxford University Press 2023. DOI: 10.1093/oso/9780192893819.003.0010

This chapter thus concerns two general threats to the neo-Humean view: (i) predictive infelicities, and (ii) the problem of changing standards of predictive utility. Sections 9.2–9.4 address the first threat, and Sections 9.5–9.6 address the second. After a brief review of neo-Humeanism in Section 9.2, in Section 9.3 I explicate a number of predictive infelicities in our best physical theories. In Section 9.4 I propose four strategies that the neo-Humean can appeal to in order to explain away predictive infelicities, and I suggest how they could be applied to the ones raised earlier. Then in Sections 9.5–9.6 I address the question of what happens to the laws when our epistemic situation changes radically. I conclude briefly in Section 9.7.

9.2 Neo-Humeanism

Nowadays the orthodox Humean view of laws of nature is David Lewis's Best System Account ('BSA'), according to which laws are the axioms of the best systematization of the totality of the particular matters of fact that obtain in the history of the universe (i.e. the 'Humean mosaic').[1] Candidate systematizations take certain claims as axioms, which jointly entail facts about the mosaic. The simplicity of a system is a function of the axioms: how many there are, and how syntactically complex they are. Other things being equal, a simpler system is better. On the other hand, the strength of a system is a function of the amount of information it provides about the particular matters of fact. Other things being equal, a stronger system is better.

These two standards conflict with one another. 'Stuff happens' would be maximally simple, but it would also be extremely uninformative about the particular matters of fact. By contrast, an exhaustive list of each of the particular matters of fact would be maximally informative, but not at all simple. So Lewis maintains that we need to strike a balance between the two. We thereby generate a ranking of candidate systematizations, and the laws are the axioms of the best system, i.e. the system that is, on balance, simplest and strongest.

Why search for systematizations that balance these two desiderata? The thought is that we want a systematization that is highly informative about what happens in the world in as compact a way as possible. Indeed, the BSA is often sketched by imagining a one-on-one conversation with God in which one has limited time to get as much information as possible.[2] Thus the guiding idea behind orthodox Humeanism is that laws of nature are *efficient summaries* of the totality of the particular matters of fact.

---

[1] See Lewis (1973, 1986, 1994).
[2] See, for example, Lange (2009, pp. 101–2), Albert (2015, pp. 23–4), and Beebee (2000, p. 574).

In recent years, several philosophers of an otherwise Humean bent have rejected this guiding idea.[3] The basis of this 'neo-Humean' approach lies in a rejection of the Lewisian presupposition that *mere* efficient summaries would be particularly helpful to creatures in our epistemic situation. More concretely: efficient summaries of the entire history of the universe might omit or obscure details that would be particularly helpful to us in our attempts to *navigate* the universe. Because of this, orthodox Humeanism leaves it mysterious why creatures like us are interested in discovering the laws in the first place.

In contrast with orthodox Humeanism, neo-Humeanism recommends that we think of the laws not as maximally efficient summaries of the entire history of the world, but as principles that are maximally effective at helping creatures in our epistemic situation predict the behaviors of physical systems. In short, the laws are *optimal predictive principles.* Consequently, neo-Humeans seek to replace the standards of simplicity and strength with a different collection of standards meant to generate principles of maximum predictive utility.[4]

It will be useful in what follows to have a sense of the sorts of standards that neo-Humeans have in mind. Suppose, then, that tomorrow an altogether unfamiliar physical system appears in your living room. We might imagine that it's a shiny green sphere, about the size of a pineapple, hovering roughly a foot off the floor. Naturally, all sorts of questions would arise: 'What will happen if I touch it? Should I get rid of it? Can I put it down the garbage disposal? *What does it taste like?'*

All of these questions are getting at the same thing: How is this system going to behave in various circumstances? To that end, suppose you phone up the Altogether-Unfamiliar-Physical-System (AUPS) Hotline to ask for advice. Imagine how this conversation might go:

AUPS ASSISTANT: Thanks for calling the AUPS Hotline. How can I help?
YOU: There's an altogether unfamiliar physical system in my living room. What should I do?
AUPS ASSISTANT: I'd be happy to help. To figure out what to do, I'm first going to need some information.

Let's pause here. What sort of information is the AUPS Assistant going to ask for?

In principle they might ask for anything. They might ask about the system's size, shape, smell, and so forth. They might ask you to bring a compass near it to see if it produces a magnetic field, or turn all the lights off and see whether it

---

[3] For example, see Hicks's (2018) 'Epistemic Role Account,' Jaag and Loew's (2020) 'Cognitive Usefulness Account,' and Dorst (2019) 'Best Predictive System Account.'

[4] What makes these views 'Humean' is that they maintain Lewis's metaphysical picture of the Humean mosaic, and merely seek to change the systematizing standards.

glows in the dark. Alternatively, they might ask for the current atmospheric conditions on Venus's north pole. Or they might want to know what this altogether unfamiliar physical system was doing four weeks ago: what color was it then, and exactly how far was its center of mass from that of Betelgeuse?

Clearly, some of these questions you are better positioned to answer than others. A better AUPS Assistant will only ask you questions that you are able to answer. In that vein, let's imagine some features of the best possible AUPS Assistant:

1. The information they request will be spatially local: it will concern what is happening nearby the system in question.[5]
2. The information they request will be temporally local: it will concern what the system is doing now, or has been doing for the last few minutes, but not information about what it was doing four weeks ago (before it appeared in your living room).
3. The information they request will not require you to locate either yourself or the system with respect to very distant objects, or to otherwise orient yourself or the system in spacetime.
4. They will be able to give you detailed information about exactly what the system will do in a variety of different circumstances.
5. They will be able to give you this information no matter what the system is like. If a different sort of altogether unfamiliar physical system had appeared in your living room, they would have been able to help just as well with that too.

I take it that these features of the ideal AUPS Assistant are relatively uncontroversial (though there may, of course, be others).

The basic suggestion is that the AUPS Assistant can be thought of as essentially a stand-in for a set of optimal predictive principles. Optimal predictive principles would help us figure out how novel systems are going to behave in a variety of circumstances. To do this effectively, they need to have the features we just attributed to the ideal AUPS Assistant: spatial and temporal locality, spatial, temporal, and rotational symmetries, informative dynamics, and wide applicability. In what follows, I'll refer to these features as the 'predictive desiderata': they are the desiderata that optimal predictive principles would satisfy.[6]

---

[5] Do we want the information to be nearby the *system*, or nearby *us*? Presumably, the latter is what we are more likely to have good information about. On the other hand, if we are gathering information about how a particular system is going to behave, oftentimes that is going to be because either (a) the system is near us or (b) we at least know a bit about the system, and derivatively about its surroundings.

[6] See Callender (Chapter 1 in this volume) for further discussion of "ideal advisor" views of lawhood.

The core neo-Humean idea is then to use these desiderata as the standards for lawhood: the laws are the axioms of the best system, where the best system is the one that best satisfies the predictive desiderata. This circumvents the worry about Lewis's proposal: whereas a systematization of the entire history of the universe based on simplicity and strength might end up being relatively unhelpful to creatures in our epistemic situation, a systematization based on the predictive desiderata is undoubtedly going to be *quite* helpful. So on this view it is no mystery why we are interested in discovering the laws.

One of the most compelling arguments for these neo-Humean theories is that putative actual laws of nature exhibit these predictive desiderata to considerable degrees (see, for example, Dorst 2019 for further discussion). For the most part, the laws of our best physical theories provide highly informative, deterministic dynamics about a wide range of physical systems; with a few exceptions, they are spatially and temporally local; and in general they exhibit spatial, rotational, and temporal symmetries. So there looks to be a powerful argument for the neo-Humean position based on the kinds of laws posited in actual scientific practice.

9.3 Predictive Infelicities

The problem is in those qualifications. There are certain respects in which putative actual laws appear to fall short of predictive optimality: respects in which they do not perfectly satisfy the predictive desiderata. And since neo-Humeanism essentially defines the laws in terms of their predictive optimality, any respects in which putative actual laws of nature fall short of this ideal are going to be inherently problematic for the view. There are four such predictive infelicities in our current best physical theories that I want to consider here: quantum indeterminacy, quantum nonlocality, special relativistic light cone restrictions, and the assumption that the fundamental laws must be exceptionless.

9.3.1 Quantum Indeterminacy

Quantum mechanics is notoriously indeterministic, at least on the canonical formulation, so it's not as informative about the dynamics of quantum systems as we might hope. If you want to predict the outcome of a measurement of the z-spin of an electron, and the electron is not in an eigenstate of the z-spin operator, quantum mechanics will only give you probabilities, not certainties.

Now, one has to be a bit careful here, as the indeterminism itself is not a property of the central dynamical principle—the Schrödinger equation, which is fully deterministic—but rather only shows up when one performs a measurement on a quantum system. One then applies the Born rule, which provides the probabilities for obtaining various measurement outcomes. So it would be slightly misleading to say that quantum mechanics involves an indeterministic dynamics. Nevertheless, the indeterminacy in measurement outcomes is still objectionable from the point of view of creatures trying to make predictions. Other things being equal, you would prefer the AUPS Assistant to tell you that this system is *definitely* not going to explode, as opposed to saying that it might not.

9.3.2 Quantum Nonlocality

Quantum mechanics is also notoriously spatially nonlocal, as Bell famously demonstrated in his (1964): how a system behaves at point a can depend on how another system behaves at point b, even if a and b are space-like separated. Obviously, this conflicts with the predictive desideratum of spatial locality. Though again, one has to be careful here: some interpretations of the theory (like Many Worlds and Many Minds) are apparently able to salvage locality.[7] Nevertheless, it would be problematic if neo-Humeanism had to tie itself to a particular interpretation of quantum mechanics in order to maintain its tenability. A better strategy would be to confront the alleged nonlocality of quantum mechanics head-on and see what can be said about it from the neo-Humean perspective.

9.3.3 Special Relativistic Light Cone Restrictions

In Special Relativity, the only events that can exert a causal influence on a given event e are within e's past light cone. Superficially, this looks like it's going to be helpful for predictive purposes, because in order to predict e, one only has to know a cross section of e's past light cone—everything else can be ignored. But as Ismael (2019) has pointed out, this actually ends up proving problematic. A genuine prediction p of e must occur at a time earlier than e.[8] And for us to talk sensibly of p occurring before e, they must be time-like separated, in which case p must occur in e's past light cone. This means that the past light cone of p is going to be 'narrower' than the past light cone of e. (Think of them as two traffic cones, one stacked on top of the other.) Thus, the set of events that can influence the prediction p is a proper subset of the events that can influence e. In other words, as long as p occurs before e, not all of the information needed to predict e is available at p. The extra information is distributed in the area between p's past light cone and e's past light cone, or, equivalently, across any time-like worldline

---

[7] See, e.g., Wallace and Timpson (2010) and Albert and Loewer (1988) for discussion.
[8] Here I'm taking 'p' to represent the event of the prediction of e.

connecting p and e. So all of the requisite information will only become available once e itself has occurred—rendering the prediction useless.[9]

9.3.4 The Laws' Exceptionlessness

Physicists treat the fundamental laws of nature as exceptionless: they apply everywhere and at all times. Again, at first blush this seems helpful for predictive purposes: if the laws have no exceptions, then we can tell with certainty exactly what will happen whenever we apply them.

However, it turns out that the requirement of exceptionlessness is actually hard to square with a conception of laws in terms of predictive utility. That's because a simple, easy-to-use principle that *almost* always generates the right predictions, but occasionally leads us wrong, might be preferable (all things considered) to a massively complex principle that always generates the right predictions.[10] At the very least, if predictive utility is the ultimate goal, such a possibility—that the best predictive principles would admit of some exceptions—cannot be foreclosed. But, pace authors like Cartwright (1983) and Braddon-Mitchell (2001), we *do* seem to foreclose the possibility of fundamental laws that have exceptions. So the laws' exceptionlessness itself looks like a potential predictive infelicity.

9.4 Strategies for the Neo-Humean

How should the neo-Humean handle these predictive infelicities in our best physical theories? One option would be to try to address each of them in a piecemeal fashion: find an explanation, consistent with their view of laws, about why the foregoing inoptimalities might arise in our physical theories—or why they are not actually inoptimalities in the first place. I think this may be possible, but it would also be short-sighted. There are other predictive infelicities in our best theories that I haven't focused on here,[11] and future physical theories may bring their own infelicities—perhaps ones that we can't even anticipate at this point. Addressing the four mentioned here in a piecemeal manner would leave other actual and potential infelicities unaddressed.

---

[9] Oftentimes we'll be able to get a pretty reliable prediction without knowing all of the information about a cross section of e's past light cone. The worry is just that we are closed off *in principle* from acquiring all of the information necessary to entail the occurrence of e.

[10] See Blanchard (Chapter 10 in this volume) for a new Humean account of laws motivated by this kind of concern.

[11] Other potential infelicities include the failure of CPT symmetry and the extreme difficulty of solving our physical equations even for relatively simple multi-body systems. Thanks to Mike Hicks for these suggestions.

A better approach would be to articulate some general strategies, consistent with the neo-Humean position, for explaining away predictive infelicities. If the strategies are general enough, then they will be able to address more than just the four infelicities I have enumerated here. I follow this approach in this section, where I outline four explanatory strategies that the neo-Humean can appeal to. These strategies may be applied in tandem or on their own, though their applicability in any given case will depend on the details of the particular infelicity in question.

Before proceeding, a word of caution: my aim here is to develop resources for the neo-Humean to defuse *modest* infelicities, not to isolate the view from all potential counterevidence. I take the actual predictive utility of the laws to be strong evidence for neo-Humeanism, but this only counts as genuine evidence if there is also the (epistemic) possibility of counterevidence. If we found, for example, that the actual laws are not even *remotely* useful for predictive purposes, this would be a major blow to neo-Humeanism, and it is not the sort of thing that could plausibly be explained away by the following strategies.

9.4.1 Historical Accidents

Sometimes the neo-Humean may be able to argue that a given predictive infelicity in the laws is the product of historical accident.

To see why the Humean can do this, it helps to contrast her position with that of the anti-Humean. For the anti-Humean, this strategy is a non-starter: the laws are what they are—necessitation relations between universals,[12] primitive entities governing the universe's temporal evolution,[13] patterns of counterfactual stability,[14] consequences of the dispositional essences of fundamental properties,[15] etc.—independently of any historical idiosyncrasies in the development of our concept of lawhood. But the Humean has a bit more flexibility here, for on her view there is no metaphysically significant difference between laws and other regularities in nature; the laws are just a subset of all the regularities. Thus, for the Humean, the distinction between the regularities that are laws and the regularities that are not laws is traceable, not to a metaphysical difference out there in the world, but to the standards inherent in our *concept* of lawhood. And this concept itself has a natural history, gradually evolving in response to seen or unseen practical demands in human affairs. So the idea here is that we need to look at the genealogy of the concept of lawhood and consider how it came to have the shape that it does today.

---

[12] Armstrong (1983), Dretske (1977), Tooley (1977).   [13] Maudlin (2007).
[14] Lange (2009).   [15] Bird (2007).

Now, I am no historian, but the following seems like a plausible story. One might imagine that various prominent historical figures—Galileos, Newtons, Einsteins, etc.—had their own idiosyncratic ideas about what a 'law of nature' was. As they promulgated their ideas, coupled with the physical theories that employed them, they were unwittingly shaping the concept of lawhood shared by physicists. They might have been shaping it in different ways: some of them made it more useful for predictive purposes, some for other purposes, and some perhaps made it less useful overall. However, for the concept to stick around, for it to find a permanent home in our conceptual scheme, presumably it had to be useful for *something*; otherwise there would not be much point in having it, and it would fall out of use. So the features that were useful—the ones that jointly contributed to the predictive utility of the principles identified as laws—stuck around as a result of a sort of 'conceptual natural selection.'

I would emphasize again that this natural history of the concept of lawhood is speculative, but it is also pitched at a general enough level that it is difficult to falsify. Now, against this backdrop one can see how there would be room for various predictive infelicities to arise in the concept, as long as they were not too extreme. For example, these infelicities might be traced to the contributions of a particularly prominent historical figure who helped to shape the concept, but who had a philosophical understanding of lawhood that was very different from what the neo-Humean recommends.

For instance, Newton arguably thought of laws as having their origin in divine creation; roughly, God creates the kinds of particles and forces, and then the laws of nature follow as logical consequences.[16] On that view, there would, of course, be no room for laws that have exceptions, since God's decrees are inviolable. Thus we might think of the laws' exceptionlessness as a predictive inoptimality arising from the historical association of the concept of lawhood with God's commands. On this view, the laws' exceptionlessness would be akin to something like a human appendix: a largely useless feature which there has been insufficient selective pressure to weed out because most of the time it is relatively harmless. Against this backdrop, it still makes sense to say that the laws are optimal predictive principles, despite this minor inoptimality, just as it still makes sense to say that the various bodily organs function to keep the organism alive, despite the uselessness (and sometimes harmfulness) of the appendix.

---

[16] See especially Newton's remarks in *Opticks*, Query 31 (1704/1952, pp. 379–80): 'Since Space is Divisible in infinitum, and Matter is not necessarily in all places, it may also be allowed that God is able to create Particles of Matter of several Sizes and Figures, and in several Proportions to Space, and perhaps of different Densities and Forces, and thereby to vary the Laws of Nature, and make Worlds of several sort in several Parts of the Universe.'

9.4.2 Reconceptualizing the Infelicity

Occasionally, something that looks initially like a predictive infelicity might actually be a prerequisite for predictive utility. To see this, often what we need to do is carefully consider our position as embedded agents with limited epistemic access to the universe. Once those limitations are brought into sharper focus, we can see that they place further constraints than we may have realized on what counts as an optimal predictive principle.

This could turn out to be the case for the laws' exceptionlessness. Obviously, creatures like us do not have access to the totality of the particular matters of fact. (Indeed, this is a precondition for our being interested in making predictions in the first place.) Now consider how this fact constrains us when we attempt to determine whether some candidate predictive system is the *best* predictive system—i.e., whether, according to the neo-Humean, that system contains the laws. Even if we are able to implement the predictive desiderata flawlessly, given our epistemic situation we are never going to be certain that the candidate system we are assessing really is the best predictive system, because we are closed off from the totality of the particular matters of fact that are the subject of systematization in the first place. Thus, even if we have a system that we currently think is best, we cannot take for granted that it will not lead us radically wrong in the future.

Clearly, then, we need to be responsive to evidence about the reliability of the candidate systems we are assessing. And, of course, if a system occasionally licenses false predictions, then on any plausible theory of confirmation that fact itself constitutes evidence against the system's reliability. So, in searching for the best predictive system, we are naturally going to have a strong preference for systems that have no exceptions—just because any such exceptions undermine our confidence in the future reliability of those systems.

We can strengthen this point by thinking more carefully about the kinds of cases in which systems might lead us wrong. Suppose that our current evidence indicates candidate system C leads us wrong 1 percent of the time. Furthermore, suppose it turns out that we can identify a commonality to all of the cases in which C leads us wrong, such that by accounting for this extra factor F, we no longer get false predictions. In that case, there is a nearby system C^* that contains all of the axioms of C, plus the rider 'except in conditions F, in which case...'. C^* is then strictly preferable to C on predictive grounds, and generates no false predictions among our evidence.

On the other hand, suppose there is no obvious commonality to all of the cases in which C has generated false predictions. Then we are essentially in the position of knowing that C usually works, but occasionally it inexplicably leads us wrong. This would, I take it, lead us to be wary of using C for predictive purposes in the future, especially if they were important predictions such that people's lives depended on their accuracy. (If we knew that it would only lead us wrong in unimportant cases, perhaps we wouldn't mind; but it's hard to see how we could know this when we can't find any commonality to all the cases in which it has led us wrong so far.) It would therefore seem eminently reasonable to refuse to accept C as the best predictive system and continue to search for a different one. Either way, then, a system like C that admits of exceptions fails to qualify as the best predictive system.

To be sure, this strategy of reconceptualizing apparent infelicities is not generally applicable. My point is only that it will sometimes be possible, especially if we are carefully attuned to the restrictions on our epistemic situation as we go about investigating the laws.

9.4.3 Tensions among the Predictive Desiderata

A third strategy to account for predictive infelicities is to appeal to tensions among the various predictive desiderata. The desiderata themselves fall into roughly two categories: (1) information maximizers, and (2) variable constraints. The information maximizers (namely, informative dynamics and wide applicability) are designed to make our predictive principles as informative as possible about the future behavior of as many physical systems as possible. By contrast, the variable constraints (namely, the various locality and symmetry desiderata) are designed to act as restrictions on the kinds of variables that our predictive principles can require as input. The purpose of these restrictions, again, is to make sure our principles do not require various sorts of information that it would be prohibitively difficult for us to ascertain.

The basic tension is that putting constraints on the kinds of variables that can figure into our predictive principles also makes it more difficult to find a principle that is maximally informative. One might think of this analogously to curve fitting. If I'm trying to fit a curve to a set of data points, then the more variables I include in my curve equation, the more data points I'm going to be able to hit with my curve, so the more 'informative' my curve will be. But if you restrict the variables I can include in my curve equation, then unless the data points are aligned in a very simple manner, the curve is no longer going to be able to hit every point, and it will be correspondingly less informative. So it's inherently difficult to simultaneously maximize the informativeness of our predictive principles and also respect the variety of variable constraints that are operative among the predictive desiderata. Unless nature is exceedingly kind to us, we should expect to have to make compromises among the desiderata.

From this perspective, it is not overly surprising that a theory like quantum mechanics fails to be both deterministic and spatially local, since those two desiderata already come into a fairly straightforward conflict. The failure of quantum mechanical laws to exhibit these features should not be read as evidence against the neo-Humean view. Rather, it is evidence that nature has not been as kind as possible to us, and that therefore we've had to make trade-offs among the predictive desiderata.

This strategy may also work to defuse the worry about special relativistic light cone restrictions. The problem there was that the total information required to generate a prediction p about an event e is not available until e occurs. In fact, however, there is something misleading about the way we stated this. The question 'What information is required to determine whether e occurs?' does not admit of a univocal answer. A better way to formulate this question is: 'What information is required *by candidate system C* to determine whether e occurs?' Different candidate systems will require different sorts and amounts of information to generate an accurate verdict about the occurrence of e.

So here again we have a conflict between the information maximizers and the variable constraints. On the one hand, the information maximizers would dictate that our systematization ought to generate a definitive prediction about whether or not e occurs. On the other hand, the variable constraints would dictate that the prediction of e ought not to require information that is prohibitively difficult for us to ascertain by the time of the prediction. It turns out that nature does not allow both of those desiderata to be satisfied. Special Relativity satisfies the information maximization desideratum at the cost of the variable constraints: it makes e's occurrence dependent on information that is not available at p. But in principle there could be other systematizations that do the reverse, i.e. that confine the facts that e depends on to those that are epistemically available at p. To do so, they likely would have to say that e's occurrence is indeterministic (since the information available at p does not fix the occurrence of e). We thus have a choice between viewing e's occurrence as (a) deterministic and dependent on information unavailable at p, or (b) indeterministic and dependent only on information available at p. Obviously we have opted for (a), but *in principle* there was a choice here—a choice that was necessitated by the tension between information maximizers and variable constraints.

9.4.4 The Instability of Predictive Optimality

The fourth strategy to account for predictive infelicities involves appealing to what we might call the 'instability of predictive optimality.' While this can help explain the appearance of certain predictive infelicities, we will also see that it introduces a rather significant problem for the neo-Humean, which will occupy us for the remainder of this chapter.

The instability of predictive optimality arises because of a tension between the form and function of a predictive principle. As we've seen, the form of an optimal predictive principle would be responsive to the epistemic situation we find our-selves in—in particular, it would not require us to input information that we are unable to ascertain. But of course, the function of a predictive principle is funda-mentally ampliative: it's supposed to help us ascertain information that we are otherwise unable to ascertain. And in so doing, it may expand our epistemic grasp in such a way that it's no longer optimal as a predictive principle, because the greater our epistemic grasp is, the fewer constraints there are on what sorts of information a predictive principle may appeal to.

More abstractly, we may picture our current epistemic grasp as a sphere, S, such that everything in the interior of S is within our grasp. An optimal predictive principle P is responsive to the types of facts that are in the interior of S, and its function is to *extend S*. Now if it extends S radically enough, say to S', we may end up in a situation where some other predictive principle would be better optimized to the sorts of facts in S'. P's ampliative success may thus undermine its own optimality.[17]

As an example, consider the desideratum of temporal translational symmetry. That was a desideratum on our predictive principles because it meant that we do not have to first locate ourselves in time before we can apply those principles. Now, it is clear why this *used* to be a predictive desideratum. After all, before we learned much about the history of the universe, we didn't know how far we were from the beginning of time, so it would've been quite inconvenient if our predict-ive principles required us to input our temporal distance from the origin of the universe in order to generate any predictions. But nowadays we have a pretty good idea of the age of the universe, so it is no longer much of a problem for us to have to plug our current temporal position into our predictive principles.

So some of the standards meant to facilitate prediction are defeasible, and their importance is likely to diminish as our epistemic grasp increases. Thus, in future physical theories we may expect to see certain violations of what we currently think of as predictive desiderata. This would not necessarily show that the neo-Humean conception of laws is incorrect, only that the standards of predictive utility—the predictive desiderata—have evolved.

But here's the problem. For the neo-Humean, laws are *defined* as optimal pre-dictive principles, and we've just seen that which predictive principles count as 'optimal' is likely to change as our epistemic grasp changes. So the question natur-ally arises: By changing our epistemic grasp, can we change the laws?

---

[17] To be clear, the concern here is not that our predictive interests may happen to change as our epistemic grasp changes. Hold the predictive interests constant. Even still, as our access to the uni-verse changes, the principles that we use to infer from what we know to what we don't know may change.

9.5 Can We Change the Laws?

Answering 'Yes' to this question is going to invite a whole host of worries. Scientists certainly do not seem to treat the laws as dependent on our epistemic grasp. It's not that, by learning more, we're changing the laws—rather, we're discovering what the laws were all along. Indeed, one might go so far as to say that thinking we can change the laws is a category mistake: the laws are simply not the sorts of things that we humans can exert any influence over.

This strikes me as quite plausible, though I have no idea how to argue for it, aside from pointing out how extreme a position it is to allow that we can influence the laws by changing our epistemic grasp. Note, for example, that the sort of influence over the laws that we are considering here is stronger than the sort of influence Lewis entertains in 'Are We Free to Break the Laws?' (1981). There, Lewis argues that we are free to act in ways such that, were we to so act, the laws would be different. Our 'power' over the laws in that case is merely counterfactual. In the present case, it is actual: by expanding our epistemic grasp significantly enough, we literally make the laws different than they previously were. Moreover, it is hard to see why this wouldn't work in both directions: if our collective epistemic grasp *decreases* dramatically enough (maybe as a consequence of widespread amnesia and cognitive decline caused by unusual cosmic rays bombarding Earth) presumably this could also change the laws. This strikes me as absurd.

So it seems to me that we *cannot* allow that changing our epistemic grasp changes the laws. Let's assume this is right. If so, the worry for the neo-Humean is this: given that throughout history our epistemic position has changed (and hopefully *will* change) quite radically, different predictive principles will count as 'optimal' in different historical epochs. Which ones are supposed to count as the laws?

Presumably it will not work to say that *all* such optimal predictive principles count as the laws. For it seems possible that the optimal predictive principles of different historical epochs could be inconsistent with each other: some might imply that a given outcome is determined, while others might imply it is stochastic, for example. And a view that countenances the possibility that the laws of nature are contradictory seems doomed from the start. So what we apparently need here is a principled way to distinguish the optimal predictive principles of a *particular* historical epoch as the ones that count as the laws.

One initially attractive candidate is this: the laws are the predictive principles that would be optimal at the end of inquiry. This proposal is at least initially plausible, because if we're going to privilege one historical epoch over all others, the end-of-inquiry epoch is the most distinctive. The question then becomes how to understand the 'end of inquiry' in this context. There are three ways we might interpret that phrase.

The first interpretation maintains that the end of inquiry is when we know all of the particular matters of fact. On this proposal, when we know all the particular matters of fact, whatever principles we then settle on count as the laws.

But this suggestion is not going to work. For when all of the particular matters of fact are known, there would be no point in trying to identify principles that are maximally predictively useful, because there would be nothing left to make predictions about. Consequently, it becomes entirely unclear what kind of standards we would use, at that point, to select optimal predictive principles. Notice, for example, that if we know all the particular matters of fact, then there are no constraints whatsoever on what kinds of variables our principles may appeal to. Nor is there seemingly any reason to prefer a dynamical principle to one that just lists every particular fact that ever obtains. Thus if we imagine ourselves knowing all of the particular matters of fact, any reasons for preferring certain sorts of principles to others seem to drop out, and we are left with no guidelines for selecting which principles count as maximally predictively useful. In short, when we know all of the particular matters of fact, the concept of 'predictive utility' fails to get any purchase.

A second interpretation of what's meant by the 'end of inquiry' is not 'the time when we know all the particular matters of fact' but, rather, 'the time when we have discovered the final, true theory of physics,' e.g. the elusive 'theory of everything.' But we can't use that understanding here either, for the 'true theory' in question would just be the one that correctly identifies the laws. Then the claim that the laws are the optimal predictive principles we would select at the end of inquiry just amounts to the claim that the laws are the optimal predictive principles we would select at the time when we have discovered the laws.

A third interpretation of 'the end of inquiry' in this context is the time at which we, as a matter of fact, would no longer revise our optimal predictive principles in light of new evidence. The laws would then be those optimal predictive principles that we settle on in the long run.[18] This suggestion avoids circularity, but nevertheless it still does not work. For as we have already seen, once all the particular matters of fact are known, the concept of an optimal predictive principle no longer gets any purchase. So in the long run the set of optimal predictive principles that we settle on is the null set. This proposal thus implies that there are no laws.

Moreover, there is something else problematic about all of these proposals. One of the benefits of the neo-Humean view was supposed to be that it makes sense of why creatures *like us* are interested in discovering the laws. But if we say that the laws are the optimal predictive principles that we would select at the end of inquiry (however we understand that phrase), it's no longer obvious why

---

[18] Thanks to Ronald de Sousa for pointing out this interpretation to me.

creatures like us *are*, in fact, interested in discovering the laws. Why should we care about principles that would be predictively optimal for creatures in a radically different epistemic position than our own?

In short, it doesn't seem like we can make headway on the proposal that the laws are to be identified with the optimal predictive principles of a particular historical epoch: the most plausible such epoch is the end of inquiry, but no understanding of that notion yields a tenable position. This throws into question whether the neo-Humean can coherently maintain that the laws really do not change as our epistemic grasp changes.

It might be helpful at this point to step back and reconsider our position. Readers may notice that we are actually in familiar territory here; Lewis himself confronted a similar problem with his infamous 'ratbag idealist':

> The worst part about the best-system analysis is that when we ask where the standards of simplicity and strength and balance come from, the answer may seem to be that they come from us. Now, some ratbag idealist might say that if we don't like the misfortunes that the laws of nature visit upon us, we can change the laws...just by changing the way we think!   (Lewis 1994, p. 479)

The ratbag idealist's suggestion was absurd to Lewis. Over his career, he proposed two different ways of handling it.

Lewis's later (1994) response was that if nature is kind, the problem needn't arise. The thought was that the standards of simplicity and strength are only partly dependent on us. That is, they are fully objective up to a point, and then beyond that point the way they are precisified depends on the way we think—on matters of our psychology. The idea, then, is that if nature is kind, there will be one system of laws that is robustly best—so much better than all others that its being the best doesn't actually depend on how we happen to understand the notions of simplicity and strength.

One might think that a similar move could work in our case. Maybe nature will be kind, and it will turn out that the optimal predictive principles do not change, even as our epistemic access to the universe increases dramatically. In this scenario, as our sphere of epistemic grasp expands, the same principles that were best at expanding S are also best at expanding S', S'', etc.

This move coheres nicely with a recent proposal by Michela Massimi (2018), who has suggested that the standards used to identify the laws not only are liable to change in the future but have already changed in the past. For example, she argues that Newtonian and Hamiltonian mechanics employ different conceptions of simplicity: the former is concerned with postulating simpler causes of phenomena, whereas the latter employs the principle of least action. To identify the laws in the face of these shifting standards, she proposes a 'novel perspectival BSA,' according to which the laws are axioms or theorems that remain constant across the temporal *series* of best systems, each of which may employ somewhat different notions of simplicity, strength, etc.[19]

There is much to explore about Massimi's proposal, but for our purposes here we need to consider whether it can help the neo-Humean identify the laws in the face of shifting standards of predictive utility. The suggestion would have to be similar to the one that we just gleaned from Lewis: even as the standards shift over time, there would be certain predictive principles that remain constant, and those are the ones that count as laws.

Unfortunately we have already seen reasons to think that this suggestion will not work. When we know all of the particular matters of fact, the concept of predictive optimality fails to get any traction, so *none* of the currently optimal predictive principles will still be optimal then. Moreover, it seems fairly implausible that as our epistemic grasp expands, the optimal predictive principles will still obey the variety of variable constraints that we initially placed on them. After all, those constraints were motivated from a highly impoverished epistemic position, comparatively speaking. It would be shocking if those were still the best to be found even after our epistemic grasp increases dramatically. And at any rate this does not strike me as the sort of thing it would be reasonable to hope for, especially if we are going to rest a theory of laws upon that hope.

So much for Lewis's later response to the ratbag idealist worry—it won't help us here. Lewis's *earlier* response was to appeal to rigidification. The suggestion was that when we consider what the laws would be if we changed our ways of thinking, we hold fixed our actual standards of simplicity and strength as constitutive of the laws. So if we counterfactually changed our standards, the laws would still be the regularities that our actual standards pick out. Changing our standards would therefore not change the laws.

The analogous move here would be to say that when we consider what the laws would be if our epistemic position changed radically (or if it had been different all along), we hold fixed the standards of lawhood that are motivated by our *present* epistemic position. This would allow us to say that the optimal predictive principles of the current historical epoch, and no others, count as the laws, and it

---

[19] One of the virtues of Massimi's account is that it offers a compelling explanation of nomic necessity. The necessity of the laws, on this picture, traces to their durability across multiple sequential best systems: 'it is the resilience of our laws (such as conservation of mass, momentum conservation, or Newton's law of gravity) despite the perspectival nature of our standards of simplicity and strength that is testimony to the nomic necessity of these laws' (2018, p. 156). Indeed, one might think this even offers a compelling explanation of the *grades* of nomic necessity: if it is true that the conservation laws have a greater degree of necessity than the particular force laws, Massimi's account might explain this by appeal to the greater resilience of the conservation laws across a broader historical swath of best systems. (Marc Lange has frequently discussed the possibility of different grades of nomic necessity in connection with the conservation laws. See, for example, his (2009) and (2012).) Of course, one worry about this account of nomic necessity is that it seems to imply that the longer we have retained an item of knowledge, the more nomically necessary it will appear. But we do not think it is at *all* nomically necessary that, say, the Earth has a moon, even though we have known this for quite a while.

would also allow us to explain, in a straightforward way, why creatures like us are interested in discovering the laws in the first place. The question we need to address, then, is whether the term 'law of nature' really employs a rigid understanding of predictive optimality.

9.6 The Rigidification Strategy

One way to try to answer this question is to imagine encountering a group of aliens whose epistemic access to the world is radically different from our own. For example, perhaps they are somehow able to perceive things at great spatiotemporal separations from their present location, and maybe they also have some sort of direct acquaintance with certain facts about the future. Given these differences, the sorts of principles that would be maximally predictively useful for them would likely be very different than the sorts of principles that would be maximally predictively useful for us.

Imagine that we meet these alien scientists who are looking for their optimal predictive principles. How would we describe what they are looking for? Are they looking for laws-for-them, or are they looking for something other than laws?

The situation here is roughly analogous to Huw Price's (2007) discussion of beings who live in a section of the universe where the thermodynamic gradient is reversed. Price uses this example to argue that the forward direction of causation is a matter of perspective:

> It remains a live empirical possibility that the universe contains regions in which the thermodynamic gradient is reversed. In such regions, it seems likely that intelligent creatures would have a time-sense reversed relative to ours... Suppose we grant that if there were such creatures, of whatever origins, then two things would follow: (i) they would think that the causal arrow is oriented in the direction that we would call future-to-past; and (ii) their perspective would be as valid as ours. Then we have all it takes to establish that causal direction is perspectival for us—whether they exist or not!   (Price 2007, p. 273)

Price's conclusion here is that the direction of causation is *perspectival*: it is relative to a given kind of creature's perspective. If we discovered beings living in a region of the universe with a reversed thermodynamic gradient, we would say that causation runs from future to past for them.

However, Ismael (2016) argues that Price is too quick in drawing this conclusion. Following Ismael, let's call the role that causal beliefs play in our practical and theoretical reasoning the *causal role*. If it is indeed possible that there are beings like the ones Price describes, what this shows is that, for them, the causal role is played by something that is time-reversed from what plays the causal role for us. Price wants to infer from this that *causation* is time-reversed for them. But this is not obvious, for we have not ruled out the possibility that the referent of 'cause' is fixed rigidly by way of the causal role. As Ismael puts it, the question amounts to 'whether "cause" just means "whatever plays the Causal Role" or whether it refers rigidly to whatever plays the role of causal relations *for us*"' (Ismael 2016, p. 251, emphasis in original).

Roughly the same issues arise in the present case. If the neo-Humean is right, the role that laws play in our practical and theoretical reasoning is essentially that of an optimal predictive principle. So our question is whether 'law of nature' just means 'whatever plays the role of an optimal predictive principle' or whether it means 'whatever plays the role of an optimal predictive principle *for us*.'

Again following Ismael, it may help to note that the question essentially amounts to whether we need to posit a suppressed parameter in the semantics for the claim 'L is a law of nature.' The two possibilities are:

1. L is a law of nature in world w for agent a iff L is an optimal predictive principle in world w for agent a.
2. L is a law of nature in world w iff L is an optimal predictive principle in world w from the perspective of actual human agents (in the early twenty-first century).[20]

In typical discussions about laws of nature, the agent-parameter in option (1) would be filled in by a human agent, so it would be quite difficult to tell the difference between these two options. The difference would only manifest in unusual cases, like if we encountered the aforementioned aliens with radically different epistemic access to the world.

I suspect that option (2) is correct; there is no suppressed agent-parameter in the semantics for 'law of nature.' The reason is that we have, at present, no need to draw this kind of distinction: we haven't encountered any aliens with radically different epistemic access to the world, so we haven't had to settle the question of whether they are trying to discover laws-for-them, on the one hand, or principles that simply are not laws, on the other. Since we haven't had to settle this question, it seems unlikely that our concept of law has this level of detail built into its structure.

The point may be viewed as a methodological one regarding theorizing about semantic structure: we ought to posit only as much semantic structure as is necessary to account for well-established linguistic practices and intuitions. At present, the well-established linguistic practices around laws of nature do not require us to posit a suppressed agent-parameter. Nor, I think, do our intuitions about cases

---

[20] Cf. Dasgupta (2018, p. 313) for a similar discussion concerning different possible semantics for the claim 'X is an elite property.'

like that of the alien scientists: when I ask myself how we would describe the optimal predictive principles of these aliens, I am unable to marshal strong intuitions either way.

This methodological principle is, I think, quite reasonable. Without it, there is nothing to stop us from positing all sorts of suppressed parameters that never manifest in ordinary circumstances. (L is a law of nature for agent a in solar system s at time t...) We ought to treat semantic structure like any other theoretical entity: posit as little as necessary to account for the data in question. And there are no data forcing us to posit a suppressed agent-parameter in the semantics for 'law of nature.'

If this is right, then laws are what play the role of optimal predictive principles *for us*. It follows that we *cannot* change the laws by radically changing our epistemic access to the world. In considering what the laws would be if our epistemic situation were radically different in various ways, we hold fixed our actual, current standards of predictive utility that are motivated by our actual, current epistemic position.

Note that this is not to say that the *extension* of 'laws of nature' is fixed rigidly, such that whatever the laws of the actual world end up being, those are the laws of every possible world. Rather, it is to say that the notion of 'predictive optimality' that figures into the analysis of laws is rigid. Changing our epistemic situation does not change the laws because we hold fixed the current standards of predictive optimality, but changing the *world* might change the laws because different principles satisfy those standards of predictive optimality in different worlds. (That is why both proposed semantics (1) and (2) include a parameter for the possible world in question.)

Of course, we might try to imagine what would happen in various hypothetical future circumstances. Suppose we *do* meet these aliens; what would happen to our concept of law, in that case? Would it change to include an agent-parameter in the manner of option (1) above, or would it behave like option (2)? I'm not sure, but I suspect the answer might depend on certain historical contingencies that actually have very little to do with our current concept of lawhood.[21] At any rate the answer really doesn't matter for our purposes. If the standards of predictive optimality that figure into the definition of lawhood are rigid, *the laws themselves will never change.* Admittedly, there is a temptation to think that if in the future our notion of predictive optimality changes, that would mean that the laws change too. In thinking this, we are trying to adopt a dual perspective—to have one foot in the language game, and one foot out of it, as it were. But there is no such perspective. If you're playing by our rules, if you're using our concept of lawhood, they will never be different than they now are.[22]

---

[21] Wilson (1982) raises some fascinating cases where minor accidents of history end up influencing the subsequent development of certain concepts.

[22] Of course, this is not to say that we can't be *wrong* about what they now are.

9.7 Conclusion

Lewis abandoned the rigidification strategy because it was in his eyes only a 'cosmetic' remedy: 'It doesn't make the problem go away, it only makes it harder to state' (1994, p. 479). I think the remedy is exactly as cosmetic as we need. We want to be able to do two things: (1) allow that the laws are responsive to our epistemic situation, and (2) disallow that the laws change as our epistemic situation changes. This is a fine needle to thread, but the rigidification strategy lets us thread it. We can admit, of course, that our epistemic situation will likely change in the future, perhaps in quite radical ways. When that happens, those future humans might not be interested in discovering the laws anymore. Perhaps they will look back on our current investigations into the laws much as we now look back on the practice of stenography: a rather idiosyncratic pursuit that makes sense only in light of our current conditions.

But of course, all of this may happen *regardless of what philosophical theory of laws is correct.* And here it seems the neo-Humean actually has a leg up. An anti-Humean would have to say that those future humans are hopelessly misguided, missing out on objectively important features of the world that only *we*, somehow, had the sagacity to latch onto. Better, I think, to have a theory that owns up to the possibility that the significance of the laws is not only practical (rather than metaphysical) but also, potentially, *transient*.[23]

References

Albert, D. Z. (2015) *After Physics*. Cambridge, MA: Harvard University Press.

Albert, D. Z., and Loewer, B. (1988) 'Interpreting the Many Worlds Interpretation,' *Synthese*, 77, pp. 195–213.

Armstrong, D. (1983) *What is a Law of Nature?* Cambridge: Cambridge University Press.

Beebee, H. (2000) 'The Non-Governing Conception of Laws of Nature,' *Philosophy and Phenomenological Research*, 61, pp. 571–94.

Bell, J. S. (1964) 'On the Einstein Podolsky Rosen paradox,' *Physics*, 1, pp. 195–200.

Bird, A. (2007) *Nature's Metaphysics: Laws and Properties*. New York: Oxford University Press.

Braddon-Mitchell, D. (2001) 'Lossy Laws,' *Noûs*, 35, pp. 260–77.

Cartwright, N. (1983) *How the Laws of Physics Lie*. New York: Oxford University Press.

[23] Thanks to Heather Demarest, Mike Hicks, and Elizabeth Miller for very helpful comments and suggestions. Thanks also to audiences at the 2019 meeting of the British Society for the Philosophy of Science and the 2019 meeting of the Florida Philosophical Association for helpful suggestions on earlier drafts.

Dasgupta, S. (2018) 'Realism and the Absence of Value,' *Philosophical Review*, 127, pp. 279–322.

Dorst, C. (2019) 'Towards a Best Predictive System Account of Laws of Nature,' *British Journal for the Philosophy of Science*, 70, pp. 877–900.

Dretske, F. (1977) 'Laws of Nature,' *Philosophy of Science*, 44, pp. 248–68.

Hicks, M. (2018) 'Dynamic Humeanism,' *British Journal for the Philosophy of Science*, 69, pp. 983–1007.

Ismael, J. (2016) 'How Do Causes Depend on Us? The Many Faces of Perspectivalism,' *Synthese*, 193, pp. 245–67.

Ismael, J. (2019) 'Determinism, Counterpredictive Devices, and the Impossibility of Laplacean Intelligences,' *The Monist*, 102, pp. 478–98.

Jaag, S., and Loew, C. (2020) 'Making Best Systems Best for Us,' *Synthese*, 197, pp. 2525–50.

Lange, M. (2009) *Laws and Lawmakers*. New York: Oxford University Press.

Lange, M. (2012) 'There Sweep Great General Principles Which All the Laws Seem to Follow,' in Bennett, K., and Zimmerman, D. W. (eds.), *Oxford Studies in Metaphysics, vol. 7*. Oxford: Oxford University Press, pp. 154–85.

Lewis, D. (1973) *Counterfactuals*. Malden: Blackwell.

Lewis, D. (1981) 'Are We Free to Break the Laws?,' *Theoria*, 47, pp. 113–21.

Lewis, D. (1986) *Philosophical Papers, vol. II*. New York: Oxford University Press.

Lewis, D. (1994) 'Humean Supervenience Debugged,' *Mind*, 103, pp. 474–90.

Maudlin, T. (2007) *The Metaphysics within Physics*. New York: Oxford University Press.

Massimi, M. (2018) 'A Perspectivalist Better Best System Account of Lawhood,' in Ott, W., and Patton, L. (eds.) *Laws of Nature*. New York: Oxford University Press, pp. 139–57.

Newton, I. (1704/1952) *Opticks: or, a Treatise of the Reflections, Refractions, Inflections, & Colours of Light*. Mineola: Dover.

Price, H. (2007) 'Causal Perspectivalism,' in Price, H., and Corry, R. (eds.), *Causation, Physics, and the Constitution of Reality: Russell's Republic Revisited*. Oxford: Oxford University Press, pp. 250–92.

Tooley, M. (1977) 'The Nature of Laws,' *Canadian Journal of Philosophy*, 7, pp. 667–98.

Wallace, D., and Timpson, C. G. (2010) 'Quantum Mechanics on Spacetime I: Spacetime State Realism,' *British Journal for the Philosophy of Science*, 61, pp. 697–727.

Wilson, M. (1982) 'Predicate Meets Property,' *Philosophical Review*, 91, pp. 549–89.

10

Best-System Laws, Explanation, and Unification

Thomas Blanchard

10.1

In recent years, an active research program has emerged that aims to develop a Humean best-system account (BSA) of laws of nature that improves on Lewis's canonical articulation of the view. Its guiding idea is that the laws are cognitive tools tailored to the specific needs and limitations of creatures like us. While current versions of this "pragmatic Humean" research program fare much better than Lewis's account along many dimensions, I will argue that they have trouble making sense of certain key features of the practice of fundamental physics. Indeed, these features seem to go against the very idea that laws are useful for agents like us. In my view, Humeans can address these issues by paying more attention to the explanatory role of laws. Following this idea, I will propose an account on which what makes a systematization the best is a kind of explanatory power, understood along the lines of the unificationist theory of explanation. The resulting view, I will argue, can make sense of those features of laws that other pragmatic accounts of laws have trouble explaining.

The guiding idea of the BSA is that the laws are the members of the systematization of the "Humean mosaic" that fare best with respect to certain theoretical standards—the "Humean mosaic" being the complete set of particular, non-modal matters of fact about the universe. The BSA is a version of Humean reductionism about laws of nature (the view that laws reduce to the Humean mosaic) as it posits no metaphysical structure over and above the Humean mosaic: the laws are nothing more than summaries of the mosaic that have certain desirable theoretical features.[1] A key question in the debates over the BSA is whether such a metaphysically lean view of laws can still make sense of their characteristic functions such as enabling induction, supporting counterfactuals, etc. Another important question—and my focus in this chapter—is what makes a systematization "the best". In Lewis's (1983) canonical version of the BSA, the best system is

---

[1] In this chapter I am using 'Humeanism' and 'BSA' more or less interchangeably. There are in fact other versions of Humeanism than the BSA, but none as plausible or popular.

Thomas Blanchard, *Best-System Laws, Explanation, and Unification* In: *Humean Laws for Human Agents*. Edited by: Michael Townsen Hicks, Siegfried Jaag, and Christian Loew, Oxford University Press. © Oxford University Press 2023. DOI: 10.1093/oso/9780192893819.003.0011

the one that best balances strength (understood as the amount of information that the system by itself provides about the mosaic) and simplicity (understood as a syntactic property of the system). Today it is widely agreed that this proposal is not quite right, and merely a first pass. Accordingly, contemporary defenders of the BSA have proposed various criteria with which to replace or supplement Lewis's standards (Albert 2015; Cohen and Callender 2009; Dorst 2019b; Hicks 2018; Jaag and Loew 2020; Loewer 1996, 2007). While they differ from one another in important ways, these proposals all converge on the idea that what makes a system "the best" is its usefulness for cognitively limited and practically oriented creatures like us.

For anti-reductionists, this pragmatic move may seem beyond the pale. Lewis himself was very concerned to avoid the charge that the BSA makes the laws relative to us.[2] But properly executed, a pragmatic take on the BSA need not yield any of the absurd consequences one may fear. (For instance, it need not entail that we can change the laws at will.) Moreover, from a Humean point of view, two considerations make it attractive to introduce pragmatic criteria into the BSA.

The first has to do with a particular challenge for the Humean—the challenge of explaining why the search for laws occupies such a central place in fundamental physics (see Hall 2012, pp. 39–41). Anti-Humeans about laws can easily explain why physicists care so much about the laws: the laws are (or are grounded in) fundamental features of reality that govern how nature behaves, and this makes them automatically worthy of physicists' attention. But, of course, on a Humean standpoint the laws are not part of fundamental reality or metaphysically privileged in any way. So Humeans must find some other explanation for our interest in the laws. And an obvious idea (in fact, perhaps the only plausible one available to Humeans) is that knowing the laws is pragmatically beneficial for agents like us. Humeans are therefore well advised to endorse a pragmatic reading of the BSA, on pain of making it mysterious why the laws matter to us.

Second, a pragmatic approach is well poised to address one of the main objections against Lewis's BSA. As shown by Hall (2012, 2015), Roberts (2008), and Woodward (2014), physicists do not value strength and simplicity as Lewis understands these notions, nor do they trade off strength and simplicity in the manner envisioned by Lewis. Thus Lewis's account doesn't match how physicists themselves think about the laws. This objection has a particular sting to it, as one of Lewis's major selling points for his account that made it attractive to many was its supposed fit with scientific practice (Lewis 1983, p. 41). But once we understand the laws as designed for agents like us, this mismatch between the BSA and scientific practice seems to largely disappear. Consider, for instance, the fact that physicists value fundamental theories that are compatible with many different

---

[2] Cf. his discussion of the 'ratbag idealist' (Lewis 1994, p. 479).

possible initial conditions and are in that respect very uninformative. (Jaag and Loew 2020 call this feature of the laws "modal latitude".) If the goal of the best system is to elegantly encode as much information about the mosaic as possible, as Lewis's BSA claims, this makes little sense. But suppose instead that the laws are designed in part to help us predict the behavior of the many subsystems of interest present in our physical environment. Given that such systems all start in very different states, it is no surprise that good candidates for the status of laws are expected to be compatible with a wide range of initial conditions. (Another reason why the laws must have modal latitude to facilitate predictions is that limited agents like us can rarely if ever know the exact initial conditions of a system (Jaag and Loew 2020, p. 16).) In a similar vein, defenders of the pragmatic approach to the BSA have shown that it can make sense of many other features of the laws that are not predicted by Lewis's account, e.g. the fact that laws can be tested independently of one another (Hicks 2018), or are expected to display certain symmetries (Dorst 2019b).

Clearly, then, a pragmatic approach to the BSA is the way to go for Humeans. But some hurdles remain to be cleared. While existing pragmatic versions of the BSA fit scientific practice considerably better than Lewis's account, they still fail to capture two key aspects of the way in which physicists evaluate candidate best systems. Those two features of fundamental physics are especially problematic for pragmatic Humeanism, as they appear to go against the very idea that the laws are tailored to be useful to limited agents like us. In fact, these features seem to make more sense on an anti-Humean view of laws, thus raising the worry that, with respect to fitting scientific practice, it is anti-Humeanism that has the upper hand.

The first feature of fundamental physics that is problematic for pragmatic Humeanism is the fact that physicists aim for a "theory of everything" (TOE): a complete, all-encompassing theoretical framework that can in principle account for every physical phenomenon in the universe. This has been an especially salient and distinct feature of fundamental physics since Newton. The ideal of a TOE plays such an important role in fundamental physics that any theory that fails to account for a certain range of physical phenomena is automatically deemed non-fundamental, even if it is otherwise empirically successful. For instance, "effective field theories" that are highly predictively accurate at a certain level are regarded as non-fundamental because they break down at certain energy scales. Moreover, this completeness requirement appears to take precedence over other criteria for laws. For instance, simplicity only comes into play as a criterion of choice between theories that hold the promise of being able to account for all physical matters of fact whatsoever (with the possible exception of the initial conditions of the universe). As Woodward (2014, p. 102) notes, this is one upshot of Einstein's remark that "the supreme goal of all theory is to make the basic irreducible elements as simple and as few as possible *without having to surrender the adequate representation of a single datum of experience*" (1934, p. 165, my emphasis). A best-system account had better incorporate such a requirement, then, on pain of failing to match a key aspect of scientific practice. But pragmatic versions of the BSA currently on the market do not include any requirement of completeness.[3] Thus, the accounts of Dorst (2019b), Hicks (2018), and Jaag and Loew (2020)—the most detailed pragmatic versions of the BSA currently available—all leave room for the possibility that the best system might fail to account for portions of the mosaic as long as it fares substantially better than its competitors in other useful aspects (e.g. by being more easily confirmable, or more computationally tractable, etc.). Yet physicists do not seem willing to consider this possibility.

It is no surprise that those accounts do not capture the premium that physics puts on completeness. From a pragmatic Humean standpoint, this aspect of the practice of physics is somewhat of an enigma.[4] To illustrate, suppose you are trying to predict the future behavior of a rock sitting on your desk.[5] Physicists expect the laws to be complete in the sense that, for any possible exact initial conditions of the rock, those laws can in principle predict the future behavior of the rock down to its minutest microphysical details. (Likewise for every other physical subsystem, of course.) But to do this the laws need information about every point of an enormous spatiotemporal region, e.g. the entire cross-section of the rock's backward light cone in relativity. This is much more information than we can ever hope to gather, and calculating the behavior of the rock on its basis would be unfeasible for us anyway. Compare with macro-generalizations like "massive objects on a stable surface stay at rest unless pushed" or "objects sitting unstably at the border of a table have a high probability of falling." These yield only coarse-grained predictions of the rock's macroscopic behavior and break down for some of the rock's possible initial conditions (e.g. thermodynamically abnormal ones). But at least they *can* actually be used by agents like us, and generally yield highly reliable predictions. More generally, when modeling the behavior of complex systems one can often achieve enormous gains in representational and computational tractability at a small cost in accuracy and predictive power (Dennett 1991). If the search for laws is driven by the need to identify generalizations useful to agents like us, the premium put on completeness by fundamental physics is therefore mysterious, at least *prima facie*, since it comes at a considerable cost in tractability and user-friendliness.[6] (Pragmatic Humeanism here seems to make

---

[3] Nor did Lewis's account. Indeed, it couldn't: given the way Lewis understands strength, a comprehensive theory would have (absurdly) to provide a complete description of all that happens in the universe.

[4] But see Dorst (Chapter 9 in this volume) for a discussion of how pragmatic Humeanism could accommodate this feature of fundamental physics.

[5] An example of Elga (2007) also discussed by Jaag and Loew (2020).

[6] Jaag and Loew (2020, pp. 11–14), who recognize the issue, offer various considerations in response. For instance, they point out that the laws are 'error-tolerant' in the sense that small errors in the specification of initial conditions usually lead only to small errors in predictions (cf. Callender 2017, ch. 7). This makes the laws applicable even if we have only incomplete information about initial conditions. While this does alleviate the issue somewhat, the point remains that the laws could be

better sense of the practice of the special sciences.[7] Since the behavior of objects of those sciences is not feasibly representable or predictable in all of their micro-physical details, the common strategy in those sciences is to sacrifice accuracy and scope by constructing generalizations and models that represent the behavior of those systems only roughly and approximately, but in a way that is tractable and usable by limited agents like us.)

Here is another feature of fundamental physics that is problematic for prag-matic Humeans. As John Roberts (2008, pp. 16–24) observes, the notion of law at work in fundamental physics is highly selective: physicists sharply distinguish between fundamental laws and regularities that are "striking and pervasive" but nevertheless not a matter of fundamental law. Roberts's examples of the second category include the second law of thermodynamics, astronomical regularities such as Kepler's rules, or the fact that all planets orbit the Sun in the same direc-tion, and global cosmological facts such as the cosmic microwave background or the large-scale flatness of the universe. Roberts goes on to argue that because those regularities are highly informative, Lewis's BSA has trouble explaining why scientists do not regard them as laws. A similar problem besets existing pragmatic Humean accounts. While they recognize a variety of uses for the laws, these accounts all put the emphasis on the laws' ability to help limited agents like us make easy and reliable predictions on the basis of the limited portions of the mosaic we can observe. (On the pragmatic Humean approach, this focus on pre-diction makes sense, as creatures like us obviously have a clear practical interest in being able to make speedy and accurate predictions.) And this makes it puz-zling why scientists do not regard the regularities cited above as fundamental laws, given how predictively useful they are. Call this the *problem of selectivity*.

For the sake of illustrating the problem, assume that the laws of our world are those of classical mechanics, as was once believed to be the case. Note that adding (say) the second law of thermodynamics to the laws of classical mechanics would yield a system much more predictively useful for agents like us than classical mechanics alone. After all, with the second law of thermodynamics in hand, one can effortlessly and reliably predict the behavior of an enormous number of phys-ical systems of interest to us, e.g. that the cup of coffee in my hand will reach room temperature within the next hour or so. True, that information can also be extracted from the laws of classical mechanics, together with information about the initial conditions of the cup and the room. But predicting the behavior of my cup of coffee on the basis of classical mechanics is far more difficult than predict-ing it on the basis of the second law in two respects. It requires much more

---

made even more user-friendly with a small sacrifice in comprehensiveness, and that from a pragmatic Humean standpoint it is mysterious why physicists are unwilling to make such trade-offs.

[7] Here I echo some remarks of Woodward (2014, p. 119), who argues that simplicity/strength trade-offs posited in Lewis's BSA are more characteristic of the special sciences than of fundamental physics. See also Frisch (2014).

information about initial conditions—namely, information about the exact initial microstate of the system formed by the cup and its environment, or at least about the probability distribution over the possible initial microstates of that system. And calculating the behavior of the cup based on that information and the laws of classical mechanics is more computationally difficult and involves more steps than extracting it from the second law. Another way to predict the cup's behavior via the laws of classical mechanics would be to first use them to derive the second law from initial conditions of the universe, and then apply the second law to predict the cup's thermodynamic behavior. But of course this also requires more information and would be more computationally challenging than predicting the cup's behavior from a system that has the second law built into it right from the start. From the standpoint of current pragmatic best-system accounts, it is therefore mysterious why physicists are not willing to regard the second law of thermodynamics as an additional fundamental law over and above those posited by classical mechanics, since such an addition would help limited agents like us make faster and easier predictions.[8] (A similar case could be made, I believe, with other generalizations that physicists do not regard as fundamental, like the law of free fall and other principles of terrestrial dynamics, or astronomical regularities such as Kepler's rules or the fact that all planets orbit the Sun in the same direction.)

Now, like Lewis's account, all current pragmatic versions of the BSA include simplicity on their list of desiderata for best systems. So an obvious suggestion is that the gain in predictive usefulness obtained by adding the relevant generalizations to the laws of classical mechanics is more than offset by the resulting decrease in the simplicity of the system, and this is why physicists do not regard those generalizations as fundamental laws. But that thought is hard to square with the way in which these accounts understand simplicity and its pragmatic benefits. In many of those accounts, the simplicity desideratum is intended to exclude systems that agents like us couldn't possibly comprehend or manipulate, such as systematizations that list all the facts about the mosaic, or Lewis's "predicate F" system (see, e.g., Albert 2015, p. 23). But that motivation doesn't apply in our example. While adding (say) the second law of thermodynamics to classical mechanics yields a slight increase in complexity, the resulting system is certainly not representationally or computationally intractable for agents like us. A pragmatic preference for simpler systems could also be motivated by the fact that simpler systems require fewer cognitive resources to be stored (Jaag and Loew 2020, p. 11). But while the system made of the second law and the laws of classical mechanics requires slightly more storage in long-term memory than classical mechanics alone, this seems a small price to pay for the resulting gain in predictive usefulness. Finally, the simplicity requirement is also often motivated based

---

[8] Frisch (2014) makes a similar point in the context of a discussion of Albert's and Loewer's pragmatic version of the BSA (Albert 2000, 2015; Loewer 2007, 2012).

on considerations of user-friendliness: simpler theories should be preferred because they are easier to comprehend or enable easier and faster computations (e.g. Dorst 2019b, pp. 896–7). But this rationale *favors* the addition of the second law (and other non-fundamental generalizations such as the law of free fall) to the laws of classical mechanics, since, as noted above, doing so greatly streamlines the inference of an enormous number of facts.

Another suggestion is that these generalizations do not count as laws because they fail to meet a pragmatic criterion of breadth. We have an obvious interest in identifying generalizations that apply to as many subsystems as possible, and accordingly virtually all pragmatic versions of the BSA include a desideratum of breadth in their recipes for lawfulness.[9] Perhaps, then, the second law of thermodynamics doesn't count as a law because there are many subsystems of the universe to which it doesn't apply. (This is also true—and even more so—of the generalizations of astronomy or terrestrial dynamics.) But this suggestion faces several objections. For one thing, the range of application of the second law—from gases and cups of water to galaxies and black holes—is still extraordinarily broad. So the proposal only works if the breadth standard that a generalization must meet to count as a law is set very high. On a pragmatic picture it is not clear why such a stringent standard would have become part of the practice of physics. Moreover, it is not entirely clear that lawful generalizations would meet that standard. The laws of classical mechanics *in principle* apply to all physical systems[10], but in practice their breadth is severely limited: there are plenty of physical systems that are too complex for agents like us to be able to predict their behavior on the basis of the laws. And, on pragmatic Humeanism, presumably it is breadth in practice rather than in principle that would seem to matter. Finally, the proposal arguably leaves it mysterious why fundamental physics draws such a sharp and rigid distinction between regularities that are fundamental laws and regularities that are not. If lawfulness is so intimately tied to predictive breadth, it would make more sense for physicists to adopt a graded and context-sensitive notion of law, so that generalizations would count as more or less lawful depending on how broad they are, and/or which generalizations are laws would vary with the context. (For instance, Kepler's rules would count as laws in contexts where we are especially interested in astronomical predictions but not in others.)[11]

---

[9] For instance, Dorst (2019b, pp. 887–9), Hicks (2018, pp. 997–8), and Jaag and Loew (2020, pp. 19–20). This requirement helps those accounts explain why certain global facts about the universe such as its total number of particles do not count as laws despite being very informative in Lewis's sense.

[10] Remember that we are working on the assumption that they are the fundamental laws.

[11] Interestingly, such a flexible picture of laws is explicitly endorsed by at least some pragmatic Humeans, namely Cohen and Callender (2009). Though that picture seems to me to be significantly at odds with how physicists themselves think of laws, it is worth noting that it seems to fit the situation in the special sciences (as Cohen and Callender note in favor of their account). In the special sciences one finds predictive models that employ generalizations of varying degrees of breadth and invariance

A final response is that the second law *is* a fundamental law after all. Albert and Loewer argue, on pragmatic Humean accounts, that the low-entropy start of the universe (the "Past Hypothesis" or PH) and the statistical-mechanical probability distribution (PROB) conditioned on the PH are fundamental laws (Albert 2000, 2015; Loewer 2007, 2012). Together with the dynamical laws, these two facts entail the second law of thermodynamics (and arguably all other special science generalizations). This proposal makes the second law a theorem of the best system and thus a fundamental law or at least a direct consequence of the fundamental laws. But while one does find scattered remarks that fit with that view in the physics literature,[12] this proposal doesn't seem to have taken much hold among physicists. Perhaps this is because physicists are still in the grip of an antiquated conception of laws that prevents them from recognizing the lawful status of PH and PROB. But those not yet firmly committed to Humeanism may not find this claim especially plausible.

To sum up: pragmatic Humeanism, in its current incarnations, has trouble making sense of the premium that fundamental physics puts on comprehensiveness, and of the fact that physicists recognize so few generalizations as laws. And the issue is an especially pressing one. For these two features of laws seem to go against the general idea that the laws are tailored to be useful to agents like us. But, as noted above, on a Humean picture it is not clear how to make sense of physicists' interest for the laws other than through their supposed pragmatic benefits. So the problem here is one for Humeanism *tout court*. Moreover, anti-Humean views of laws (some of them at least) seem better poised to make sense of the features of laws under consideration. On a governing view of laws, for example, it makes sense to expect the laws to apply to every part and parcel of the universe. One can perhaps make sense of a scenario in which the laws govern only some portions of natural reality, though some complications would arise in developing that picture. (For example, what happens in interactions between those components of reality that are governed by laws and those that are not?) But in any case that hypothesis seems more contrived and unnatural than the scenario on which the laws govern all of nature, so that it is no surprise that physics takes the latter as working hypothesis. A governing view can also easily explain why the second law of thermodynamics is not a fundamental law in addition to those of classical mechanics: the second law is an enormously useful and informative generalization, but it is the laws of classical mechanics that do the real work of governing thermodynamic phenomena. If right, this undercuts one of the main considerations that Humeans routinely offer in favor of the BSA, namely that it fits scientific practice better than competitors. So the issue here is one that Humeans have to take seriously.

(see Woodward 2003), which play the functions characteristic of laws (explanation, counterfactual support, etc.) in some contexts but not others, and with little effort being made to rigidly separate those generalizations into 'fundamental' and 'non-fundamental' ones.

[12] For instance, Feynman (1965) envisions the possibility that the PH may be a law.

10.2

Fortunately, further reflection on the problem of selectivity suggests the beginning of the solution. I have argued that current pragmatic BSAs do not get the distinction between laws and non-laws right, and count too many generalizations as laws. But even if they did get the right results, it would be for the wrong reasons. Physicists do not decide whether a generalization is a law based on how useful it is for limited agents like us to make predictions. Instead, I suggest, they rely primarily on *explanatory* considerations. Consider the fact that the second law of thermodynamics, which was generally regarded as a fundamental law in early nineteenth-century theories of heat, had lost that status among physicists by the end of the century (see Brush 1976). The reason was that by that time it had become clear that entropy increase could be explained as a consequence of the laws of classical mechanics. (Not, of course, as a consequence of those laws alone, but as a consequence of the laws together with facts about the initial state of the universe, as became clear through Boltzmann's work.) By contrast, the principles of classical mechanics *were* regarded as fundamental laws at that time because it was widely thought that, while being able to explain a great number of phenomena, they could not themselves be explained in terms of deeper principles. Further support for that conjecture comes from other examples of striking but non-lawful generalizations mentioned in the previous section. Quite evidently, the reason why physicists do not regard, say, Kepler's rules as fundamental is that they can be explained as consequences of classical mechanics and the initial conditions of the solar system (including among other things the absence of any body at least as massive as the Sun in the vicinity of the solar system). Likewise, the uniformity in planetary orbital directions, the cosmic microwave background, and the flatness of the universe can all be explained as consequences of generalizations we already regard as fundamental and facts about initial conditions (respectively, the origins of the solar system as a gigantic gas of dust and cloud, the Big Bang, and the inflationary period).

These considerations suggest the following hypothesis. No matter how informative, broad, or useful a generalization is in itself, physicists will not deem it fundamental if it can be explained through generalizations that they already regard as fundamental laws. And conversely, they will deem it fundamental if it is not itself explainable in terms of further facts. In other words, physicists take the fundamental laws to be the *ultimate explainers*—the principles that explain other facts about the universe, while not themselves being explainable in terms of deeper principles. Hence a proposal for Humeans: an adequate BSA should take as starting point the idea that the chief virtue that makes a system best is a kind of *ultimate explanatory power*.

As I explained, the key motivation for this proposal is that it seems to get the distinction between laws and other regularities right—at least more so than other Humean accounts. (It also fits tightly what some physicists have to say about laws. Weinberg, for instance, speaks of the quest for a final theory as "the ancient search for those principles that cannot be explained in terms of deeper principles" (1992, p. 18).) But questions and concerns immediately arise. Explanation is an elusive notion, and using it as starting point for a Humean account of laws raises a number of issues. Many Humean-friendly theories of explanation—most obviously the deductive-nomological account (Hempel 1965)—presuppose a distinction between laws and non-laws already in place and hence cannot be used to elucidate what laws are. Furthermore, it is unclear how Humeans can make sense of the laws as *ultimate* explainers, since Humeanism also holds that laws do have explanations—they are metaphysically explained by their instances.[13] Finally, even if the proposal fits with how physicists themselves think of laws, it is unclear whether it can do a better job than other Humean accounts at motivating puzzling aspects of scientific practice, such as the ones discussed in Section 10.I. On an explanationist best-system account, the search for a comprehensive theory is naturally understood as a search for a theory that can (in principle) explain everything, but what exactly is the value of such a theory? And granting that physicists' distinction between laws and other striking regularities is based on explanatory considerations, what exactly is the payoff of drawing such a distinction? The underlying issue is that whereas making good predictions has an immediate and tangible value for agents like us, the point or value of explanation is rather opaque, at least from a Humean perspective.[14] In Salmon's words, 'Why ask, "Why?"?' (Salmon 1978). One answer is that explanations provide us with a sense of understanding, but this raises more questions than it answers. So, however well it fits actual scientific practice, an explanationist take on the best-system approach runs the risk of making it mysterious why physicists care so much about the laws.

In the remainder of the chapter I will sketch how an explanationist BSA can be developed to address these questions and concerns. In the next section I will propose a way to spell out the account in terms of explanatory unification. In Section 10.4 I will argue that the proposal can make good sense of why physicists care so much about the laws, and can also make sense of the features of laws discussed above that other pragmatic Humean accounts of laws leave mysterious.

---

[13] This issue is related to the well-known problem of "explanatory circularity" for Humeanism: if laws are just regularities, they are explained by their instances and hence cannot themselves figure in explanations of those instances, on pain of explanatory circularity (see, e.g., Lange 2013). In my view the best answer to that objection is that it runs together two distinct kinds of explanation, scientific and metaphysical (Loewer 2012; see also Dorst 2019a and Bhogal 2020). I will briefly return to this issue in the next section.

[14] Non-Humeans may claim that explanation has an obvious value: its function is to track metaphysically robust relationships of dependence in the world. But, of course, Humeans deny the existence of such relations.

10.3

Several considerations suggest unification as a key notion to develop an explanationist best-system account. The idea that explanation involves unification has a long pedigree in the philosophy of science (where it is associated mainly with Friedman 1974 and Kitcher 1981, 1989). And there is something deeply intuitive to the idea. Unification involves drawing together seemingly unrelated phenomena under a single cohesive theoretical framework, and such an achievement seems explanatorily valuable. In addition, even a cursory glance at the history of physics reveals a prominent role for considerations of unification. Many theories that physicists take or once took seriously as fundamental gained that status largely because they unified previously disparate domains: e.g. celestial and terrestrial dynamics for Newtonian mechanics, electricity and magnetism for Maxwell's equations and special relativity, and electromagnetism and weak nuclear theory for the electroweak theory. And contemporary physics' search for a TOE is fundamentally a search for unification—first, of the electroweak and strong forces into a Grand Unified Theory (GUT), which is then itself to be unified with gravity. This strongly suggests that the ultimate explanatory power that physicists expect from fundamental laws should be understood as unification. Finally, unificationism about explanation and the best-system account of laws go very well together. What makes a system best for the BSA is its ability to cover a wide range of phenomena via a suitably simple set of principles, and this is also what explanatory unification involves. Accordingly, several defenders of the BSA appeal to unification to make sense of the explanatory role of laws (e.g. Loewer (1996, p. 113) and Bhogal (2020)[15]), though they do not go as far as claiming that explanatory unification is what *makes* a system the best, as I am proposing here.[16]

Before explaining further how this goes, let me clarify that I do not wish to claim, as for instance Kitcher does, that *all* explanation is unification. Such an imperialist view of unification faces several objections. For instance, it entails that causal explanation can be reduced to unification. But it is hard to see how unification can account for the asymmetry of causal explanation (Barnes 1992). And Kitcher's attempt to fit causal explanation within a unificationist framework forces him to endorse an implausible 'winner-takes-all' on which only the most unifying theories are explanatory at all (Woodward 2017). I want to be a pluralist about explanation, and claim only that the explanatory role *of*

---

[15] Bhogal uses the idea that laws explain by unifying to distinguish scientific explanations from metaphysical ones (which do not aim at unification) and to thereby answer the explanatory circularity objection to Humeanism. This response fits very well with the account I propose in this chapter.

[16] Psillos (2002, ch. 10) defends a similar view, though with motivations different from mine. Kitcher (1989) also seriously considers the possibility that lawfulness can be understood on the basis of a prior notion of explanatory unification, and offers an interpretation of Kant's theory of laws along these lines (Kitcher 1994).

fundamental physical laws can be understood in terms of unification. Explanation in the special sciences, on the other hand, is mostly causal and does not aim at unification.[17] This modest position escapes the objections against unificationism just mentioned.

(In fact, I do not even wish to claim that the explanatory role of fundamental physical laws is exhausted by unification. Physicists presumably rely on a diverse range of considerations when assessing the explanatory value of a fundamental theory. A fully developed explanationist best-system account is thus likely to include additional criteria on good systematizations besides unification, though I will not explore what these might be here.)

My way of understanding unification is inspired by Friedman's (1974) account, though different from it in important respects. The main idea is as follows. One starts with a set S of facts to be unified. A theory T consists of a subset of those facts. We can think of the facts included in T as the basic theoretical principles or laws of the theory. The degree of unification of S by T is a function of how sparse T is (i.e., how many laws it contains), and of how many facts in S can be derived from T in a certain way. (I will explain what counts as a proper derivation shortly.) T is maximally unifying when it can properly derive all the facts in S, and no sparser theory exists that does the same. When a theory is maximally unifying, all facts in S can be derived from its laws, and those laws themselves cannot be derived from deeper, more encompassing principles. As should be clear, this fits closely with the idea of laws as ultimate explainers. My proposal, then, is that what makes a systematization the best is that it maximally unifies the mosaic, or at least comes closest to the ideal of maximal unification. (The reason for that caveat will appear later.)

To make this more precise, we need to specify the set S of phenomena that the laws are intended to unify. That set includes all the fundamental physical facts that compose the Humean mosaic. We need not take any specific stance on the nature of those fundamental physical facts here.[18] We must also include in S all the facts that supervene on the Humean mosaic. This is in line with the idea that physics aims to explain not only physical phenomena but also all other phenomena that depend on the physical. (Also, Humean laws are regularities that supervene on the mosaic. Since a fundamental theory is supposed to be a subset of the facts in S that set better include the laws.)

We also need to explain what it means for a fact f in S to be properly derivable from a theory T. Requiring f to follow from T alone would be too strong. True,

---

[17] Of course, on my proposal considerations of unification come into play in determining what the laws are and hence also in (partly) fixing the causal facts that work as explanantia in causal explanations. But this impact of unification on causal explanation does not entail that causal facts themselves do explanatory work by unifying.

[18] Lewis, of course, takes them to be instantiations of perfectly natural properties at spacetime points, but that characterization raises a number of issues (Loewer, 2007).

some particular facts about the universe directly follow from a fundamental theory. For example, the fact that a particular magnetic field has a net flux equal to zero can be deduced directly from Maxwell's second equation. But generally the laws entail a particular fact only given additional information—information that we would naturally describe as being about 'initial conditions'. Consider, for instance, a proposition p_2 describing the position of a planet at a given time. To derive p_2 from the laws of Newtonian mechanics, one needs a further proposition p_1 describing positions, velocities, and masses of the Sun and planetary bodies at an earlier time. I propose that we understand proper derivation as follows:

A fact f in S can be properly derived from T just in case there exist facts $f_1, \ldots, f_n$ in S distinct from f such that $f_1, \ldots, f_n$ and T together entail f.[19]

When I say $f_1, \ldots, f_n$ must be *distinct* from f, I mean that f is conceptually and metaphysically independent of $f_1, \ldots, f_n$ (Lewis 1986). Without this qualification, we would get the disastrous result that the best system is the empty one, since such a system would be maximally sparse and could 'account' for every fact in S by deriving it from itself.[20] Note that this conception of derivation includes no built-in constraint that f be derived from T and facts at other times. But, as Hicks (2018) notes, it is a plausible conjecture that, given the way that the contents of the universe are *de facto* distributed in spacetime, any theory that aims to derive a large number of physical facts while remaining tractable by agents like us will have to take the form of a dynamical theory that derives facts at a time from facts at other times.

Finally, to fully flesh out the proposal we need to explain when a theory counts as sparser than another. This raises a number of issues. For example, compare a system T_1 consisting of Galileo's laws of terrestrial dynamics and Kepler's astronomical rules, with the system T_2 obtained by conjoining those laws into a single statement. Obviously T_2 is not sparser than T_1 in any meaningful way, but explaining why is not easy.[21] Friedman's solution to this problem appeals to the idea that genuine identification works by reducing the total number of 'independently accepted' regularities, but that notion faces severe difficulties (Kitcher, 1976).

---

[19] This conception of derivation as a kind of entailment works only in the case of deterministic theories. There are natural ways to extend the account proposed here to encompass indeterministic theories. But I will not discuss them here.

[20] The distinctness requirement also helps address a related objection. If a system T contains a law L, it seems that the system T' obtained by eliminating L from T always counts as better. For T' is certainly sparser than T, and can derive L (and hence also all the facts that L helps derive) by listing all of L's instances. But since L's instances are not distinct from L, this doesn't count as a proper derivation. This fits well with the idea that scientific and metaphysical explanations are importantly different, as many Humeans have argued in response to the explanatory circularity objection (see note 13). While T's derivation of L from its instances might count as a proper metaphysical explanation of L, it doesn't count as a scientific explanation of L.

[21] See Hempel's discussion of the "problem of conjunction" (1965, p. 273).

A better suggestion is that the statement obtained by conjoining Galileo's and Kepler's laws cannot account for any phenomena not already covered by these laws, and hence does not genuinely unify them into a single law.[22] Compare with the genuine unification of terrestrial and celestial dynamics provided by Newton's law of gravitation: that law can account for all phenomena covered by Galilean and Keplerian laws, as well as other phenomena, such as meteorites and other objects that lie at the boundary of those laws (Douglas 2009, p. 456). (Interestingly, on that proposal replacing several laws by a single law yields a genuinely sparser theory only if it also increases in the number of phenomena derivable from the resulting theory.) These remarks certainly do not amount to a full account of sparseness, but giving such an account goes beyond the scope of this chapter.

To explain the logic of my account, let us assume that the laws of our world are those of Newtonian mechanics. On my view, what makes them the laws is, first, that all the facts about the mosaic can be properly derived from Newtonian mechanics, generally together with further information about other parts of the mosaic, in the manner that p_2 can be derived from p_1 (and p_1 itself would be derivable from even earlier facts about the solar system, and so on). Second, there is no sparser set of principles which can also derive all the facts in S. If there was one, then Newton's laws (which are among the facts in S) could themselves be properly derived from this deeper, more encompassing set of principles. It is because they cannot be derived in this way that they are the ultimate explainers. (The relevant sense of 'ultimate', note, is compatible with laws being grounded in their instances and hence not *metaphysically* ultimate. Replacing a law with a list of its instances would not yield a more unifying system. For instance, a system that simply listed all instances of Newton's laws with a list of their instances would be able to derive both those laws and all of the facts that Newtonian mechanics can properly derive, but would not be in any reasonable sense sparser than Newtonian mechanics.)

I will close this section by mentioning one complication for my proposal and explaining how to address it. (Readers pressed for time may safely skip to Section 10.4.) The complication is that if the universe had a beginning, its very initial conditions may not be explainable in terms of the laws, so that no fundamental theory can be the best system in my sense. Call this *the problem of initial conditions*. (One wrinkle is that many of the theories that physicists take seriously are time-symmetric in the sense that they permit the derivation of past from future facts. Those theories therefore can 'properly derive' initial conditions of the universe—namely, from *later* time slices of the universe. However, physicists do not seem to regard such later-from-earlier derivations as explanatory. Perhaps there is in fact no privileged direction of explanation in physics and this is mere prejudice on their part. Or (more likely in my view) it is evidence that physicists rely on additional considerations besides unification when assessing explanations.

---

[22] Here I am indebted to Psillos (2002, p. 272), who makes what I take to be the same suggestion.

At any rate, the time symmetry of physical theories does not offer a clear way out of the problem of initial conditions.)

The proper response to the problem, in my view, is that one can still hope that nature is kind enough to let a clear winner emerge from the competition between systems. In particular, suppose there is a system T from which every fact in S can be properly derived except facts about initial conditions of the universe. T, then, is a theory of nearly everything. And suppose also that serious competitors to T (i.e. alternatives to T that can explain a reasonable amount of facts in S) cannot explain those initial conditions either, and either fail to account for some other facts in S or can account for all remaining facts but in a less sparse way than T. T would then clearly come closest to the ideal of a maximally unifying theory, and thus deserves the title of best system. Modern physics, while willing to consider the possibility that initial conditions may not be explainable, does seem predicated on the hope that a theory with the features just described exists.

Initial conditions raise another issue for my account, however. Assume again that the laws of our universe are those of Newtonian mechanics, and let p be a proposition describing the initial conditions of the universe. Supposing that p cannot be properly derived from Newtonian mechanics, my proposal would seem to require making p a fundamental law as the system made of Newton's laws and p would account for more facts than Newtonian mechanics alone. This is at odds with how physicists think of initial conditions. A sensible response to this worry, however, is that p is likely to be too complex a proposition for us to ever be able to represent it, so that a system that includes it as a law would be cognitively intractable by agents like us. What if p describes just one of the small, localized facts that together compose the initial conditions of the universe? However, counting p as a law goes very much against the spirit of the unificationist approach, since this would do little to increase the unification of the mosaic. The resulting law would be an *ad hoc* addition designed to explain just one little fact rather than drawing together seemingly disparate facts under a common umbrella, as *bona fide* laws do. A unificationist best-system account can therefore address the issue by requiring laws to earn their keep: adding a new law to a system is permissible only if that new law helps the system account for a substantial amount of facts that could not be derived from it otherwise. I leave further development of this idea for future work.

10.4

I have argued that a proper best-system account should take as point of departure the idea that laws play a kind of ultimate explanatory role, and sketched a way to spell out such an account in terms of unification. I now want to discuss how the resulting account can help pragmatic Humeans make sense of the two problematic features of fundamental physics discussed at the outset of this chapter. These, remember, were the fact that physicists aim for a theory of everything, and count only a few select generalizations as laws. These two aspects of scientific practice are built into my account, on which the best system must be able to account for all or nearly all phenomena in the mosaic, and must do so by positing as few laws as possible. Of course, that my account recovers these aspects of scientific practice doesn't mean that it explains them. The question remains of why physicists spend so much time and effort finding a theory with these features, despite its lack of user-friendliness for predictive purposes.

To answer that question, it helps to return to an issue for my account already mentioned at the end of Section 10.2. This is the fact that the point of explanation is rather opaque, so that an explanationist version of the BSA runs the risk of leaving it mysterious why physicists care about the laws. One answer to that worry is that explanations are valuable because they provide understanding. Friedman (1974), for instance, endorses such a view in the context of defending his unificationist account of explanation. According to him, we gain understanding by reducing the number of independent facts we have to posit as brute. Assuming that physics seeks understanding, it therefore makes sense that physicists seek to identify the sparsest set of principles from which everything else can in principle be derived. Still, the question arises of why understanding itself is valuable. What exactly would be lost if we stopped seeking understanding?

A more promising line of response, in my view, is that explanations are instrumentally valuable, and that seeking and identifying explanations has tangible benefits for agents like us. In ordinary cognition, explanatory reasoning has been shown to have wide-ranging effects on a variety of cognitive tasks such as discovery, confirmation, and learning (Lombrozo 2011; 2016). My suggestion is that Humeans can provide a story of this kind to explain why laws with explanatory/unificatory power are valuable: searching for such a system is an efficient and fruitful way to explore the universe.[23] To see this, note that many striking facts about the ground of physics have been discovered via explanatory inference—that is, by noting that these facts together with the laws explain certain observations. Consider for example Laplace's discovery that the solar system originated in a gigantic gas cloud. Laplace arrived at this 'nebular hypothesis' by reasoning that, together with Newton's laws, this hypothesis would explain the puzzling fact that in the solar system all planetary orbits lie on a single plane and follow the same direction (a fact that Newton himself took to be the result of the direct intervention of God).[24] Likewise, striking facts about our early universe such as the Big Bang, the inflationary period, or the low entropy start of the universe have been established on the ground that together with the laws these facts explain various cosmological regularities—respectively, the cosmic microwave background,

---

[23] Dorst (2019a) has a similar proposal about how Humeans should understand the value of scientific explanation, though he develops it in a different way than I do here (partly because his account of explanation is less tied to unification than mine).

[24] See Roberts (2008, pp. 17–19).

the large-scale flatness of the universe, and the second law of thermodynamics. So we have a pervasive pattern in physics where some important and interesting fact E about the mosaic on the ground that E together with the laws L enables the derivation and hence the explanation of a seemingly puzzling observation O. (This amounts to showing that given E one can 'properly derive' O from L in the sense of the phrase introduced earlier.) This pattern suggests that by playing their explanatory role the laws serve an important function for us: they help us extract information from our observations. An explanationist version of the BSA can therefore vindicate the pragmatic Humean idea that what makes a system the best is its usefulness for agents like us. Limited agents like us face a particular epistemic predicament: we can observe only a very small portion of the universe, and need to figure what the rest is like on the basis of those very restricted observations. The laws help us solve that predicament by enabling the kind of explanatory inference just described. No wonder, then, that physicists are so interested in discovering the laws.

Now, it remains to be shown that this line of thought is still plausible when explanation is understood as unification, as I have proposed. That is, we need to show how a system that is unifying in my sense—a system that allows the derivation of everything in the mosaic from a set of principles as sparse as possible—favors the exploration of the mosaic in the way just described. If such a story can be told, we would finally have a way to justify the premium that physicists put on comprehensiveness and sparseness in pragmatic Humean terms.

I cannot offer a fully fleshed-out story of this form in this chapter, but here is a sketch of how it may go. Return to the example of Laplace's nebular hypothesis. Before Laplace's work, it was not yet known whether and how the uniformity in planets' orbital directions (call that fact U) could be made to fit with the laws of Newtonian mechanics. Faced with this question, an eighteenth-century physicist could have adopted one of three strategies. The first would be to try and provide a Newtonian explanation of U by figuring out what kind of initial conditions could have led to this striking uniformity via Newton's laws. This was, of course, Laplace's strategy. A second strategy would be to posit that U does not admit of a natural explanation in terms of physical laws, thereby 'reconciling' U with Newtonian mechanics by placing the former outside of the latter's purview. This was, in effect, Newton's move. A third strategy would be to explain U by positing it as a new basic law in addition to those of Newtonian mechanics: if it is a fundamental law that all planets orbit the Sun in the same direction, that immediately entails that they all do, and there is no further explanation to be sought. (Pragmatic versions of the BSA that emphasize the predictive role of laws validate that third strategy, since adding U to the book of laws would yield a substantial increase in predictive power.)[25]

---

[25] Perhaps U is not sufficiently predictively useful to deserve inclusion. But the point would still hold if we were to replace U with the second law of thermodynamics, or another extremely broad but still non-fundamental generalization.

But now note that, if one wants to hold on to Newtonian mechanics as fundamental theory, the requirements of comprehensiveness and sparseness in effect
privilege the first strategy. The rule that every physical phenomenon be properly derivable from the laws prohibits the second strategy. And the requirement
that the best system be as sparse as possible rules against the third strategy, by
prohibiting positing new laws lightly. It makes postulating a new law a solution
of last resort, to be avoided unless there really is no hope of accounting for that
phenomenon on the basis of principles already recognized as laws.[26]

Note, moreover, that it is only by pursuing the first strategy that one could be
led to discover the facts about the origins of the solar system that explain
U. (Indeed, our best reason to believe that the solar system started out as a rotat-
ing gas cloud is still that this together with the laws explains uniformities in
planetary orbits.) Had the scientific community in the eighteenth century not
pursued that strategy, and been content to either restrict the explanatory scope of
Newtonian mechanics or posit U as a new fundamental law, that fact may well
have never been discovered, or only much later.[27] Other examples with the same
moral are easy to find. Think of other striking physical regularities that physicists
do not regard as a matter of fundamental law—e.g. the red shifting of galaxies, the
large-scale flatness of the observable universe, or the constant increase in entropy.
In each case, attempts to explain those regularities on the basis of current fundamental physical theories led, via inference to the best explanation, to the discovery of further striking facts about the early history of the universe (the Big Bang,
the inflationary period, and the Past Hypothesis). Again, had scientists been content to restrict the scope of their fundamental theories or posit those regularities as
laws, they would likely have missed out on those discoveries.

Putting these considerations together suggests a hypothesis about the value of
the comprehensiveness and sparseness requirements on the best system. Those
requirements act as norms on the kinds of moves that physicists can make when
it is unclear how a preferred fundamental theory T can accommodate some
observation O. In effect, they encourage one possible reaction to that situation—
namely, searching for further facts through which T can explain O—at the
expense of other possible reactions. And the history of physics shows that in our
world searching for such explanations is fruitful, and constitutes an excellent way
to discover significant facts about the universe. It is therefore no surprise that the

---

[26] I note in passing that this proposal makes sense of the fact that physicists have yet to endorse
Albert and Loewer's proposal that the Past Hypothesis is a law: the jury is still out on whether the PH
can be explained in terms of the dynamical laws and other facts about the early universe.

[27] Roberts (2008, pp. 21–2) argues that Lewis's BSA warrants positing U as a law and hence cannot
explain why Laplace devoted so much effort to providing an explanation of U. This is part of his
broader argument to the effect that Humean accounts of laws cannot make sense of physicists' distinction between laws and other regularities (cf. Section 10.1). In effect, the account offered in this chapter
provides a response to this argument.

practice of fundamental physics, which aims at the discovery of such facts, is structured by such norms.[28]

Here is an objection against this line of argument. It is clear that investigation of the world must obey norms requiring scientific theories to be reasonably comprehensive and simple. If it were acceptable to restrict the scope of one's theory (or postulate a new principle *ad hoc*) whenever one encounters some recalcitrant phenomenon, scientific theorizing would likely not go very far, and certainly wouldn't lead to the kind of significant discovery exemplified by Laplace's nebular hypothesis. But why insist on *maximal* comprehensiveness and sparseness? A worry similar to the one raised in Section 10.1 for other pragmatic Humean accounts arises here. After all, maximally comprehensive and sparse theories of the sort one finds in fundamental physics can *in practice* be applied only in a very few select contexts, so that there are many significant facts about the universe that we would want to know but could never hope to discover through those theories. For example, it would be ludicrous to hope to derive any interesting fact about, say, the actual evolutionary history of life on earth by doing fundamental physics.

But that objection can be put to rest if we pay attention to the fact that fundamental physics isn't all of science, and that the special sciences deploy a quite different strategy to investigate the world. More precisely, while the special sciences also display a concern for unification, each of them seeks to unify only one specific domain of reality (e.g. life, the mind, or the economy). And scientific theorizing in these disciplines is also heavily influenced by considerations of user-friendliness, tractability, and practical relevance to agents like us. These further considerations lead to trade-offs with considerations of unification. For instance, there is no expectation that a good special scientific theory be able to account for *all* phenomena even within its domain (because a fully comprehensive theory would likely be too complex to help us efficiently navigate the world). And likewise, there is generally little pressure to make the theory as sparse as possible (since often a demand for sparseness interferes with considerations of user-friendliness and tractability). By contrast, the strategy adopted by fundamental physics to investigate the world involves pushing the concern for unification to its maximum, while other considerations take a back seat.

By adopting these different strategies, physics and the special sciences nicely complement each other. By yielding theories limited in range but easily tractable and usable by agents like us, the special sciences help us discover many significant facts about the Humean mosaic that we could never uncover by doing

---

[28] A similar explanation can be given of the requirement that a scientific theory can properly explain a fact by deriving it from other facts only if the latter are distinct from the former. (This requirement, remember, is built into my account of unification.) One could "explain" U by deriving it from a list of its instances, but such a "metaphysical explanation" would not yield any interesting scientific discovery of the kind Laplace made. The distinctness requirement in effect prohibits cheap explanations of that sort.

fundamental physics. For instance, we can uncover the most important milestones in the history of life by using the generalizations of evolutionary biology along with those of anatomy, molecular biology, and other areas of biology. But there are also significant facts about the universe that, in all likelihood, limited agents like us could discover only by adopting the investigative strategy characteristic of fundamental physics.

One vivid illustration is the discovery of the Higgs boson. Its existence was first postulated in the context of the Weinberg–Salam–Georgi electroweak theory. That theory seeks to provide a unified account of an enormous number of physical interactions (including at energy scales that we never encounter) in terms of a few basic principles, and thus nicely exemplifies physics' obsessive pursuit of unification. The theory entails that at very high energy regimes the W and Z bosons (which carry the weak interaction) and the photon (which carries the electromagnetic interaction) are all massless. This raises the question why at lower energy regimes the W and Z bosons become massive while photons remain massless. Postulating the Higgs boson provides an explanation of that fact by supplying a mechanism that breaks the symmetry between photons and W/Z bosons as temperatures decrease. This is a dramatic example of a significant discovery falling out of physics' strategy of searching for a maximally unified theory of the universe, and it is hard to imagine how we could have made that discovery other than by following this strategy.

In short: what justifies the norms of maximal comprehensiveness and sparseness is not that any fruitful inquiry into the natural world must abide by them. Instead, it is the fact that an investigative strategy guided by these norms enables the discovery of phenomena that no other investigative strategy available to us could uncover.

I will close by noting an interesting consequence of the argument presented in this section for the way in which Humeans should think of the value of laws. Most versions of the BSA assume that physicists seek to know the laws because the end product of that search is valuable. That is, we want to discover the laws because they contain information that is extremely useful to successfully navigate the world around us. In this picture the benefits of the laws kick in only once we know them. But consider Laplace's discovery of the nebular hypothesis. It would be wrong to say that Newtonian mechanics was known to be the true fundamental theory. For one thing, it isn't. And the question whether it could account for all phenomena in nature was still very much in dispute. In that context, the discovery of the nebular hypothesis served as a major piece of confirmation for Newtonian mechanics by showing that the theory could account for puzzling phenomena such as R without resorting to divine intervention. So the example holds an important lesson. On the view proposed in this chapter, searching for the laws is a valuable scientific aim not only because the end product of that search is valuable but also because the very *process* of searching for a maximally unifying account for the world leads to the discovery of significant facts about the mosaic. In other words, in our quest for the laws, it is not just the destination, but also the journey that matters.[29]

References

Albert, D. (2000). *Time and Chance*. Cambridge MA: Harvard University Press.

Albert, D. (2015). *After Physics*. Cambridge MA: Harvard University Press.

Barnes, E. (1992). 'Explanatory Unification and the Problem of Asymmetry', *Philosophy of Science*, 59, pp. 558–71.

Bhogal, H. (2020). 'Nomothetic Explanation and Humeanism about Laws of Nature', in K. Bennett and D. Zimmerman (eds.), *Oxford Studies in Metaphysics,* 12. Oxford: Oxford University Press, pp. 164–202.

Brush, S. (1976). *The Kind of Motion We Call Heat*. Amsterdam: North-Holland Publishing Company.

Callender, C. (2017). *What Makes Time Special?* Oxford: Oxford University Press.

Cohen, J., and Callender, C. (2009). 'A Better Best System Account of Lawhood', *Philosophical Studies,* 145, pp. 1–54.

Dennett, D. (1991). 'Real Patterns', *Journal of Philosophy*, 88, pp. 27–51.

Dorst, C. (2019a). 'Humean Laws, Explanatory Circularity, and the Aims of Scientific Explanation', *Philosophical Studies*, 176, pp. 2657–79.

Dorst, C. (2019b). 'Towards a Best Predictive System Account of Laws of Nature', *British Journal for the Philosophy of Science*, 70, pp. 877–900.

Douglas, H. (2009). 'Reintroducing Prediction to Explanation', *Philosophy of Science*, 76, pp. 444–63.

Einstein, A. (1934). 'On the Method of Theoretical Physics', *Philosophy of Science*, 44, pp. 248–68.

Elga, A. (2007). 'Isolation and Folk Physics', in H. Price and R. Corry (eds.), *Causation, Physics and the Constitution of Reality*, Oxford: Oxford University Press, pp. 106–19.

Feynman, R. (1965). *The Character of Physical Law*. New York: Modern Library.

Friedman, M. (1974). 'Explanation and Scientific Understanding', *Journal of Philosophy*, 71, pp. 5–19.

Frisch, M. (2014). 'Why Physics Can't Explain Everything', in Wilson, A. (ed.), *Chance and Temporal Asymmetry*. Oxford: Oxford University Press, pp. 221–40.

Hall, N. (2012). 'Humean Reductionism about Laws of Nature', unpublished manuscript, https://philarchive.org/archive/HALHRAv1.

---

[29] Thanks to Michael Hicks, Jenann Ismael, Siegfried Jaag, Christian Loew, and audiences at the January 2016 Workshop in Philosophy of Physics at the University of Arizona and the 33rd Conference in the History and Philosophy of Science at CU Boulder for extremely helpful comments and discussion.

Hall, N. (2015). 'Humean Reductionism about Laws of Nature', in Loewer, B., and Schaffer, J. (eds.), *A Companion to David Lewis*. Oxford: Blackwell, pp. 262–77.

Hempel, C. (1965). *Aspects of Scientific Explanation*. New York: Free Press.

Hicks, M. (2018). 'Dynamic Humeanism', *British Journal for the Philosophy of Science*, 69, pp. 983–1007.

Jaag, S., and Loew, C. (2020). 'Making Best Systems *Best for Us*', *Synthese*, 197, pp. 2525–50.

Kitcher, P. (1976). 'Explanation, Conjunction and Unification', *Journal of Philosophy*, 73, pp. 207–12.

Kitcher, P. (1981). 'Explanatory Unification', *Philosophy of Science*, 48, pp. 507–31.

Kitcher, P. (1989). 'Explanatory Unification and the Causal Structure of the World', in Salmon, W., and Kitcher, P. (eds.), *Minnesota Studies in the Philosophy of Science*, vol. 13. Minneapolis: University of Minnesota Press, pp. 410–505.

Kitcher, P. (1994). 'The Unity of Science and the Unity of Nature', in Perrini, P. (ed.), *Kant and Contemporary Epistemology*. Dordrecht: Kluwer, pp. 253–72.

Lange, M. (2013). 'Grounding, Scientific Explanation, and Humean Laws', *Philosophical Studies*, 165, pp. 255–61.

Lewis, D. (1983). 'New Work for a Theory of Universals', *Australasian Journal of Philosophy*, 61, pp. 343–77.

Lewis, D. (1986). 'Events', in *Philosophical Papers*, Vol. II. Oxford: Oxford University Press, pp. 241–69.

Lewis, D. (1994). 'Humean Supervenience Debugged', *Mind*, 103, pp. 473–90.

Loewer, B. (1996). 'Humean Supervenience', *Philosophical Topics*, 24, pp. 101–27.

Loewer, B. (2007). 'Laws and Natural Properties', *Philosophical Topics*, 35, pp. 313–28.

Loewer, B. (2012). 'Two Accounts of Laws and Time', *Philosophical Studies*, 160, pp. 115–37.

Lombrozo, T. (2011). 'The Instrumental Value of Explanations', *Philosophy Compass*, 6(8), pp. 539–51.

Lombrozo, T. (2016). 'Explanatory Preferences Shape Learning and Inference', *Trends in Cognitive Sciences*, 20, pp. 748–59.

Psillos, S. (2002). *Causation and Explanation*. Routledge.

Roberts, J. (2008). *The Law-Governed Universe*. New York: Oxford University Press.

Salmon, W. (1978). 'Why Ask, "Why?"? An Inquiry Concerning Scientific Explanation', *Proceedings and Addresses of the American Philosophical Association*, 51: pp. 683–705.

Weinberg, S. (1992). *Dreams of a Final Theory*. New York: Pantheon Books.

Woodward, J. (2003). *Making Things Happen*. Oxford: Oxford University Press.

Woodward, J. (2014). 'Simplicity in the Best Systems Account of Laws of Nature', *British Journal for the Philosophy of Science*, 65, pp. 91–123.

Woodward, J., and Ross, L. (2017). 'Scientific Explanation', in Zalta, E. (ed.), *Stanford Encyclopedia of Philosophy*, Fall 2017 Edition, https://plato.stanford.edu/entries/scientific-explanation/.

11

A Discourse on Methods; or, Humean Metaphysics of Science without Best Systems

John T. Roberts

11.1 Introduction

The best-known Humean theory of laws of nature is David Lewis's best-system analysis.[1] Almost all of the other Humean theories of laws in the literature are also "best-system theories": Like Lewis's account, they say that to be a law is to belong to a system of statements or propositions that maximizes some desirable feature.[2] In fact, in conversation, philosophers commonly use "the Humean account of laws" and "the best-system account of laws" interchangeably. But Humeanism about laws and the best-system approach to laws are not the same thing: Humeanism imposes a metaphysical constraint on an acceptable theory of laws, and the best-system approach is one strategy for meeting that constraint. There are other theories of lawhood that also meet the constraint.[3]

In this chapter, I aim to present a sketch of a new Humean theory that eschews all reference to best systems, and to make it clear that this theory is plausible enough to be worth looking into further. The theory is meant to include accounts of all the items in the same well-loved metaphysical neighborhood as lawhood—including counterfactuals, singular and general causation, powers and dispositions, and objective chance. But in this chapter I will focus just on laws and counterfactuals.

The central innovation of this theory is to use the concept of an *effective method* as the key primitive: the project is to give informative and illuminating truth conditions for counterfactuals, law-statements, causal claims, and so on that

---

[1] Lewis (1973, 1994).

[2] For example, Braddon-Mitchell (2001), Cohen and Callender (2009), Dorst (2019), Hall (unpublished), Hicks (2018), Loewer (2012), Roberts (1999).

[3] E.g., Ayer (1956), Roberts (2009), and Skyrms (1980), as reinterpreted by Armstrong (1983). Of course, the naive regularity theory is also a Humean theory without best systems, though it is perhaps not the best advertisement for this type of theory. See Roberts (2008, ch. 1) for some reasons why Humeans might be skeptical of the best-system approach.

John T. Roberts, *A Discourse on Methods; or, Humean Metaphysics of Science without Best Systems* In: *Humean Laws for Human Agents*. Edited by: Michael Townsen Hicks, Siegfried Jaag, and Christian Loew, Oxford University Press.
© Oxford University Press 2023. DOI: 10.1093/oso/9780192893819.003.0012

presuppose no non-logical concept other than that of an effective method. Effective methods are very closely related to what Nancy Cartwright (1979) calls "effective strategies"; roughly speaking, an effective method is a way of trying to promote some end, which has at least some chance of actually promoting that end, and is therefore worthy of being seriously considered by a rational agent interested in promoting that end. An important part of my task here is to show that it is consistent with Humeanism to invoke the concept of an effective method in the course of giving a theory of laws.

I will proceed in three stages: First I will show how to give truth conditions for counterfactuals, assuming that it is fair to take for granted the notion of an effective method. Second, I will show how to define lawhood on the basis of the same assumption. Of course, this assumption will not go unchallenged: which methods are the effective ones might well depend on which counterfactuals are true and on what the laws are, in which case my account will be circular. Perhaps more importantly (given the theme of this volume), it's far from obvious that the resulting account of laws and counterfactuals qualifies as Humean. In fact, someone could endorse my analyses of laws and counterfactuals, while adopting a non-Humean conception of effective methods, resulting in a kind of non-Humean mirror image of the theory I present here. The pressing question for me is whether a *Humean* version of this theory is really possible: This would require a way of conceiving of effective methods in a way that not only makes them more fundamental than laws and counterfactuals (so that it is permissible to appeal to them in a theory of laws and counterfactuals) but also makes them acceptable by Humean lights. In the third stage of this chapter, I will turn to this problem.

11.2 Stage One: Counterfactuals

11.2.1 Strategy: Counterfactuals from Semifactuals

My strategy for treating counterfactuals is to do all the heavy lifting with my account of *semifactuals*. A *semifactual* is a counterfactual whose consequent is true in the actual world—for example, 'If I were a decent billiards player, then I would (still) be less famous than Barack Obama.'

Suppose we have a theory of the truth conditions of semifactuals, by which I mean a theory of the following form:[4]

SF schema: Where b is true in the actual world, $(a \mathbin{\Box\!\!\rightarrow} b)$ is true just in case, and in virtue of the fact that: ...

---

[4] I use the symbol $\Box\!\!\rightarrow$ as the counterfactual-conditional connective.

Then we can extend this theory into a general theory of the truth conditions of counterfactuals by adding the principle FC (which stands for 'False Consequent'):

FC: Where c is false, $(a \,\square\!\!\rightarrow c)$ is true just in case, and in virtue of the fact that, there is some true b such that $(a \,\square\!\!\rightarrow b)$ is true and $(a \,\&\, b) \vDash c$.[5]

This is just what I propose to do.

Why should anyone adopt this strategy?[6] The only really good answer I know is that it is vindicated by its fruits; see below. But it is worth pointing out that a big part of the content of FC is uncontroversially true; in particular:[7]

FC [biconditional part]:   Where c is false, it is necessary that: $(a \,\square\!\!\rightarrow c)$ is true if and only if there is some true b such that $(a \,\square\!\!\rightarrow b)$ is true and $(a \,\&\, b) \vDash c$.

The part of FC that this leaves out is the 'in virtue of the fact that' clause. In effect, FC [biconditional part] tells us that the counterfactuals supervene on the semi-factuals; the rest of FC adds that the counterfactuals with false consequents depend on the semifactuals. If you endorse grounding-talk, you may paraphrase FC as saying that the false-consequent counterfactuals are grounded (in a particular way) in the semifactuals, and FC [biconditional part] as the supervenience thesis that follows from this grounding thesis. FC [biconditional part] should be uncontroversial, because it follows from the soundness of an obviously correct rule of inference. Namely:

Extended Weakening the Consequent: If $(a \,\square\!\!\rightarrow c)$ is true, and $(a \,\&\, c) \vDash d$, then $(a \,\square\!\!\rightarrow d)$ is true.

The task before me is to fill in the SC schema.

11.2.2 Means, Ends, and Semifactuals

Towards that end, consider a famous counterfactual:

SomeoneElse: If Lee Harvey Oswald had not assassinated John F. Kennedy, then someone else would have.

[5] Please understand the symbol $\vDash$ as standing for the relation of semantic entailment as defined in standard first-order logic; in other words, no special logical axioms or rules from the logic of counterfactuals (such as Extended Weakening the Consequent—which I am about to introduce) count.

[6] It might be worth noting that Goodman (1983) adopts much the same strategy; an account of which semifactuals are the true ones is almost equivalent to an account of which truths are 'cotenable' with which counterfactual antecedents.

[7] FC [biconditional part] contains an occurrence of 'it is necessary that' which has no counterpart in FC, but I take this necessity to be implicit in FC's 'in virtue of' clause.

The alert reader[8] will notice that **SomeoneElse** is not a semifactual, but it is obviously equivalent to one, namely:

StillKilled: If Lee Harvey Oswald had not assassinated John F. Kennedy, then John F. Kennedy would (still) have been assassinated.

Who believes **SomeoneElse** and **StillKilled?** Answer: Conspiracy theorists. They believe that Oswald was merely a pawn used by a vast and powerful group of conspirators determined to get rid of the President, and if Oswald had not done the deed, then the conspirators would simply have used a different pawn. Who disbelieves **SomeoneElse** and **StillKilled?** Those who accept the conclusion of the Warren Commission that Oswald was a "lone wolf," and there was no powerful group operating behind the scenes.[9]

Why do conspiracy theorists think that **StillKilled** is true? Well, one other thing that they believe is that getting to Oswald and preventing him from pulling the trigger was not, under the circumstances, a good way of saving the President's life. Had it been your goal to prevent Kennedy from being assassinated, then stopping Oswald from doing the deed might have been a necessary condition, but it would not have been enough. On the other hand, those who think Oswald was a lone wolf presumably also think that stopping Oswald might well have been an effective way of saving Kennedy. They need not think that it definitely would have been; given that they deny **StillKilled**, all that consistency demands of them is that they affirm:

MightNot: If Oswald had not assassinated Kennedy, then Kennedy might not have been assassinated at all.

And **MightNot** expresses a thought that goes hand in hand with the thought that, under the actual circumstances, stopping Oswald might have been an effective means of saving Kennedy.

There does not seem to be anything unique about all these conditionals about Kennedy and Oswald; the crucial point here should generalize: When a is false and b is true, we are disposed to accept and believe the semifactual $(a \,\Box\!\!\rightarrow\, (\text{still})\ b)$ just in case we are disposed to accept and believe that the (merely possible) act of

---

[8] More carefully: the alert reader *who is not a certain type of conspiracy theorist* will notice this— namely the type that not only denies that Oswald acted alone but also denies that Oswald was the trigger-man at all. I will ignore this possibility in what follows.

[9] Of course, things are more complicated than I suggest here: There are any number of other reasons why a person might accept or reject **SomeoneElse**. But I doubt that this makes any important difference to the following discussion; judge for yourself.

making it the case that a does not (or did not), under the actual circumstances, count as an unused but potentially effective means to the end of making it so that $\neg b$; we are disposed to accept and believe the negation of that semifactual just in case we are disposed to accept and believe that the act of making it the case that a does, under the actual circumstances, count as an unused but potentially effective means to the end of making it so that $\neg b$. Less carefully but more briefly: We tend to believe ($a \,\square\!\!\rightarrow$ (still) b) just in case we think that a does not represent a *missed opportunity* to make it the case that $\neg b$.

I suggest that our judgments about which semifactuals are true tend to line up in this way with our judgments about which unused means were potentially effective ones for a very good reason: Namely, because the truth conditions of these judgments are identical. That is:

Semifactual–Means Equivalence: Necessarily, whenever b is true, the semifactual ($a \,\square\!\!\rightarrow b$) is true iff making a true is not an unrealized but potentially effective means to the end that $\neg b$.

Equivalently: Necessarily, whenever b is true, ($a \,\square\!\!\rightarrow b$) is false iff making a true is an unrealized but potentially effective means to the end that $\neg b$.

There is a very strong appearance that **Semifactual–Means Equivalence** is at least extensionally correct, and it is very plausible that it is necessarily so. The case of **StillKilled** seems representative. One thing you might worry about is that **StillKilled** is a rather special kind of conditional, namely one whose antecedent asserts the absence of a cause that is actually present. But things seem to work the same with conditionals in which the antecedent asserts the presence of a cause that is actually absent; consider this example:

Shattered: If I had dropped this glass, it would have shattered.

(Assume that **Shattered** is uttered in a context in which the salient glass is very fragile, I did not actually drop it, and the context makes it clear that "If I had dropped this glass" means "If I had dropped this glass at time t," where at time t I was in a room with a very hard stone floor, gravity is normal, there are no wizards around, and other conditions are what we would happily call "normal.") **Shattered** evidently stands or falls with this semifactual:

StillNotBoth: If I had dropped this glass, then it would (still) have been the case that the glass is not both dropped and unshattered.

According to **Semifactual–Means Equivalence**, **StillNotBoth** is equivalent to:

BadMeans: Getting me to drop this glass does not qualify, under the actual circumstances, as an unused but potentially effective means to getting the glass to be both dropped and unshattered.

And this equivalence seems plausible: If you wanted the glass to be both dropped and unshattered, then getting me to drop it, under the actual prevailing circumstances, would have been a terrible means of doing so. On the other hand, if we wanted to modify the circumstances in such a way as to make **BadMeans** false, then a potential modification (such as, say, making the floor a lot softer) would fill the bill just in case it also made **StillNotBoth** and **Shattered** become true.

Things go just as well in cases where the antecedent is not causally related to the consequent at all. For example, **Semifactual–Means Equivalence** implies that this conditional:

If I were a decent billiards player, then I would (still) be less famous than Barack Obama.

stands or falls with this proposition:

Getting me to become a decent billiards player does not, under the actual circumstances, qualify as an unused but potentially effective means to the end of making me more famous than Barack Obama.

which, again, seems just right: In the real world, both of these are obviously true, and to describe a possible circumstance in which one of them is false is, evidently, to describe a possible circumstance in which the other one is false, too.

But it is not enough for my purposes to convince you that **Semifactual–Means Equivalence** is extensionally correct, or even that it is necessarily so. You and I might agree that it is a necessary truth while disagreeing about why it is true. One possible explanation is that what we mean by calling something an "unrealized but potentially effective means" to some end just is that a certain semifactual is false; another is that something's having the status of an unrealized but potentially effective means is grounded in the negation of that semifactual, or perhaps in the absence of whatever it would take to ground that semifactual. In either case, the arrow of explanation points from the truth value of the semifactual to the status of the means as a potentially effective one. My proposal here is to turn this arrow of explanation around. On my theory, various merely possible events stand there—in a manner of speaking[10]—as unused but potentially effective means to

---

[10] I do not mean to say that I am ontologically committed to merely possible events; the point is just that there are truths that can be aptly stated in ordinary language by sentences like: "Stopping Oswald was an unused, but potentially effective, way of saving Kennedy."

various unrealized ends, and others don't, and this is what accounts for why the true semifactuals are true and the false ones false. If you agree with me about this, then you might say that semifactuals are grounded in, or metaphysically explained by, unrealized-but-potentially-effective-means facts. (Or more briefly: semifactuals are grounded in missed opportunities.) Personally, though, I think it's more plausible that the explanatory dependence here is straightforwardly semantic: our natural-language counterfactual locutions are a handy syntactic device for reporting facts about what missed opportunities there are.[11] Be that as it may, for the purposes of this chapter, what's important is just the thought that, in the **Semifactual–Means Equivalence**, the direction of explanation runs from the right-hand side (about potential effectiveness) to the left-hand side (the truth of the semifactual).

11.2.3 A Preliminary Official Formulation

To give a kind of official formulation, the theory of counterfactuals I am proposing can be expressed as the conjunction of two principles:

SF: Where b is true, $(a \,\square\!\!\rightarrow (\text{still})\ b)$ is true just in case, and in virtue of the fact that, a is not an open, potentially effective means to $\neg b$.

FC: Where c is false, $(a \,\square\!\!\rightarrow c)$ is true just in case, and in virtue of the fact that, there is some true b such that $(a \,\square\!\!\rightarrow b)$ is true and $(a \,\&\, b) \vDash c$.[12]

These principles require a bit of clarification.

First, I speak of a proposition being a "means" or an "end"; more precisely, the means here is the possible act of rendering the means-proposition true, or of bringing about the event whose occurrence the means-proposition states; the end is the outcome that makes the end-proposition true.

Second, by an *open* potentially effective means to an end, I mean a potentially effective means that does not actually get used, and in this sense is "left open" (more on this in just a moment). What I call "means" and "ends" are propositions, occurrences, or states of affairs that could be true, occur, or obtain, even if no agent is employing them as means, or aiming at them as ends. For example, one good means to getting a pot of water on the stove to boil more quickly is to add a little salt; in my way of talking, this means that the proposition *that a little salt is added to the pot of water* is a means to the proposition *that the water comes to a*

---

[11] Again, please don't take this to be a statement of serious ontological commitment: In an obvious sense, there are plenty of missed opportunities in the world, whether or not the basic furniture of the universe includes some entities aptly called 'missed opportunities.'

[12] Please continue to understand the symbol $\vDash$ as standing for the relation of semantic entailment as defined in standard first-order logic; in other words, no special logical axioms or rules from the logic of counterfactuals count here.

boil sooner (than otherwise similar unsalted water would do). The means-proposition here could be true even if I wasn't thinking about boiling times when I added the salt to the pot; it could even be true even if neither I nor any other agent rendered it true[13]: A bizarre accident might result in salt being added to the pot. When I call a proposition a "means to some end," I just mean that it is a sort of proposition that someone who had the power to render true might render true as a means to that end.

It is even possible for a proposition to be a "potentially effective means" in my sense even if there is no agent capable of employing it as a means: being potentially effective is not at all the same thing as being remotely feasible. A means-proposition corresponds to a "possible act" only in the very weak sense that it is not logically or *a priori* ruled out that some agent does something to render the proposition true.[14] A "potentially effective" means is one that might actually get the job done, should someone implement it; a means can be like that even if there is no one in the universe with the ability to implement it. So, for example, if Big Bang cosmology is right, then *that the initial mass-energy of the universe was B* (where *B* happens to be much greater than its actual value) might be a potentially effective means to *that the universe collapsed back in on itself before any galaxies got a chance to form*. No agent was ever in a position to employ this means—so it represents a "missed opportunity" only in a thin sense—but still, it seems non-vacuously true that any agent who could have employed it, and wanted to head off galaxy formation, might have achieved that end via this means.

Finally, for a potentially effective means to be "open", it is not enough that the means-proposition is false. To see why, consider an example:

An Aspirin and a Prayer.   At time t, I had a mild fever. A potentially effective means to the end *that my fever go down shortly after t* is *that I take a dose of aspirin at t*. Another potentially effective means to the same end is *that I take a dose of aspirin and say a prayer at t*. I assume that my taking aspirin and saying a prayer is exactly as potentially effective, *qua* means of getting a fever to go down, as just taking aspirin; the prayer adds nothing to the efficacy of the aspirin, and takes nothing away from it either.[15] As it happens, I did take a dose of aspirin, I did not say a prayer, and my fever did not go down.

---

[13] As I am using the words here, "*S* rendered it true that *p*" just means that *S* acted in a way whereby they made it so that *p*. I use "rendered" rather than the more colloquial "made" in order to avoid giving the impression that I am talking about what contemporary metaphysicians call "truthmakers."

[14] So, it's *almost* true that the proposition *p* is a possible act in my sense iff it is not *a priori* impossible that *p*. Only almost, because of cases like the proposition *that the cat is on the mat for reasons not owing to the activities of any agent*, which could be true, but couldn't be rendered true by an agent. Thanks to Nina Emery for helpful discussion about this point.

[15] Another possible view is that something can count as a potentially effective means only if it 'contains no fat,' so that in this case getting me to take an aspirin and say a prayer does not count as a

If falsity of the means-proposition alone were a sufficient condition for a potentially effective means to count as "open," then getting me to take an aspirin and say a prayer would count as an open, potentially effective means to getting my fever to go down; hence, by SF, the following semifactual should be false:

1. If I had taken a dose of aspirin and said a prayer at t, then my fever would still have failed to go down.

And so the following might-conditional should be true:

2. If I had taken a dose of aspirin and said a prayer at t, then my fever might have gone down.

But this is wrong; 1 is true, and 2 is false. I *did* take an aspirin, and my fever *didn't* go down, and adding a prayer could not, *ex hypothesi*, make any difference. Cases like this show that falsity is not a sufficient condition for a potentially effective means to count as an open one.[16]

The right thing to say about this case is clear enough: The possible act of getting me to say a prayer is an unused means, but not a potentially effective one; the possible act of getting me to take an aspirin is a potentially effective means, but not an unused one. Combine the two into one possible act, and you get a potentially effective means that was not in fact used. But it would be misleading to call this an *open, potentially effective* means—and it would be just as misleading to call it a "missed opportunity"—because it contains no single act that is *both* unused *and* potentially effective. So here is how we need to define what it is for a potential effective means to be open:

m is an *open* potentially effective means to e just in case: (i) m is a potentially effective means to e, (ii) m is false, and (iii) for any natural way of breaking up m into a conjunction m_1 & m_2, such that m_1 is false and m_2 is true, m_1 alone is a potentially effective means to e.

---

potential effective means after all. The problem I am about to talk about in the text could be given a different solution than the one I'm about to give by someone who took this alternative. But I think that to solve the problem in that way would just be to move a bump in the rug, which will have to be dealt with sooner or later.

[16] Or at least, it's not a sufficient condition for it to qualify as an *open, potentially effective means* in the sense we are interested in here—the sense in which it is roughly equivalent to a thin sense of 'missed opportunity,' and in which it is necessarily linked with the truth values of semifactuals. There may be other perfectly reasonable ways of using the same phrase.

What counts as a "natural" way of breaking up a means is, alas, beyond the scope of this chapter.[17]

11.2.4 Means, Ends, and Effective Methods

This still leaves us with the question: what is it for something to be a "potentially effective means" to some end? Presumably, no singular event or state of affairs can be an effective means to some end just as a one-off thing; if pouring a little salt into the pot of water presently boiling on my stove is an effective means to getting that water to boil sooner, then this must be because it is an instance of a general schema, all of whose instances count as effective means to analogous ends. Such a general schema is simply an *effective method*. We can represent a *method* (whether it is potentially effective or not) by an ordered triple of predicates or propositional functions $\langle M, K, E \rangle$, where M can be called the *means* in this method, K the *enabling condition*, and E the *end*. $\langle M, K, E \rangle$, in plain speech, is the method of getting E to happen, by means of getting M to happen, while conditions K hold. Of course, M and E can both happen even in cases where nobody makes M happen, and they can happen when someone does make M happen, but not for the end of getting E to happen. It can also happen that M and K are both true on some occasion, but E is false; this is a case where the method failed to work; a method need not be sure-fire in order to be reliable.

Some methods, but not others, are effective. I will return to the question of just what effectiveness is supposed to be below. For now, though, there are several good ways of indicating its extension and intension. Roughly speaking, the method $\langle M, K, E \rangle$ should qualify as one of the effective ones just in case events of type M have the power to cause events of type E under conditions K. Also roughly speaking, $\langle M, K, E \rangle$ should qualify as effective just in case M raises the objective chance of E under any condition $K+$ that entails K and also entails a complete specification of whether X holds for every condition X such that $\langle X, K, E \rangle$ is itself effective.[18] From my point of view, it would be a mistake to consider either of these characterizations as definitions or analyses of effective methods, because I think the arrow of dependence points the other way, and I hope to show how to reductively analyze causation, causal powers, and objective chance in terms of effective methods. (But not in this chapter, which gets only as far as laws and counterfactuals.)

---

[17] I think it might suffice to require that it is logically possible that $\sim m_1 \,\&\, \sim m_2$.

[18] This characterization is meant to be based on Cartwright's (1979) characterization of effective strategies, with effective methods replacing causation.

11.2.5 The Final Official Formulation, and the Big Picture

With the notion of an effective method in hand, we can spell out what it means for something to be an "open, potentially effective means", and thereby clarify the principle **SF**:

m is a *potentially effective means* to the end *e* just in case there exists an effective method $\langle M, K, E \rangle$ and an index *r* such that $M(r) = m$, $E(r) = e$, and $K(r)$ is true.

m is an *open, potentially effective means* to *e* just in case *m* is a potentially effective means to *e*, *m* is false, and for any natural way of analyzing *m* as a conjunction $(m_1 \,\&\, m_2)$ where m_2 is true, m_1 is also a potentially effective means to *e*.

These definitions make it possible to rewrite the theory of counterfactuals I am proposing as follows:

SF: Where *b* is true, $(a \,\square\!\!\rightarrow\, \text{(still)}\ b)$ is true just in case, and in virtue of the fact that: (i) *a* is false, and (ii) there is no *M, K, E, r* such that:

- $\langle M, K, E \rangle$ is effective;
- $M(r) = a$, $E(r) = {\sim}b$;
- $K(r)$ is true;
- and there is no *M1, M2* such that *M* is logically equivalent to (*M1 & M2*), *M2(r)* is true, and $\langle M1, K, E \rangle$ is not effective.

FC: Where *c* is false, $(a \,\square\!\!\rightarrow\, c)$ is true just in case, and in virtue of the fact that, there is some true *b* such that $(a \,\square\!\!\rightarrow\, b)$ is true and $(a \,\&\, b) \models c$.

The only substantive, non-logical concept taken for granted here is that of an effective method. Written in this way, **SF** can look disturbingly complicated, but remember that it is simply a careful formalization of a very simple informal idea: A semifactual is true just in case its antecedent does not represent a missed opportunity (in a rather thin sense of this phrase) to make whatever its consequent says not hold.

   This gives us the following big-picture view of what counterfactual discourse is all about: The semifactuals, in effect, package information about what sorts of effective methods there are; in particular, the content of a particular semifactual $(a \,\square\!\!\rightarrow\, b)$ packages the information that there exists no effective method in virtue of which, under actual circumstances, *a* is (though unused) a potentially effective way to promote the negation of *b*. It is very handy to have such a device in our language, because it provides a very simple way of stating a piece of information with a very complex logical form—namely, a quantification over effective methods. Furthermore, there are many cases where this very piece of information is practically important to us. For example, when deciding whether to pin blame for some undesirable outcome on someone, it is important to know whether there were any missed opportunities to prevent that outcome. Hence, it is quite useful to have a small and easy-to-handle syntactic package for the sort of bundle of information that semifactuals convey.

But what about the rest of the counterfactuals, the ones with false consequents? What is their significance? A true counterfactual that is not itself a semifactual typically entails (via such principles as **Extended Weakening the Consequent**) many semifactuals;[19] it also entails the negations of very many false semifactuals;[20] the negation of a semifactual, in turn, amounts to the negation of a generalization about effective methods. Hence, the true non-semifactual counterfactuals package a great deal of information of an even more complex form.

Of course, since we are not omniscient, we do not always know which counterfactuals are the semifactuals and which are not. So we cannot always tell a priori which sort of information is packaged by a particular counterfactual. But even in cases where we cannot, information about which counterfactuals are true and which are false can be very useful, because it still conveys lots of indicative conditionals about what sorts of effective methods there are, if this or that is actually the case. For example, even if we are ignorant about whether Kennedy was assassinated, we can be sure that if he was assassinated, then **StillKilled** implies that stopping Oswald from killing him is not among the open, potentially effective means of saving him from assassination, and that if he wasn't assassinated, then **StillKilled** implies certain other facts about what open, potentially effective means that there are.

11.3 Stage Two: Laws of Nature

The theory of laws I propose is this:

Laws: It is a law of nature that L just in case, and in virtue of the fact that, L is logically contingent[21], and it belongs to a member of an *unviolatable* set of propositions.[22]

---

[19] For example: **SomeoneElse** entails **StillKilled**.

[20] For example, this counterfactual with a false consequent: "If I had dropped the glass, then it would have broken" entails (given the obvious auxiliary premises) the negation of this semifactual: "If I had dropped the glass, then our set of fine stemware would (still) have been complete."

[21] The truths of pure logic, of course, belong to every stable set, and to every unviolatable set. You might want to classify them as laws of nature 'by courtesy,' in which case the account given in the text characterizes those truths that are *merely* laws of nature.

[22] It might be a good idea to add that L has the form of a universal or statistical generalization, since laws of nature are traditionally thought to be general rather than singular or particular. But I won't explore this issue further here.

where "unviolatable" is defined thus:

Unviolatability: A set of propositions $\mathcal{U}$ is *unviolatable* $=_{def}$ $\mathcal{U}$ is a deductively closed, proper subset of the set of all true propositions, and for every effective method $\langle M,K,E \rangle$, and every index r, either $E(r)$ is consistent with $\mathcal{U}$, or $M(r)\,\&\,K(r)$ is inconsistent with $\mathcal{U}$.

In other words, a set of truths is unviolatable just in case there is no effective method for rendering one of them false—except possibly for a method whose means itself is inconsistent with the set (and therefore can be implemented only by an agent who is already independently capable of violating the set), or whose enabling condition is itself inconsistent with the set (and therefore is available to be used only under conditions under which the set has already been violated). So, what it would mean for (say) the laws of nature to form an unviolatable set is that, unless you are some kind of being already gifted with a power to violate a law of nature, or else you find yourself in a universe where the laws are already violated, it is in principle impossible for you ever to discover an effective method for producing violations of any law. This seems a pre-philosophically plausible explication of what makes the laws special; to defer to the authority of *Star Trek*: The laws of physics are those things that not even Scotty can change, even when Captain Kirk orders him to do so. (However, if you spot Scotty a temporary relaxation of some of the laws, you can count on him to work miracles.)

Readers familiar with Lange's (2005, 2009) work on laws and counterfactual stability will notice a close family resemblance between this account of laws and Lange's. This is no accident; I was led directly to this view of laws by thinking about Lange's notion of a counterfactually stable set. In fact (though I won't try to prove this here), I believe that it follows from my account of counterfactuals that a set is unviolatable in the sense defined above just in case it is stable in Lange's sense.[23],[24] The theory I am presenting here is importantly different from Lange's; Lange takes "subjunctive facts" as a primitive, whereas I reduce counterfactuals to something I take to be more basic. (Not to mention the fact that Lange's view is resolutely non-Humean, whereas—as I'll argue in the following section—mine is compatible with Humeanism.) But I obviously owe much to Lange here.

---

[23] With one exception: Lange's definition counts the set of all truths as stable; since I define an unviolatable set to be a proper subset of the set of all truths, the set of all truths is not unviolatable in my sense.

[24] For this reason, much of what Lange shows to be true of stable sets carries over to unviolatable sets. For example, there can be nested families of unviolatable sets (which is possibly the key to understanding the idea of there being stronger and weaker degrees of nomic necessity); moreover, if there exists more than one unviolatable set, then they must form a nested hierarchy. See Lange (2005).

11.4 Stage Three: Effectiveness for Humeans

I propose that the truth conditions for propositions of the form "M is an effective method" and "M is a non-effective method" are what Peirce's theory of truth would imply they are: $\langle M, K, E \rangle$ is effective just in case it is one of the methods we would classify as effective, in the limit of scientific inquiry, assuming that conditions are optimal.[25] We might call this a "Quasi-Peircean" theory of effectiveness, because unlike Peirce's theory of truth it applies only to propositions of one narrow class.[26] This formulation is a useful initial characterization of the account, but it is not acceptable as an official formulation; for one thing, it is stated in the form of a counterfactual. Let me now work towards a better formulation.

Let our *E-opinions* be the set of our opinions of the form "M is (is not) an effective method," and let the *E-methodology* be the methodology we use for updating our E-opinions in response to new empirical evidence. (Or better: the way in which, in calm and lucid moments, we think we should revise our E-opinions in response to new empirical evidence.) The E-methodology is not a method for going from some body of evidence about actual occurrences and regularities in them to a hypothesis about which methods are the effective ones. Instead, it is a method for taking a previous set of E-opinions and updating them. This is not because of the role of the priors in Bayesian confirmation theory. Instead, it is because of what Cartwright (1994) calls the "No Causes In, No Causes Out" principle. Cartwright (inter alia) argues that premises about events and statistical correlations imply nothing about what causes what unless they are supplemented with some background assumptions about causal relations. Without such background assumptions to appeal to, we are typically unable to distinguish between a case of causation and a case of two effects of a common cause. Under favorable conditions, though, background hypotheses can help us make this distinction, by telling us where to look for cases of "screening off."

These familiar points are, of course, about the methodology for investigating general causal relations, rather than effective methods. But it seems obvious that there is a close relation between general causal relations and effective methods.[27] Some readers might be inclined to think that what I am calling "effective methods" are really just general causal relations under a different name; in particular, that when I say $\langle M, K, E \rangle$ is effective, that's just my unusual way of saying that under condition K, Ms cause Es. I think this is almost right: The extension, and the intension, of "effective method" as I use it might be more or less the same as those

---

[25] Peirce (1935): "Truth is that concordance of an abstract statement with the ideal limit towards which endless investigation would tend to bring scientific belief."
[26] Hall's (unpublished) version of the best-system analysis is a Quasi-Peircean theory of laws, in this sense.
[27] See Cartwright (1979).

of "general causal relation."[28] But the concepts *effective method* and *general causal relation* are different, and we characterize a triple $\langle M,K,E \rangle$ differently when we classify it as an effective method than when we classify it as a general causal relation. My view is that "effective method" is a more fundamental way of classifying it, both conceptually and metaphysically. Nonetheless, because of the similarity between their extensions, we should expect the methodology for discovering effective methods to look very much like the methodology for discovering causal relations.

Thus, we cannot infer directly from data about actual occurrences and correlations to conclusions about which methods are the effective ones; we need some background hypotheses that tentatively classify some methods as effective and some as non-effective. The role our previous E-opinions play in the E-methodology is to provide the needed background assumptions.

This is one reason why the theory I am presenting here is not a "best-systems" theory: best-system theories take the laws to be specified by whatever possible theory or system maximizes a set of theoretical virtues, among all those consistent with the Humean mosaic; there is no place in this scheme for the hypothesis we start out with to play a role. By contrast, I identify the correct hypothesis about which methods are effective as the one reached by starting from our current hypothesis and revising it in a certain way; there is no guarantee that the hypothesis this process leads to will maximize any particular virtues.

I suggest that a method $\langle M,K,E \rangle$ is effective if and only if it is classified as effective by the theory that results from starting with our actual, current E-opinions, taking a specification of the entire Humean mosaic to be a new batch of empirical evidence, and updating those E-opinions via the E-methodology. This statement in effect identifies a function from possible Humean mosaics to possible extensions of the predicate "is an effective method." This function is here characterized in terms of our actual, current E-opinions, and the E-methodology itself. But like all functions this one can be defined in many different ways, and what is essential to its identity is simply the way it maps inputs to outputs. My official account of effectiveness is that a method is effective in a given possible world if and only if it is in the set to which this function maps that world's Humean mosaic. The result is plainly a Humean account of effectiveness.

Quasi-Peircean accounts like this one face a distinctive challenge. In the case of this account, the challenge is that it seems to make it mysterious why we should be at all interested in finding out which methods are the effective ones. One

---

[28] One possible difference: As I mentioned above, I think that if taking an aspirin is the means in an effective method for the end of reducing a fever, then so is taking an aspirin and doing something irrelevant, like saying a prayer. But I suspect it would be a mistake to say that instances of taking aspirin and saying a prayer are causes of fever reduction; it seems to me that causes cannot include excess, irrelevant content, whereas effective means can. But the theory of causation is beyond the scope of this chapter.

reason for this is that it seems to make the effective methods a very unnatural kind; its boundaries are drawn by an enormously complex function that we cannot even specify directly, but must characterize in terms of its relation to a methodology we employ and a set of opinions we happen to hold. To see another reason, consider the question:

"Why bother carrying out any inquiry at all that involves using the E-methodology?"

The natural and obvious answer is:

"Because that's our best way of gaining knowledge about which methods are effective, and we have an interest in gaining such knowledge."

But this answer is not available to me. For given my theory, that answer is equivalent to this one:

"Because using the E-methodology is our best way of gaining knowledge about what opinion we would reach in the long run if we applied the E-methodology under favorable conditions."

But this answer assumes that we have an interest in where the E-methodology leads—which is just what we wanted explained. Thus, my Quasi-Peircean account of effectiveness is plausible only to the extent that it is plausible that we serve our interests by employing the E-methodology, and that the interests in question do not include a direct interest in knowing which methods are effective.

This is so if there are ways in which we use our E-opinions, which are such that things will tend to go better for us (other things being equal) insofar as we keep updating our E-opinions using the E-methodology. I think it is pretty clear that this is so. A full argument for this claim would take more space than I have here, but let me present one consideration that makes it plausible: The E-methodology will direct us to classify method $\langle M, K, E \rangle$ as effective (tentatively), when our available evidence supports the hypothesis that $P(E|M\&K) > P(E|K)$, AND this evidence does NOT support the hypothesis that this correlation of M with E is asymmetrically "screened off" by another factor N, such that we judge $\langle N, K, E \rangle$ to be effective. (Notice that this illustrates the way in which the advice of the E-methodology depends not only on the evidence we receive but also on the opinions about effectiveness of methods that we started out with.) It seems fair to assume that our total evidence will tend to be a good guide to whatever statistical correlations hold within our local environment. So, this part of the E-methodology will tend to lead us to classify as effective lots of methods with the property that, in our local environment, the end occurs more frequently with the means and the enabling condition than it does with the enabling condition alone.

One of the most important ways in which we use our E-opinions is in deciding which methods to seriously consider employing in order to promote our practical ends: We assume it makes sense to employ only those methods which are likely to be effective. Hence, maintaining our E-opinions using the E-methodology will predictably have the result that in the near future we will tend to use methods in which the means and the end are (given the enabling condition) positively correlated in our local environment. It is reasonable to expect that this will increase the frequency with which our practical ends come true, since it is reasonable to expect that it will increase the frequency with which conditions that are positively correlated with our practical ends will come true.

Importantly, this benefit of using the E-methodology is independent of whether we will ever discover the truth about which methods are the effective ones. For example, it might be that the statistical correlations that hold in our local environment are a rare fluke, and they do not hold within the Humean mosaic at large. Still, as long as these correlations hold within our local environment, our use of methods that look effective in the light of these correlations will tend to serve our interests. So this is the kind of benefit that I need in order to defend my account: It shows that we have an interest which we serve by using the E-methodology, even if we never come close to discovering which methods are really the effective ones. Hence, quite apart from whatever direct interest we might have in learning which methods are effective, we have an interest in updating our E-opinions by means of the E-methodology. This is just what it takes to answer the challenge to my Quasi-Peircean proposal.

11.5 In Closing: Ratbag Idealist Judo, Revisited

One common objection to Lewis's analysis of lawhood charges it with a kind of chauvinism: The standards used to pick out the best system are supposed to be our standards, but what is so special about us that our standards should occur in the correct analysis of what it is to be a law of nature? Lewis himself considers a related objection, involving a character called "the ratbag idealist": This is someone who accepts the best-system theory, notes that it makes lawhood depend on our standards of strength, simplicity, and the balance between them, supposes that these standards are ultimately up to us, and then concludes that we have the power to change the laws of nature—indeed, to change *what they have always been*—simply by changing our standards. The objection to the best-system theory here is: The ratbag idealist seems to be right—we really can change our standards, and if this theory of laws is right, that means we really can change the laws. But obviously that's wrong: We cannot change the laws. Ergo, we should forget about the best-system theory.

Obviously, there are analogous objections to the theory I have presented here: This theory characterizes effectiveness of methods in terms of some things about us, including the opinions about the effectiveness of methods that we hold right now. Hence, it makes the counterfactuals and the laws depend on these opinions. But why should our current opinions about which methods are effective play such a central role in our metaphysics? And doesn't it follow that we could change the laws and the counterfactuals just by changing our current opinions, as the ratbag idealist might say?

Lewis himself gave one reply to these worries: In his account, our standards are picked out rigidly, so that even in other possible worlds where we employ different standards, the laws are still picked out using our actual standards. Likewise, if we were to change our standards, that would make no difference to the laws, because the best-system account picks them out using the standards we used before the change. Similar replies can be given on behalf of my account.

But while this reply might be strictly correct, it doesn't seem to get to the heart of the matter. The laws are picked out by a function from possible mosaics to possible sets of laws—but why this function, rather than some other? Lewis's motivation in choosing this function depends on its relation to our actual standards. But there are other possible standards, and corresponding to them there might be other functions. What is so special about us, that the function that determines the laws of nature should be the one that is thus related to our standards, rather than one of the ones thus related to some other set of standards? An analogous question can be directed at me: why should the function that maps possible Humean mosaics onto possible sets of effective methods be the one that is related in a certain way to our initial opinions about which methods are reliable, rather than the one thus related to some other possible initial opinions?

As Hall points out, an interesting reply is available to a best-system theorist; an analogous reply is open to me. Hall characterizes it as a "judo move," because it take an alleged problem for the best-system analysis and flips it into a problem for the Non-Humean. To quote Hall:

> In effect, the [anti-Humean] who endorses the ratbag idealist challenge to the [best-system analysis of laws] is committing herself to the following position: "I believe there are facts about the world that logically go beyond the sort of facts that even in principle could serve as our evidence. Still, we can have good reason to think that we have got these facts right, if we construct theories about them that, among other things, score well on our standards of simplicity. Of course, what counts as 'simple' is a matter having to do with highly idiosyncratic features of human psychology, features that we could change if we wanted to. All the same, I am committed to treating simplicity of a theory as an epistemically good guide to its truth." (Hall unpublished, pp. 38–9)

And what the Non-Humean is thus committed to is (Hall says) "crazy."

I interpret Hall's judo move as a dilemma for the non-Humean critic who wields either the original chauvinism objection or the more vivid objection featuring the ratbag idealist. Suppose that the standards of simplicity etc. that are implicit in the practice of (human) science are indeed biasing, in a way that makes a significant difference to the system or theory chosen on the basis of those standards—a significant difference that will persist even after all of the data about the mosaic are in. Then science itself has an anthropocentric bias, which makes an ineliminable difference to its ways of evaluating hypotheses about what the laws are. If the laws themselves have such an anthropocentric bias—say, because lawhood is determined by a function that is characterizable in terms of human standards of simplicity etc.—then there is no worry here: Human science uses an anthropocentrically biased epistemology in order to study an anthropocentrically biased subject matter. This is as it should be: If the trail of the human serpent lies over that which you want to study, then your method of studying it had better take account of the way that serpent slithers. By contrast, if lawhood has nothing anthropocentric about it at all, as the non-Humean suggests, then there is a great mismatch between our epistemology and our metaphysics: Science unavoidably relies on all-too-human standards, in a way that will continue to exert a distorting bias on its best guess about what the laws are, even in the long run. Hence, there is a difficulty for the non-Humean theorist of laws (unless they are happy to embrace skepticism about our ways of investigating the laws), but none for the best-system theorist.

On the other hand, suppose the standards of simplicity etc. that are implicit in the practice of (human) science are not biasing in this way; in other words, once all of the data specifying the Humean mosaic are taken account of, the parochially human aspects of these standards will not end up making any difference to the system or theory that they favor; any anthropic bias in those standards gets washed out when enough evidence is considered. In this case, there is no problem about biased standards either for the best-system theorist or for the non-Humean critic.

Therefore, in either case (i.e. whether or not the all-too-human character of our standards of simplicity etc. imposes a bias that cannot wash out no matter how much evidence we consider), there is no difficulty for the best-system analysis, but in one case there is a grave difficulty for its non-Humean critic. Thus, when we consider the roles that our standards of simplicity etc. play in the epistemology of science and in the best-system analysis of lawhood, we find no disadvantage for the best-system analysis vis-à-vis its non-Humean competitors.

I can give an analogous reply to the analogous objection to my Humean account of effective methods. Our best methodology for investigating the effectiveness of methods involves a process of revision that begins where we begin—with our current opinions to the effect that certain methods are effective and others are not. We cannot start with a blank slate, because of the same considerations that motivate Cartwright to say "No causes in, no causes out." So where our reasoning leads us will depend not only on the empirical evidence we receive but also on the opinions we start out with. Perhaps this feature of this methodology means that it suffers from a bias that cannot be remedied, no matter how much empirical evidence we take into account. If that's so, then on a non-Humean conception of effectiveness, science's way of trying to find out which methods are effective, is fatally flawed. By contrast, my Quasi-Peircean, Humean account of effectiveness has no such skeptical implication, since it implies that the method of science and the truth it seeks are both anthropocentrically biased in the same way; the epistemology fits the metaphysics. On the other hand, if our best methodology for studying effectiveness is not irredeemably biased (say, because once sufficient evidence is taken into account, there are no more traces of our idiosyncratic starting point), then the way in which my account of effectiveness appeals to our starting point does not inappropriately bias its classification of methods as effective or non-effective—so there is no worry after all. In neither case is the way my account gives a role to our initial opinions about which methods are reliable a liability.[29]

References

Armstrong, D. M. (1983). What is a Law of Nature? Cambridge: Cambridge University Press.

Ayer, A. J. (1956). 'What is a Law of Nature?', Revue Internationale de Philosophie, 10, pp. 144–65.

Braddon-Mitchell, D. (2001). 'Lossy Laws', Noûs, 35(2), pp. 260–77.

Cartwright, N. (1979). 'Causal Laws and Effective Strategies', Noûs, 13(4), pp. 419–37.

Cartwright, N. (1994). Nature's Capacities and Their Measurement. New York: Oxford University Press.

Cohen, J., and Callender, C. (2009). 'A Better Best System Account of Lawhood', Philosophical Studies, 145(1), pp. 1–34.

Dorst, C. (2019). 'Toward a Best Predictive System Account of Laws of Nature', British Journal for the Philosophy of Science, 70(3), pp. 877–900.

Goodman, N. (1983). Fact, Fiction and Forecast, 4th ed.. Cambridge, MA: Harvard University Press.

[29] I am very grateful to Alison Fernandes, Mike Hicks, Siegfried Jaag, and Christian Loew for enormously helpful comments on an earlier draft of this paper. For helpful discussion of these ideas, I am grateful to John Carroll, Luke Elson, Nina Emery, and Alastair Wilson.

Hall, N. (unpublished). 'Humean Reductionism about Laws of Nature', https://philarchive.org/archive/HALHRAv1 (accessed Aug. 2020).

Hicks, M. T. (2018). 'Dynamic Humeanism', British Journal for the Philosophy of Science, 69(4), pp. 983–1007.

Lange, M. (2005). 'Laws and their Stability', Synthese, 144, pp. 415–32 .

Lange, M. (2009). Laws and Lawmakers. Oxford: Oxford University Press.

Lewis, D. K. (1973). Counterfactuals. Oxford: Blackwell.

Lewis, D. K. (1994). 'Humean Supervenience Debugged', Mind, 103(412), pp. 473–90.

Loewer, B. (2012). 'Two Accounts of Laws and Time', Philosophical Studies, 160(1), pp. 115–37.

Peirce, C. S. (1935). 'Truth and Falsity and Error', in Hartshorne, C., and Weiss, P. (eds.), Collected Papers of Charles Sanders Peirce (Volumes V and VI). Cambridge, MA: Harvard University Press, pp. 394–8.

Roberts, J. (1999). '"Laws of Nature' as an Indexical Term: A Reinterpretation of Lewis's Best–System Analysis', Philosophy of Science, 66(3 Supplement), pp. 502–11.

Roberts, J. (2008). The Law-Governed Universe. Oxford: Oxford University Press.

Skyrms, B. (1980). Causal Necessity. New Haven and London: Yale University Press.

12

Humean Reductionism about Essence

Ned Hall

12.1 Introduction

Here are some claims about black holes:[1]

- Black holes are regions of spacetime from which nothing—not even light—can possibly escape.
- Every black hole has an 'event horizon': a boundary such that things on the outside of it can escape from the black hole, whereas things on the inside cannot.
- Every black hole at equilibrium is completely physically characterized by just three parameters: its mass, angular momentum, and charge.
- A massive body that is collapsing in upon itself will form a black hole if its constituent matter ever falls within a certain distance—the 'Schwarzschild radius'—that is proportional to the total mass of that body.

Here are some more claims about (or at least partially about) black holes:[2]

- There is a supermassive black hole at the center of the Milky Way galaxy.
- The first detection of the merger of two black holes was observed in February 2016 by the LIGO Scientific Collaboration.
- The by now widely entrenched use of the phrase "black hole" to refer to black holes was significantly helped along by the noted physicist John Wheeler's early adoption of it.
- There are exactly seventeen black holes within 20 light years of our solar system.

Let's take for granted that all of these claims are true. (Okay, not the last one, which I just totally made up. Still, it will help focus attention on the key issues if we *pretend* that it is true.) You will have noticed a difference between these claims.

---

[1] Taken more or less at random from the Wikipedia article on black holes (Wikipedia contributors 2022) and the *Stanford Encyclopedia of Philosophy* article 'Singularities and Black Holes' (Curiel 2021).

[2] Same sources, except for the last one, which I just made up out of whole cloth.

Ned Hall, *Humean Reductionism about Essence* In: *Humean Laws for Human Agents*. Edited by: Michael Townsen Hicks, Siegfried Jaag, and Christian Loew, Oxford University Press. © Oxford University Press 2023.
DOI: 10.1093/oso/9780192893819.003.0013

In some good sense, the first set of claims is more significant—vastly more significant—than the second.

What sort of philosophically illuminating account can we give of this difference?

I am going to set aside approaches to this question that look to semantics to provide the key. That may strike you as hasty: after all, you will likely have noticed that the first four claims are *generalizations* about black holes, whereas the second four are not. But it's easy enough to state generalizations about black holes that fall on the wrong side of the 'vastly more significant' divide: 'There are no black holes with a mass of exactly 5 kg', for example. More importantly, one thing we should hope that an illuminating answer to our question would provide is an *explanation* for the truth (if it is a truth) that existential or singular claims about black holes are thereby not 'significant', in the intended sense.

Of course, there are other semantic avenues of approach. For example, perhaps we can make out some illuminating and purely semantic sense in which the first four claims are properly *about* black holes, whereas the second four are not—presumably, because they are, to too great an extent, about other things as well. Fair enough; something like that might work. But here I wish to focus on a very different proposal—or rather, a very interesting choice between two very different kinds of proposal.

The first kind of proposal aims to draw the distinction we want by means of a suitable appeal to metaphysics. Perhaps what is going on here is that each of the first four claims tells us something about the *essence* of the kind 'black hole'; or about *what it is* for something to be a black hole; or about facts that either display or are explained by the *grounds* of black holes; or (somewhat more anemically) simply what is *necessary* of black holes. Here it is important to understand that the key expressions—'necessary', 'ground', 'what it is for', 'essence'—are all to be taken in a distinctively metaphysical sense.

Admittedly, that key term 'metaphysical' is not exactly the clearest piece of technical philosophical vocabulary ever introduced, so we will have to return to what it might mean. But hopefully it is clear enough to allow you to detect an important difference between these sorts of 'metaphysics-first' approaches and a rival style of approach that foregrounds *epistemology*. Here, the key idea is that what distinguishes the first four claims from the second is simply their *epistemic utility*: In some sense that (of course) will need to be made more precise, knowing the first four claims facilitates inquiry (specifically, into black holes) to a vastly greater extent than does knowing the second four claims. Suppose you want to explain phenomena involving black holes, or answer important questions about black holes—or for that matter, be able to spot important questions about black holes that are as yet unanswered. Then the first four claims and others like them will be highly useful, whereas the second four claims will not. It is that difference in epistemic utility—and not some difference at the level of metaphysics—that we are responding to when we judge the first four claims vastly more significant than the second four.

I do not want to settle here which approach is correct. (And in case you're wondering, I don't myself have a firm view on the matter. That's because I subscribe to a regrettably minority position within our discipline, which is that we ought to remain agnostic about questions that are extraordinarily difficult to settle, unless and until we find arguments that finally settle them. Which *does* happen—but not that often, and certainly not, yet, in this case.) What I want to do instead is to draw what I think are some illuminating parallels between the question we face here—'Go metaphysical, or go epistemic?'—and a thriving philosophical debate about laws of nature.

To help bring the debate into focus, I'm going to assume that the right language to use to characterize the distinction that we are after between the two sets of claims is the language of 'essence': The first four claims tell us something about the essence of the kind black hole, whereas the second four do not. Now for the parallel: In the case of laws of nature, one fundamental philosophical question is whether we should give a *metaphysical* or rather an *epistemic* account of what they are; that is the core issue that divides 'Humeans' from 'anti-Humeans'. I think that, in much the same way, a key question we face with essences is whether to give a metaphysical or rather an epistemic account of what they are. (So note well: in choosing to deploy the term 'essence' I am *not* taking sides on the 'Go metaphysical, or go epistemic?' question. Just as we can be anti-metaphysical and still freely use the expression 'law of nature', so too we can be anti-metaphysical and freely use 'essence'.) And the advantage of connecting these two topics is that we have made enough progress on the former for the lessons learned to provide valuable guidance for investigating the latter. Or so, at any rate, I'll try to show!

Despite my agnosticism about which approach is correct, I'll mostly be investigating the prospects for the 'go epistemic' option—what I will call 'Humean reductionism about essence'. And that is because I think it has been relatively neglected, with the consequence that both its potential virtues and its potentially interesting consequences have, perhaps, not been adequately appreciated. Before diving in, here's a preview of some consequences that may fall naturally out of such an approach:

- The *importance* of the concept of 'essence' can be readily vindicated—that is, Humean reductionism about essence does not simply devolve into Humean eliminativism about essence.
- At the same time, essences turn out to be interest-dependent and capable of vagueness.
- The view offers the resources to answer an important background question: Which kinds *have* essences?

- The view helps clarify the distinction between essences of *kinds* and essences of *individuals*.
- The view offers an important revisionary account of certain key examples, such as Fine's (1994) famous Socrates/{Socrates} case.
- The view naturally motivates a strikingly deflationary treatment of essentiality about origins.
- The view neatly solves a curious puzzle about the essences of arbitrary fusions.

You may have noticed that I labelled these as consequences that *may* fall out of a Humean reductionist approach to essence. That's partly, of course, because all I'm going to be able to do in this essay is sketch the broad contours of such an approach, drawing inspiration from what I take to be the best version of Humean reductionism about laws of nature. But it's also partly because the viability of the approach (and, for that matter, the viability of the parallel treatment of laws) crucially hinges, I think, on a question that I at least don't know how to answer: Can the core epistemic aims that such concepts as 'law' and 'essence' subserve *themselves* be characterized without any appeal to metaphysics? Modern-day Humeans, as far as I can tell, tend to blithely assume that the answer is 'yes' (that is, if they pay heed to the question at all). I think that's hasty, and at the very end of this chapter I will return to this topic with some tentative suggestions about ways forward.

12.2 Preliminaries: Essence of Kinds vs. Essences of Individuals

When we philosophize about 'essence', we might focus on either *individuals* or *kinds*. I think it pays to keep these two possible foci distinct. To see this, consider that each of the following three metaphysical positions is perfectly intelligible.

12.2.1 Individuals Only

The fundamental ontological ingredients of reality are individual substances, together with the properties and relations those substances instantiate. In addition, however, there is a metaphysically fundamental distinction, pertaining to each individual, between those of its properties that is has *essentially*—that are part of its *essence*—and those that is has only accidentally. Put another way, the essence of an individual x fills in the blank in this formula[3]:

What *makes* x the individual that it is is that it has—

---

[3] Compare Fine (1994), p. 1: 'What is it about a property which makes it bear, in the metaphysically significant sense of the phrase, on what an object is? It is in answer to this question that appeal is naturally made to the concept of essence.'

Filling in this position will require giving an account of *which* of an individual's properties belong to its essence. And other questions will arise: For example, we might want to say that an individual that is essentially F *also* has the property 'being essentially F'; in which case we can ask whether it has *this* property essentially. We might wonder whether two distinct individuals can differ *only* in their essences,[4] and if so under what general conditions this happens. We might wonder whether an individual's essence grounds any *de re* modal truths pertaining to it (e.g. via the thesis that if an individual is essentially F, then that individual could not have existed without being F). And so on. There are many ways to develop an account which answers such questions. But the key point is that in giving such an account we don't *need* to draw on the concept of a 'kind'. In fact, we could augment the position being sketched here by insisting that there *are no* 'kinds', at least in any metaphysically interesting sense of that term. There are properties, to be sure; and so if we wanted to, we could use 'kind' interchangeably (if misleadingly!) with 'property'. Or we could restrict a bit, and say that to each property F there corresponds a 'kind' whose members are all and only those individuals that have F essentially. Or we could restrict a bit more, and count a property F as corresponding to a 'kind' in this way only if it is a property that is part of the essence of every individual that has it.

Even if we introduce talk of 'kinds' in one of these ways, though, there need be no interesting sense in which kinds, so understood, *have* essences. On this metaphysical picture, the concept of 'essence' applies *only* to individuals.

12.2.2   Kinds Only

The fundamental ontological ingredients of reality are individual substances, the properties and relations those substances instantiate, and *kinds*. Kinds have essences: bundles of properties that distinguish each kind from each other kind, and which collectively explain, for each individual that is a member of a given kind, *why* it is (or *what it is for it to be*) a member of that kind. Put another way, the essence of kind K fills in the blank in this formula:

What *makes* an individual a member of K is that it has—

Thus, we might say that part of the essence of the kind *gold* is the property of having atomic number 79, since part (all?) of what *makes* something gold is that it has this atomic number.

---

[4] Or a little bit more carefully, can share in common all 'categorical' properties, or properties *not of* the form 'being essentially F'.

As with the 'individuals only' view, plenty of questions need answering if we are to fill out this position. For example, we might want a specification of the essence of a kind solely in terms of properties that neither are nor presuppose other *kinds*. If so, we won't be happy with the typical specification of the essence of the kind *gold* as 'being an atom with atomic number 79'. Never mind that we *call* things gold that are not atoms (but, rather, fusions thereof). The trouble is that 'atom' itself is a kind term; so a fan of this view should want a substantive answer to the question of what makes an individual an *atom*. Absent an answer to that question, we haven't yet adequately specified the essence of the kind *gold*, after all. More generally, we might hope that a full philosophical theory of essences of kinds would illuminate what constraints (if any) there are on which properties can belong to which essences, how we might be able to discover which kinds there are and what essences they have, what the connections are between a kind's essence and the laws governing it or causal powers it has, etc.

Still, such a philosophical theory need not recognize *any* sense in which *individuals* have essences. Offhand, you might have thought otherwise. For example, you might have thought that when an individual *a* belongs to some kind *K*, then *a* not only *has* the properties that make up *K*'s essence but has them *essentially*. But that extra bit of doctrine is totally optional. There is no obstacle, for example, to pairing whatever theory of essences of kinds we come up with with the view that, for every individual *a* and kind *K*, *a* could have *belonged* to *K*, and could have *failed* to belong to *K*. On the usual assumption (noted above) that if a property *F* is essential to an individual *a*, then *a* could not have existed without being *F*, it follows that even when *a* belongs to some kind *K*, not every property that is part of the essence of *K* is essential to *a*. But look, the crucial point is that nothing in the sketch given in the last two paragraphs requires that the concept of 'essence' *have any application* to individuals. Here is Fred, who happens to be an atom of gold, and so also happens to have atomic number 79. So what? For all that, Fred *could* have been an atom of silver, or a duck, or a prime number.... But more centrally, on the view I'm imagining the fact that Fred has atomic number 79 tells us nothing special about the nature *of Fred*, nothing that would answer the question 'what makes Fred the individual that it is?' (Indeed, one might hold that this question never has a correct and non-trivial answer.) For all that, we might learn quite a lot about the nature of the kind *gold* by learning that the property of having atomic number 79 is essential to *it*.

12.2.3 Individuals and Kinds

As with the 'kinds only' view, the fundamental ontological ingredients of reality are individual substances, the properties and relations those substances instantiate, and kinds. But this time, the concept of 'essence' has significant application to *both* individuals *and* kinds. And the ways in which it applies to each category are related. To begin with, the kinds themselves exhibit a hierarchical structure, with some kinds related to others as species is to genus. More exactly, suppose that what it is for an individual to belong to kind *K1* is to have the properties in set *S1*. And what it is for an individual to belong to kind *K2* is to have the properties in set *S2*. And, finally, set *S2* is a proper subset of set *S1*. Then we will say that *K1* is *subordinate* to *K2*. The relation of subordination partially orders the kinds.

We could, if we like, add some more doctrine. Some kinds are *maximally general*: they are subordinate to no other kinds. Some kinds are *maximally specific*: no other kind is subordinate to them. For every kind *K* that is not maximally general, there is exactly one kind *L* such that *K* is subordinate to *L*, but for no intermediate kind *M* is *K* subordinate to *M* and *M* subordinate to *L*. In such a case, we'll say that *K* is *immediately subordinate* to *L*. For every kind *K* and kind *M* where *K* is subordinate to *M*, there is a finite (and possibly empty) set of kinds {*L1*, *L2*,... *Ln*} such that *K* is immediately subordinate to *L1*, *L1* is immediately subordinate to *L2*,..., *Ln* is immediately subordinate to *M*. Finally, if *S* is a set of kinds such that for every pair of kinds *K* and *L* in *S* either *K* is subordinate to *L* or *L* is subordinate to *K*, then *S* is finite.

These pieces of doctrine ensure that the kinds as a whole form a collection of trees, with each kind featuring in exactly one tree, the maximally general kinds at the top of the trees, and the maximally specific kinds at the bottom.

Next, we add some doctrine concerning individuals. First, each individual belongs to exactly one maximally specific kind. Second, the properties that make up the essence of this kind *also* constitute the essence of the given individual.

And that's it.

As before, much remains to do to fill out the picture. What are the most general kinds? Can anything be said in general about when one kind is immediately subordinate to another? Why are the key pieces of doctrine true? (For example, why can't there be more and more specific kinds, without end?) Never mind. What matters for our purposes is that, on this picture, the notion of 'essence' has substantive application to *both* kinds and individuals: each kind has an essential nature, with these natures related to each other in a hierarchical fashion; and each individual has an essential nature, given by the (one) maximally specific kind to which it belongs.

12.2.4 Focusing Our Question

I said at the beginning of this section that it pays to keep distinct these two possible foci (individuals vs. kinds) for philosophical theorizing about 'essence'. Why? Well, it's never a good idea to conflate two things that *are* distinct. But more importantly, I think that very different philosophical interests motivate concern for these two varieties of essence. Each finds expression in the same form of question: 'What is it?' But the 'it' in question differs. One interest is in asking this question *of individuals*; the other is in asking it *of kinds*. Thus you might ask 'What is gold?' Or you might pick out a particular thing—the ring on my finger, for example—and ask 'What is that?' In each case, of course, the question needs to be understood in a particular way, lest you get an irrelevant answer: 'Philosophers' second-favorite example of a natural kind, next to "water"', 'Ned's wedding ring.' And so we tend to opt for slightly specialized terminology, so as to pose our questions unambiguously: 'What is it that makes something count as gold?', 'What is it that makes Ned's ring the thing that it is?'

It's worth observing that in the recent-ish literature, what I take to be these core questions sometimes get replaced by questions about *necessity*. Instead of focusing on what makes a given individual the thing that it is, for example, we might focus on *de re* modal truths concerning that individual, and more generally on what marks the distinction between possible worlds in which that very individual exists and ones in which it doesn't.[5] Or, instead of focusing on what makes something qualify as belonging to a given kind K, we might ask about metaphysically necessary truths concerning K. This shift of focus may seem to carry advantages, especially if you think that the notions of metaphysical possibility and necessity are much clearer than the notion of the essence of an individual or kind; perhaps we should use the former to analyze or even replace the latter. Myself, I think that's a mistake, although I won't argue the case here. (Except very indirectly, in that the approach I'll end up recommending for consideration doesn't in any way attempt to use metaphysical modality to define or replace the concept of essence.) I'll simply proceed under the assumption that we're right to organize our philosophical investigations around the 'What is it?' questions.

I'm also going to privilege the question about *kinds*. That's not for any deep reason: it's just that I haven't myself been able to get a clear fix on what the question about individuals is *asking* except (as we'll see in Section 12.6) by way of a detour through an account of essences of kinds. Consequently, I confess that I don't really understand 'individuals only' views. But look, that may just be my own shortcoming. At any rate, my primary interest will be in developing a 'go epistemic' approach to essences of kinds, in light of which I'll suggest some derivative lessons for our talk of essences of individuals.

---

[5] Here, for example, is the first sentence of the *Stanford Encyclopedia* article 'Essential vs. Accidental Properties': 'The distinction between essential versus accidental properties has been characterized in various ways, but it is currently most commonly understood in modal terms: an essential property of an object is a property that it must have while an accidental property of an object is one that it happens to have but that it could lack' (Robertson and Atkins, 2013, p. 1).

12.3 Metaphysical vs. Epistemic Approaches

One way to put the core contrast between the 'individuals only' and 'kinds only' views is this: Each thinks that just *one* of our two kinds of questions is apt to yield correct and non-trivial answers. (Our sample 'individuals and kinds' view thinks both do, with the answers closely connected to each other.) Correspondingly, each illustrates one way to be anti-metaphysical about the notion of 'essence': just reject one form of the 'What is it?' question as meaningless, or capable only of trivially correct answers. And, of course, you might reject both forms.

But that's not the only way to be 'anti-metaphysical'. You might also *accept* one or both kinds of question as important and capable of substantive answers but think that providing these answers in detail needs ultimately to appeal to epistemology and not to metaphysics. *Very* roughly, some feature counts as part of the essence of a given kind *K* or thing *x* because it is especially worth knowing about.

Now, depending on the background you bring to this discussion, that suggestion may strike you as Euthyphro-backwards or even oxymoronic.

As for Euthyphro-backwards: You might agree that a feature counts as part of the essence of a given kind *K* or thing *x* *if and only if* it is especially worth knowing about (at least, for certain important purposes), but think that it's worth knowing about *because* it is part of the essence, not the other way around. I suspect this view is often just taken for granted, in the background. Here for example is Fine (1994, p. 1):

> For one of the central concerns of metaphysics is with the identity of things, with what they are. But the metaphysician is not interested in every property of the objects under consideration. In asking 'What is a person?', for example, he does not want to be told that every person has a deep desire to be loved, even if this is in fact the case.
>
> What then distinguishes the properties of interest to him? What is it about a property which makes it bear, in the metaphysically significant sense of the phrase, on what an object is? It is in answer to this question that appeal is naturally made to the concept of essence. For what appears to distinguish the intended properties is that they are essential to their bearers.

Presumably, Fine didn't think it needed saying that the metaphysician's question is *important*, and the answers to it therefore worth knowing.

As for oxymoronic: You might think that while our questions *can* be given epistemic answers, doing so deprives them of their significance; so there can be no *philosophically serious* epistemic approach to 'essence'. For example, suppose you claim that the essence of any individual simply comprises those properties of the individual that, in a given conversational context, would strike the participants as particularly interesting to know about for reasons local to that conversation and/or those participants. (You're careful to characterize these reasons in a way that blocks the Euthyphro-style objection of the last paragraph.) That's not a crazy view: something like this seems to govern the way we answer questions about *who someone is*, for example.[6] But if it's the right view, then it's hard to see what *philosophical* interest (never mind metaphysical interest) there might be in the concept of the essence of an individual. Doesn't such interest in the essences of kinds (or individuals) presuppose that 'essence' be understood in a distinctively metaphysical sense?

No, as we'll shortly see: it's possible to give an epistemic treatment of the concept of essence in light of which that concept preserves its philosophical significance. It will help us to see how to arrive at this result if we first stop to consider a bit more carefully just what might be *meant* by insisting on a 'metaphysical' sense—either for the term 'essence' or for other terms ('necessity', 'ground', etc.) by which it might be defined.

12.3.1 What Is a 'Metaphysical' Approach?

We touched on this question before, in the introduction (Section 12.1). There, I cheated. In characterizing the 'go metaphysical' approach, I tried to get away with this line: 'Here it is important to understand that the key expressions—"necessary", "ground", "what it is for", "essence"—are all to be taken in a distinctively metaphysical sense.' I trusted you to understand—well enough, anyway—what I meant by 'distinctively metaphysical'. To be fair, philosophers writing about metaphysics almost always (it seems to me) presume this kind of trust. In the paper quoted above, for example, Fine clarifies the 'metaphysically significant' sense of the phrase 'what an object is' simply by appealing to the concept of essence, without pausing to worry that this concept itself might be given an epistemic analysis. As another example, here is Schaffer's (2009, p. 351) characterization of metaphysics itself:

> ...the neo-Aristotelian will begin from a hierarchical view of reality ordered by priority in nature. The primary entities form the sparse structure of being, while the grounding relations generate an abundant superstructure of posterior entities. The primary is (as it were) all God would need to create. The posterior is grounded in, dependent on, and derivative from it. The task of metaphysics is to limn this structure.

Any of the key terms—'priority', 'grounded in', 'dependent on', 'derivative from'— could be given a defensible *epistemic* reading. The context makes it obvious that Schaffer means to be ruling such readings out. But he doesn't stop to tell us how.

---

[6] It is also reminiscent of the strikingly deflationary approach to *de re* modality that Lewis (1986) endorses in *On the Plurality of Worlds*; see especially section 4.5, 'Against Constancy'.

I don't mean to be raising some serious problem here—as if none of us *really* understands what that qualifier 'metaphysical' is supposed to mean. I take it we do, at least well enough. Still, in any context in which we're trying to draw a contrast between metaphysical and epistemological approaches to some key bit of philosophical terminology ('essence', in our case), it would help to have something concrete to say by way of clarifying each side of that contrast. And here I think we can draw on an interesting and useful idea of Ted Sider's, developed in his 2011 *Writing the Book of the World*.[7] That book defends and explores a comprehensive view of metaphysical inquiry as inquiry into the *structure* of reality. But for our purposes, the key component of Sider's view is that this structure gives rise to a central normative requirement on inquiry: roughly, inquirers are *epistemically better-off* to the extent that they possess concepts that track reality's structure. As Sider (2011, p. 61; emphasis added) puts it,

> The goal of inquiry is not merely to believe truly (or to know). Achieving the goal of inquiry requires that one's belief state reflect the world, which in addition to lack of error *requires one to think of the world in its terms*, to carve the world at its joints. Wielders of non-joint-carving concepts are *worse inquirers*.

It will be helpful to put the idea slightly differently. Suppose we find ourselves equipped with some concept X, and with some way of expressing this concept in our language. Then we can ask: What *value* does our grasp of this concept, and our ability to communicate with one another about it, have for us? What sort of *reflective rationale* can we give for its deployment in our thought and talk?

Now, in some cases the best answer may reveal that a concept that we *thought* had distinctive philosophical or even metaphysical significance in fact doesn't. Take our ordinary, everyday concept of 'physical object'—the one that distinguishes between genuine physical objects and mere arbitrary aggregates. Take for granted that it is valuable for us to think in terms of this concept. Okay, but why? Well, suppose the best answer is that structuring our thought about the world around us by means of this concept makes it much easier for us to navigate our surroundings. But, so this story goes, serious theoretical inquiry into the nature of physical reality will just bog down unless we replace this concept by one that is more appropriately flexible. (For example, the physicist's concept of a 'physical system', a concept which draws no distinction between 'genuine' physical systems and mere arbitrary aggregates of particles.) In that case, it would presumably be a

---

[7] To be sure, Sider is definitely not the *only* philosopher to formulate this idea. But his expression and defense of it are particularly clear and engaging.

serious mistake to treat 'object' (in its ordinary sense) as a *philosophically* important concept.

On the other hand, the question about value may have an extraordinarily simple answer: it may be that the best reflective rationale we can give for deploying some concept X is that, without it, *we couldn't think about Xs.* To introduce the example we'll take up in more detail in the next section, consider the concept 'law of nature', and consider the view of someone like Tim Maudlin, who thinks of laws as metaphysically fundamental features of reality that govern the way earlier states of the world generate later states (see in particular Maudlin 2007). On Maudlin's view, the question 'What's so great about having the concept of "law of nature"?' has a totally obvious answer: 'Because without it, we couldn't think and talk about laws of nature.'

Now, that's not *quite* right: after all, we could surely concoct some weirdly gerrymandered language that allowed us to think and talk about laws of nature in a highly indirect manner, via some gruesome terminology. But if Maudlin is right about what laws *are*, then Sider's strictures apply: it would surely be *better* for us to think and talk about them *not* by way of some highly convoluted conceptual vehicle ill-suited to the task. So we should modify our answer slightly: Without our concept of laws of nature, we couldn't think and talk about laws of nature nearly as effectively as we can.

We've now seen examples of two positions one could take towards some putatively philosophically important concept X:

Modest eliminativism: The best reflective rationale for our deployment of X shows that X does not have a role to play in any serious philosophical theorizing. (X may still be valuable to us on *other* grounds; hence the qualifier 'modest'.)

Robust realism: X has a role to play in at least some serious philosophical theorizing because it marks out or closely corresponds to a distinctive kind of metaphysical structure; the best reflective rationale for our deployment of X is simply that it optimally enables thought and talk about this structure.

In the logical gap between modest eliminativism and robust realism lies our third option:

Respectful deflationism: X has a role to play in at least some serious philosophical theorizing, but not because it marks out any distinctive kind of metaphysical structure. The best reflective rationale for our deployment of X must be found elsewhere.

Respectful deflationism about some concept X comes with a promissory note: provision of a rationale that vindicates both the 'respectful' and 'deflationism' parts of the label. In Section 12.4 I will defend the view that contemporary 'Humean' approaches to laws of nature are best understood as a species of respectful deflationism about laws of nature. But first let's consider a much less controversial example, as a sort of proof of concept.

12.3.2 Respectful Deflationism: Proof of Concept

Suppose we have decided that arithmetic is not just a useful fiction: numbers really exist, and arithmetical statements are statements about them. We also hold that distinct arithmetical truths express distinct arithmetical propositions: the proposition that $2 + 3 = 5$ is not the same as the proposition that there is no largest prime number, for example.[8] In addition, the true arithmetical propositions exhibit an internal structure given by the facts about which of them logically imply which others of them. Perhaps we will want to label this entailment structure 'metaphysical'. But whether or not we do, we hold (say) that there is no *additional* metaphysical structure to the set of true arithmetical propositions. They are all, as it were, completely metaphysically on a par with one another.

This is a perfectly sensible view about the metaphysics of arithmetic (modulo, perhaps, its realism about numbers). If anything, someone who wants to insist that the true arithmetical propositions exhibit a metaphysical structure that goes beyond what can be captured by logical entailment takes on the burden not just of justifying that claim but of rendering it clear and coherent.

But for all that we are *metaphysically* egalitarian about the arithmetical truths, we need not be egalitarian tout court: there is nothing in the position we are sketching that should lead us to deny the central importance, for both mathematical practice and philosophical analysis and reflection on that practice, of the concept *axiom of arithmetic*. Here, it seems clear, respectful deflationism about 'axiom of arithmetic' is the way to go.

The next question to answer is the one about rationale. What is the point of our shared deployment of the concept *axiom of arithmetic*? The best answer will appeal to epistemic aims. For whatever reason, we are interested in figuring out which arithmetical statements are true. Some arithmetical statements strike *all* of us as *obviously* true—in fact, so obviously true that we could not become more confident of their truth by deriving them from other arithmetical statements. Among all such obvious-to-all-of-us arithmetical truths, some have proven to be particularly logically fecund, capable (at least collectively) of serving as the basis for logical derivations of a wide range of other truths. These truths thus appear especially well suited to play a regulative role within mathematical practice, so that, for example, if I claim to have discovered some novel arithmetical truth, it is appropriate for you to demand a proof of that truth from these statements. Finally, we deploy the concept of *axiom* simply as a way of collectively picking out these truths.

---

[8] For a contrary view, according to which there is only *one* true arithmetical proposition, see, for example, Lewis (1986), especially section 1.4.

This is only meant as a preliminary sketch of the best reflective rationale for deployment of the concept *axiom of arithmetic*. It may well be badly incomplete. For example, perhaps one important constraint on the choice of axioms is that they yield proofs of other arithmetical truths that are *elegant* or *illuminating* in some way. Never mind. The important point is that *epistemology* is driving the show: the rationale for having a shared concept of *axiom of arithmetic* is that doing so facilitates certain epistemic aims, and it is in reference to our pursuit of these aims that we become clear on why it makes most sense to select as axioms those arithmetical statements that we do. Metaphysical structure has absolutely zero role to play, beyond (perhaps) providing the network of entailment relations among the arithmetical propositions.

That's our first working example of respectful deflationism about some concept. There are two ways of misunderstanding it that you should work hard to avoid.

The first misunderstanding is to think that what we are offering here is a *semantics* for the expression 'axiom of arithmetic'. We are not. To bring this point out as clearly as possible, imagine two distinct mathematical communities. In key respects, the communities are exactly alike: in each, there are detailed, universally shared views about the good-making epistemic features that a set of arithmetical truths must have to optimally play the role of axioms. (Obviousness, logical fecundity, explanatoriness, etc.) In each community, there is a special book in which, by collective agreement, are written a small number of arithmetical statements. Each community calls this book the 'Book of Axioms'. But there is one difference between the communities: in the first community, the word 'axiom' just *means* 'sentence written in the Book of Axioms'. In the second, it means, by contrast, 'member of a set of arithmetical truths that optimally satisfies our shared epistemic desiderata'. In each community, as mathematical knowledge and understanding evolve, statements get added to and subtracted from the Book. In the first community, historians of mathematics describe this evolution by saying— *truthfully*—things like 'What used to be axioms are no longer axioms.' In the second community, the historians say—again, *truthfully*—things like 'We used to think that certain statements were axioms, but we now think we were wrong about that.' So there is a perfectly sharp difference between these communities, and it resides exactly in the meaning each attaches to the expression 'axiom of arithmetic'. But I submit that there is no philosophically interesting difference between them at all. In particular, the rationale for deployment of the concept of axiom is *identical*. For a respectful deflationist about X, what matters is the best reflective rationale for deployment of X. Semantics of our 'X'-language is an afterthought.

The second misunderstanding is to think that respectful deflationism is really just a different kind of metaphysics in disguise. This mistake is easy to fall into, because of a regrettable tendency within certain parts of contemporary philosophy to be hegemonic about the use of the word 'because'. Here is what I have in mind. Imagine that we express the view about axioms just sketched in the following way: 'An arithmetical proposition counts as an axiom because it is part of a set of such propositions that optimizes for such-and-such epistemic desiderata.' That use of the word 'because' might invite the following argument.

'You say that something counts as an axiom *because* it is part of a collection of arithmetical truths that optimizes for some epistemic value. But that use of the word "because" shows that you're giving a *metaphysical explanation* of axiomhood— you are providing an account of what *grounds* a proposition's status as an axiom, or perhaps an account of the *essence* of the kind "axiom". You might succeed thereby in metaphysically "deflating" axioms (as compared, say, to someone who thinks that certain arithmetical truths are somehow intrinsically metaphysically special). But there's a crucial sense in which your account is *not* itself metaphysically deflationary.'

Since my ultimate aim here is to defend the coherence and philosophical interest of a certain respectfully deflationist approach to 'essence', it would be bad news to accept the conclusion of this argument. Fortunately, it's easy to resist, simply by noting that 'because' can be used to *give a rationale*. To make this perfectly explicit, consider an expanded and more careful statement of our proposed view about axioms: 'Certain arithmetical truths are worth singling out for such-and-such a role in our collective mathematical practice because doing so optimizes for such-and-such epistemic desiderata—oh and by the way, for ease of reference we happen to call these truths "axioms".' Any suggestion that a 'metaphysical explanation' of axiomhood is on offer here vanishes.

12.3.3 Metaphysics Lurking behind the Epistemology?

The observations of the last paragraph notwithstanding, respectful deflationism about some philosophically important concept X does *not* guarantee a resolutely 'anti-metaphysical' account of X. To bring the reason into view, consider the abstract structure of which our 'axioms' example was an instance: We provide a *rationale* for drawing the distinction that X allows us to draw by explaining why doing so facilitates (perhaps optimally) certain epistemic ends. Well, what if a proper characterization of these epistemic ends *itself* involves appeal to some sort of metaphysical structure? Not, to be sure, the distinctive metaphysical structure picked out by X itself—for, by hypothesis, *that* doesn't exist. But some other structure. Something like that arguably happened in our arithmetical example, given that the very setup presupposed realism about numbers. But in the next example we'll look at—Humeanism about laws of nature—we'll see this question arise in a more urgent form.

12.4 Case Study: Laws of Nature

12.4.1 The Basic Schism and 'Best System' Accounts

I will mostly presume familiarity with Humean accounts of laws of nature, and with the broader debate about the nature of laws to which they are a contribution.[9] Still, a quick orientation is in order.

What unites Humeans is the view that laws of nature are, in some sense, nothing over and above *mere patterns* in the phenomena—where 'the phenomena' can themselves be characterized entirely in non-modal terms. Humeans happily sign on to the view that the sciences—especially fundamental physics—aim at discovering laws; they just think that to characterize *what* is being discovered we need invoke no real necessities in nature, or powers, or causes, etc. They thus stand together on one side of the basic schism between philosophical views, with their opponents invoking one or another metaphysical posit as necessary to a proper account of laws. (Notable examples are Maudlin's (2007) primitivism about laws, mentioned above; and Armstrong's (1983) view that laws are grounded in higher-order necessitation relations between universals.)

Since the groundbreaking work of Lewis (1973, 1983), by far the most popular form of Humeanism has been a 'best system' account. In broad outline, an account of this form invokes certain desiderata on a set of true sentences about the world (a 'candidate system'). That set of truths which optimally satisfies the desiderata fixes what the laws are. Fixes how? There is mild disagreement on this point, with Lewis (1983) asserting that the laws are the *generalizations* that appear in the 'best' system, and Hall (2009) arguing that *every* claim should count as nomologically necessary that is entailed by the sentences of the best system. There is additional and more serious disagreement about the language in which the sentences of candidate systems are to be written, with Lewis famously insisting that it must be a language whose primitive predicates express 'perfectly natural' properties and relations, and Loewer (2021) and Callender and Cohen (2009) arguing for a more permissive view (see also Bhogal, Chapter 7 in this volume, and Schrenk, Chapter 8 in this volume).

And, of course, there are differences of opinion about the best characterization of the desiderata on candidate systems (as this volume will attest). Lewis's original proposal is still probably the most cited: the best system is the one that optimizes for *simplicity* and *informativeness*. But by now the literature features many alternatives.[10] And the important point to emphasize is that given the Humean's broad aims, Lewis's particular choice of desiderata is *optional*. In fact, the Humean's anti-metaphysical aims give her wide latitude in proposing desiderata on

---

[9] For a fuller treatment of the ideas in this section, see my Hall (forthcoming).

[10] See Dorst (2018), Hicks (2018), and Jaag and Loew (2020) for recent proposals.

candidate systems: she just needs to make sure not to propose criteria that themselves invoke or presuppose the kind of metaphysical structure she seeks to dismiss. For example, it obviously won't do to say that the 'best' system must be one whose statements are all nomologically necessary. We'll see below that metaphysics might intrude in more subtle ways; but for now let's take a cue from Lewis's proposal, and take for granted that a best system account will propose broadly *epistemic* desiderata on candidate systems.

12.4.2 Correcting a Mistake about Humean Accounts of Law

That last remark suggests a way of thinking about what all Humeans are up to—a way of thinking that, alas, has yet to become mainstream. And that is to think of best system accounts, restricted to those that put forth purely epistemic desiderata on candidate systems, as versions of *respectful deflationism* about the concept *law of nature.*

Channel, for the moment, your inner Humean. Here is the idea. You think that as far as the *kinds* of truths that nature serves up, there are no interesting metaphysical distinctions to be drawn. (Well, not many: perhaps the distinction between general truths and singular truths counts as 'metaphysical'; perhaps entailment relations between truths count as 'metaphysical'.) To that extent, you view the truths that are up for investigation in much the same way that our respectful deflationist about 'axioms of arithmetic' views arithmetical truths. But your egalitarian attitude towards the *metaphysical* status of these truths need not motivate an egalitarian attitude towards the truths themselves. Just as we can ask, 'How might the basic epistemic interests that motivate arithmetical inquiry provide a rationale for singling out certain of the arithmetical truths for special attention, and having them play a distinctive role in structuring that inquiry?', so too we can ask an exactly analogous question about the non-modal truths that nature serves up. And now we can see the literature on Humeanism as containing proposals for answering precisely this question. Here, for example, is Beebee (2000, p. 574; emphasis added):

> So the idea is something like this. Suppose God wanted us to learn all the facts there are to be learned. (The Ramsey-Lewis view is not an epistemological thesis but I'm putting it this way for the sake of the story.) He decides to give us a book—God's Big Book of Facts—so that we might come to learn its contents and thereby learn every particular matter of fact there is. As a first draft, God just lists all the particular matters of fact there are. But the first draft turns out to be an impossibly long and unwieldy manuscript, and very hard to make any sense of—it's just a long list of everything that's ever happened and will ever

happen. We couldn't even come close to learning a big list of independent facts like that. Luckily, however (or so we hope), God has a way of making the list rather more comprehensible to our feeble, finite minds: he can axiomatize the list. That is, he can write down some universal generalizations with the help of which we can derive some elements of the list from others. *This will have the benefit of making God's Big Book of Facts a good deal shorter and also a good deal easier to get our rather limited brains around.*

A rationale-focused Humean knows exactly how to read this passage: the final sentence provides a reason why, given our limited cognitive abilities together with our interest in knowing as much about the world as possible, it makes sense for us to single out certain truths for special attention, and so denominate them 'laws'.

But that is not Beebee's *own* reading, as witness her parenthetical remark that the view she is describing 'is not an epistemological thesis'. That's a mistake. It precisely *is* an epistemological thesis! Namely, a thesis about the epistemic value of drawing a certain distinction between the non-modal truths, and using that distinction to structure empirical inquiry in a certain way. The fact that the thesis is *motivated* by a background deflationary metaphysical view—namely, that there *are no* modal facts, or at least none that are relevant to empirical inquiry—should not distract us from this point.

The literature, alas, has been highly distracted, right from the beginning. For what Lewis (mistakenly) foregrounded, in defending his own best system account, was a *supervenience* thesis: facts about the laws metaphysically supervene (along with all other facts, in his view) on the totality of non-modal facts. A respectful deflationist about laws won't deny this thesis, but will view it as a trivial after-thought; for her, the *real* philosophical action is in giving a detailed, illuminating account of the epistemic rationale for structuring empirical inquiry around a law/non-law distinction.

The mistake persists, albeit in a different and frankly more damaging form. Consider this claim from a recent paper by Loewer; he is offering it as a rebuttal to an argument advanced in Maudlin (2007) and elsewhere, that the Humean's laws can't explain particular non-modal facts because those laws *are themselves explained by* such facts:

> I claim that this objection rests on failing to distinguish metaphysical explan-ation from scientific explanation. On Lewis' account the Humean mosaic meta-physically determines the L-laws [L-laws = laws, as a Humean understands them]. It metaphysically explains (or is part of the explanation together with the characterization of a Best Theory) why specific propositions are laws. This meta-physical explanation doesn't preclude L-laws playing the usual role of laws in scientific explanations.   (Loewer 2012, p. 131)

Loewer's response to the explanatory circularity challenge has since spawned a mini-literature (see, for example, Lange 2013), almost all of which simply takes for granted that the Humean means to be putting forth a *metaphysically* explanatory thesis.

I think this is all just a tissue of confusions, stemming from (i) a failure to spot the ambiguity in the question 'Why are certain specific propositions laws?'; and (ii) a failure to opt for the right disambiguation. To be sure, the question *can* be read as asking for a metaphysical explanation of the status of lawhood. But it can *also* be read as a request for the rationale for drawing a distinction between the non-modal truths, and deploying that distinction in a certain way within empirical inquiry. That is the right reading; the other one should simply be rejected. Having done so, the Humean need not spare a moment's thought for the alleged circularity problem.

Here I imagine resistance, coming from a peculiar but stubborn source. I have frequently encountered in conversation, especially with grounding enthusiasts, the view that *every* fact that is not itself metaphysically fundamental must *have* a metaphysical explanation in terms of more fundamental facts. So, since Humeans certainly grant that there are facts about what the laws are, they *must* view these facts as metaphysically explained by something more fundamental. And surely part of what metaphysically explains law-facts is, as Loewer notes, the Humean mosaic. So the circularity challenge arises, after all.

No, it doesn't. Or at least, we should not think that it does until we are given a compelling reason to accept what is, after all, a wildly ambitious claim about the scope of 'metaphysical explanation'. Consider *any* case in which we agree that our interests will be well served by drawing a certain distinction within a certain subject matter. For ease of communication, we then introduce some expression X to let us mark the distinction. Is that really all it takes to generate a demand for a *metaphysical* explanation of X-hood—a demand that must be met, on pain of having to treat X-hood itself as metaphysically fundamental? Please.

But there is, I think, another quite urgent problem to which the Humean should devote serious thought.

12.4.3 Unfinished Business for Humeanism about Laws

Can the epistemic aims that thinking in terms of a concept of 'law' optimally serves themselves be characterized without appeal to the kind of metaphysical structure the Humean abjures? Yes, I think, *if* those aims are just the ones built into Lewis's original formulation in terms of simplicity and informativeness. But I also think that by now we are in a good position to see how implausible it is that

Lewis has accurately captured the key epistemic aims in question. I won't rehearse all the reasons in detail (though see my Hall 2009, along with the other chapters in this volume). To me the most pressing worry concerns *explanation*.

One of the things we want from empirical inquiry is not just a bigger and bigger pile of facts—as if the quest for fundamental laws of nature was like a search for the SparkNotes version of the Book of the World—but *understanding*. It's a truism that what drives scientific investigation is not just the desire to answer 'Is it the case that *p*?' questions, but the desire to answer '*Why* is it the case that *p*?' questions. Suppose we try to build a version of Humeanism that fully respects the centrality of that epistemic aim. We add a requirement on candidate systems, say, that one of the dimensions along which they are to be assessed is *explanatoriness* (cf. Blanchard, Chapter 10 in this volume). Then the obvious question that arises is whether that dimension itself can be accurately characterized without any appeal to the kinds of metaphysical structure whose rejection motivated Humeanism in the first place.

Showing that it *can* would require, I think, an account of explanation very different from most of those currently on offer, the most popular of which appeal to counterfactual structure in a way that seems fundamentally opposed to the anti-metaphysical spirit of Humeanism.[11] Humeans occasionally hint at the desirability of such a non-standard account. Loewer (1996, p. 113), for example, notes in passing that the Humean's laws 'do explain. They explain by unifying.' But he says very little about what exactly that means, and certainly does not attempt to provide a 'unificationist' account of explanation that would satisfyingly answer the question raised in the last paragraph. As far as I know, no one pushing for a Humean account of laws has done so.

To my mind, getting clear on the nature of explanation, and so on the exact character of the epistemic aims that knowledge of laws is supposed to subserve, is (in addition to being a fascinating philosophical topic in its own right!) the biggest piece of unfinished business for Humeanism about laws of nature. In what follows, where I attempt to sketch some forms that a broadly Humean approach to 'essence' might take, we'll find a similar piece of unfinished business concerning the nature of *inquiry*, and the kinds of information that enhance our ability to carry it out.

---

[11] If not, perhaps, opposed to the *letter*. That is, once we settle—in whatever way—that a certain set of non-modal truths qualifies as 'the laws', we can turn around and define counterfactual conditionals in terms of them in familiar ways. The underlying question doesn't go away, though. For why should counterfactuals, *so understood*, be the kinds of claims that provide ingredients for epistemically valuable explanations? Why should knowledge of *them* count as the sort of knowledge worth having because of the way it enhances understanding?

12.5 Possibilities for a Broadly Humean Approach to 'Essence'

This section sketches two approaches a Humean might take toward the concept of *essence* (that is, a Humean who doesn't simply wish to dismiss the concept outright). Drawing on the lessons of the last two sections, I will present both approaches as varieties of respectful deflationism about essence. One approach is fairly clear and concrete—but in the end, I think, unsatisfactory. The alternative I will offer strikes me as much more promising, but will also be much more schematic.

Before getting to the sketches, it is worth noting that respectful deflationism about essence may have appeal to more than just the Humean. Consider Maudlin-style primitivism about laws of nature. Someone who endorses that view obviously feels comfortable with robust metaphysical posits. But it doesn't follow that he will think he should include *essences* among these posits. At the same time, he may think that the notion of essence has useful philosophical work to do. Respectful deflationism about essence is thus an option for him. But it will likely take a profoundly different form from the form a Humean would give it, for at least two reasons. First, our primitivist about laws thinks there is much more to reality than the Humean recognizes, and his commitment to this extra structure is bound to make a difference to his views about the basic aims of inquiry, and so will likely make a difference to his more specific views about how thinking in terms of the concept of essences of kinds subserves those aims. Second, one powerful motivation for believing in metaphysically robust laws is that one is committed to a conception of explanation that requires them. (Maudlin 2007 is very clear on this point.) For reasons indicated at the end of the last section, that is a conception the Humean probably cannot share. This difference in conceptions of explanation will likewise ramify into differences of opinion about the basic aims that a respectful deflationism about essence should invoke.

So there is, I think, interesting work to be done spelling out what a *non*-Humean, respectfully deflationist, account of essence might look like. But I will leave that for another occasion. Let's turn now to some possible shapes of a Humean account.

12.5.1 A Traditional 'Best System' Account

The most straightforward approach simply adapts Lewis's account of laws. Just as there are 'systems' of truths about reality as a whole, there are, for a given kind *K*, systems of truths about *K*s. So we apply the very same standards of simplicity and informativeness, and read off the 'essence' of kind *K* from the 'best'—i.e. optimally simple and informative—*K*-system. Perhaps we focus on those generalizations within the system that have the form 'every K is a —'; then we could say that what makes an individual a member of kind K is that it has every feature that fills in one of these blanks. Or there might be some other way of extracting the essence of kind K from the best K-system.

Whatever variation we end up choosing, the epistemic rationale for carving out 'essences' in this way can simply adapt Beebee's rationale for the best system account of laws: We have, we may suppose, a basic epistemic interest in knowing as much as possible about the Ks. But we suffer from cognitive limitations that reasonably lead us to want our K-information packaged in an optimal-for-us way: i.e. in the form of a collectively simple set of truths. We thus have a rationale for privileging the information contained in the best K-system; we reserve the phrase 'essence of kind K' as a way of picking out this information.

Despite the elegance and simplicity of this approach, we have reason to hesitate. For why suppose that our sole epistemic interest is in having lots of information (with the appeal to simplicity brought in merely as a compromise, to deal with the fact of our cognitive limitations)? This was already a worry with respect to Lewis's account of laws; but in the present context I think it is even more acute. Here is why. Presumably, our practice of dividing the objects of our inquiry into kinds *itself* has a rationale. Now, one could go full Sider here, and insist that the rationale is simply that we are trying to conform our thought and talk as best we can to the objective metaphysical joints in nature—all of them. But while that may be plausible as an account of our aims in distinguishing kinds at the level of fundamental physics, I think it is not plausible as a fully general account. At any rate, it is not an approach a Humean drawn to respectful deflationism about essence should want to take.

Instead, she should want to say that when we distinguish some kind K *as a* kind, we do so in order to try to meet some important epistemic aim (where, as usual, that aim can be characterized without appeal to any *verboten* metaphysical structure). Fine; but then shouldn't that very aim influence *which* information about things of kind K matters to us? I think that it should. But the account we are considering offers no provision for such influence: as it stands, the function of the kind K is merely to demarcate a certain subject matter, after which a purely *generic* epistemic interest in having information takes over as the crucial desideratum. That strikes me as a mistake.

12.5.2 A More Tailored 'Best System' Account

Adequately correcting this mistake will require a much more nuanced account of the nature of and norms governing empirical inquiry than we now possess. (More nuanced, that is, than the account implicit above, according to which inquiry aims solely at increasing our store of knowledge, and succeeds exactly to the extent that it does so.) Absent such an account, I am simply going to point, by means of an example, to a phenomenon that I think is of central importance in this context. We might call it the 'inquiry-enhancement' value of certain kinds of information. Here is a simple illustration.

Consider the following game. Two people play. You start with a pile of coins. On each player's turn, they remove either one or two coins from the pile. The player to remove the last coin wins.

Here is one way you could teach someone to play this game well: give them a long list of instructions. 'If the pile contains sixteen coins, take one'; 'if the pile contains seventeen coins, take two'; 'if the pile contains eighteen coins, take either one or two'; etc. Against an unskilled opponent, that list will give them a very good shot at winning. *Why*, though? Here is an explanation (which will also tell you exactly how to construct the list of instructions, should you want to).

Suppose it is your turn in this game, and you face a pile of just three coins. Then, provided your opponent doesn't make a silly mistake, you are guaranteed to lose. So a three-coin pile is a 'losing situation'. And that is because of a simple mathematical fact: since three is one more than the maximum number of coins that can be taken, and since any number of coins less than three can be taken, any move you make can be countered by a move that reduces the pile to zero. But then a six-coin pile is a losing situation: for if that is what you face, then no matter what you do your opponent can guarantee that on your next turn you will face a three-coin pile. And so on: any pile of coins that is a multiple of three is a losing situation. Correspondingly, if it is your turn and the pile of coins is *not* a multiple of three, then (with proper play) you are guaranteed to win if you take enough coins that the remainder is a multiple of three.

Some observations about the explanation in the last paragraph. To begin, it really *is* an explanation, and moreover one that answers a why-question.[12] But it does not succeed as an explanation by providing causal information, or indeed any information of a metaphysically special kind. So it and its ilk may be just the place a Humean should look, if she wants to fill out an account of explanation that will dovetail neatly with her deflationary account of laws. Second, and more importantly for our present purposes, someone equipped with the explanation in the last paragraph is in a *vastly better position to conduct inquiry* about our simple game than someone who only has the list of instructions at her disposal. And this in two ways. First, she can reason much more efficiently about the game itself. Second, she is in a position to spot generalizations to *other* games, for example this one: in any game in which the players may take a number of coins up to N, the optimal strategy is (if possible) to reduce the pile to a number of coins that is a multiple of N.[13] Finally, the inquiry-enhancing character of the explanation

---

[12] Which, remember, not all explanations do.

[13] As an exercise, you might see if you can leverage the explanation in the previous paragraph into an answer to a slightly harder question: Suppose there are two piles, and on a player's turn she may

hinges crucially on the use it makes of a certain *category*: 'one more than the maximum number of coins that can be taken'. Thinking in terms of this category is essential if you want to reason effectively about this and related takeaway games.

Now I am going to go out on a limb, with a fair amount of hand-waving, and suggest that the lessons from our toy example generalize in ways that point to a Humean-friendly account of essence very different from the account sketched in Section 12.5.1. Here is the idea. Just as our ability to reason effectively about takeaway games gets enhanced quite a lot by thinking in terms of certain appropriate categories, so too our ability to inquire effectively in *any* domain will hinge on the categories we use to organize our understanding of that domain. Suzy, an aspiring chemist, wants to reason about and answer questions concerning chemical compounds and their interactions. Good; then she had better know about the periodic table. What's more—and here we are going beyond the simple lessons highlighted by our takeaway game—her ability to conduct chemical inquiry effectively requires that she know certain things *about* each element (orbital structure, number of protons in the nucleus, etc.). As with Suzy, so too with anyone conducting inquiry about any reasonably sophisticated subject matter: you need to have the right categories to think in terms of, and you need to have the right sort of information about things that fall under those categories.

These observations suggest a deflationary account of essences of kinds quite friendly to the Humean: What is essential to Ks as such is that set of truths about Ks that are, collectively, most K-related-inquiry-enhancing. K-related-inquiry includes, of course, inquiry strictly about Ks; but it also typically includes a broader range of questions such that reasoning about them effectively itself requires, in part, deploying the category of Ks. Returning to the example that began this chapter, an astrophysicist needs to know certain things about black holes not only in order to be able to answer questions properly about black holes but in order to answer questions about a range of other phenomena that partly involve black holes. And if we think about what will enhance her ability to conduct such inquiries, it's obvious that the four items on the first list will be vastly more important to her than any of the items on the second list. Finally, and crucially, this profound epistemic difference need not be underwritten by any *metaphysical* distinction.

12.5.3 Why 'Essence' May Still Be Philosophically Important

The proposal just sketched is way too schematic to serve as a serious alternative to metaphysically heavyweight accounts of essence, and at the end of this paper I

take any number of coins up to N from just one of them; suppose as before that the player who takes the very last coin wins. What is the optimal strategy?

will highlight what seem to me to be the most important bits of unfinished business. Here I just want to emphasize two points. First, the proposal really does mean to treat the concept of 'essence' as first and foremost an *epistemic* concept, as the concept of information about a kind that is *optimally useful in enhancing inquiry* involving that kind. Second, for all that, the concept of 'essence' may still be philosophically important. Strange as this may sound, whether it *is* hinges on what are ultimately empirical questions about human psychology. Let me explain.

Consider the concept 'prime number'. Any number theorist worth her salt will, of course, think about the numbers partly in terms of this concept. Is that because this concept 'carves the numbers at the joints', à la Sider? I doubt it, and at any rate let's suppose not. So why is it so important for number theorists to deploy this concept? Presumably because of some deep facts about human cognition, and what does and does not enable creatures like us to think in an organized and effective manner about the numbers.

How deep? For all we know, *quite* deep: Indeed, maybe it will turn out that any finite being built more or less like us that was capable of reasoning about numbers at *all* would have to view multiplication as one of the most basic functions on the numbers, from which it would be a short step to viewing 'prime number' as one of the most basic distinctions that need to be drawn. Then again, for all we know at present, maybe organizing our understanding of the numbers in this way reflects a fairly shallow and contingent feature of the way human cognition works. More daringly, maybe it reflects a shallow and contingent feature of the way *some* human cognition works—maybe there are potential number theorists out there disposed to organize their understanding of the numbers along completely different lines, and we don't know about them simply because the social structure of academic mathematics marginalizes them.

For my money, I think the reasons run deep. And I think the same when it comes to the categories we deploy in mature empirical sciences, and what we need to know about things that fall under those categories in order to be effective inquirers. Finally—and this is the punchline—if the reasons *do* run deep, then it is of course important for philosophers interested in the nature of inquiry to deploy the concept of essence, just as it is important for them to deploy the concept of law.

12.6 Upshots and Unfinished Business

Even with respectful deflationism about 'essence' in the schematic state in which we've left it, we can discern a number of interesting potential upshots. Collectively, they bring out very vividly how profoundly different one's orientation to a philosophically central concept can be, once one abandons robust realism in favor of respectful deflationism. I will give just brief sketches.

First, what counts as part of the essence of a kind will be both vague and interest-dependent, precisely because what counts as maximally inquiry-enhancing is itself vague and interest-dependent. These results actually strike me as quite unsurprising, at least if we think about more examples than just the usual suspects, 'water' and 'gold'. Consider *wood*. On the one hand, it seems to be perfectly sensible for someone to want to know what is essential to wood as such. On the other hand, what we should tell her seems obviously to depend on why it *matters* to her—what kinds of wood-related inquiries she plans to be engaging in. And it seems equally obvious that there will be no bright line between facts about wood that should be included in the answer, and facts that shouldn't.

Second, plenty of kinds simply won't have essences—and the distinction between those that do and those that don't will itself be vague. Consider the kind *thing owned by Ned*. There is just not some compact set of facts about things owned by Ned such that someone with the peculiar desire to conduct inquiry related to this category will be especially helped by knowing those facts.[14]

Third, on the conception of essences of kinds being advanced here, there is no obvious connection between essence and metaphysical necessity. For example, is water necessarily H_2O? Well, we can say this much: given the characteristic reasons that chemists have for deploying the category 'water' at all, it makes sense to count any substance in *any possible world with the same laws as our own* as water if and only if it is composed of H_2O. But that's it; the rationale certainly doesn't extend to other metaphysically possible worlds, if such there be. And at any rate, the only salient question here is the following: To what extent is it *contingent* that certain information about a given kind has the sort of inquiry-enhancing value needed to qualify it as information about the kind's essence? As far as I can see, that will have to be settled on a case-by-case basis. (Not very contingent, in the case of water. But in other cases, who knows?). So there will be no general thesis in the offing relating essence and metaphysical necessity.

Fourth, extending the main idea to essences of *individuals* can make sense, *if* there is a sufficiently salient kind to which the given individual belongs, and if in addition we have a clear fix on distinctively K-related questions *about* that individual. Then we can say that the essence of the individual, *qua* member of kind K, is that set of truths about the individual that optimally enhances K-related inquiry concerning that individual.

This extension itself has some interesting upshots. Here are three. First, I think it can explain the intuitive pull of Fine's famous Socrates/{Socrates} example, albeit not in a way that licenses the conclusions Fine himself wants to draw from the example. Here is Fine (1994, pp. 4–5):

---

[14] And don't let yourself be distracted here by a purely semantic point, which is that we might decide we want to reserve the term "kind" for those categories that *do* have essences.

Consider, then, Socrates and the set whose sole member is Socrates. It is then necessary, according to standard views within modal set theory, that Socrates belongs to singleton Socrates if he exists; for, necessarily, the singleton exists if Socrates exists and, necessarily, Socrates belongs to singleton Socrates if both Socrates and the singleton exist. It therefore follows according to the modal criterion that Socrates essentially belongs to singleton Socrates.

But, intuitively, this is not so. It is no part of the essence of Socrates to belong to the singleton. Strange as the literature on personal identity may be, it has never been suggested that in order to understand the nature of a person one must know to which sets he belongs. There is nothing in the nature of a person, if I may put it this way, which demands that he belongs to this or that set or which demands, given that the person exists, that there even be any sets.

On the approach we are exploring here, there is nothing at all puzzling, or even particularly noteworthy, about this example. When we consider the set {Socrates}, the salient category to which it belongs is the category *set*. And if we ask, 'What information about this individual will optimally enhance our ability to conduct set-related inquiry concerning it?', the answer is quite obviously 'its membership'. By contrast, when we consider Socrates, the salient category is probably *human* or *philosopher*. But whatever this category K is, it obviously won't help *at all* to conduct K-related inquiry concerning Socrates to know that he belongs to {Socrates}.

Second, and much more speculatively, the approach we are considering to essences of individuals makes it puzzling why, if that approach is correct, we should expect any object's *origins* to be essential to it. Here is Suzy, a human. Suppose you want to conduct inquiry into Suzy-*qua*-human. How important is it to know her parentage? Well, on certain mistaken and frankly disreputable views about heredity, it might be *very* important. (You know, 'Does she have noble blood?' etc.) But those views *are* mistaken (and disreputable). So if this approach to essences of individuals is correct, then standard Kripkean intuitions about essentiality of origins likely need an error theory, and not a philosophical vindication. (Conversely, if there *are* good philosophical reasons for respecting those intuitions, then that will show that the deflationary approach to essences of kinds sketched in this chapter likely *can't* be extended neatly to essences of individuals—itself an interesting result.)

Finally, consider a certain kind of tangle we can get into when thinking about the essences of arbitrary fusions. Here is a motley assortment of objects. (My sandals, the dogs that live across the street, the Eiffel Tower, etc.) Grant that it has a mereological fusion, and ask what is *essential* to this fusion. A very common answer is that it is essential to this object that it be composed of just those parts. But to endorse that answer about *every* fusion leads either to mereological essentialism or to the view that some objects must be distinguished from at least some of the fusions of (some of) their parts. (For example, I am not identical to the mereological fusion of my cells, since it is essential to that fusion that it be composed of just those cells, but not essential to *me* that I be so composed.) On the other hand, if we say that only some fusions have their parts essentially, then we owe an account of which ones do and which ones don't (and why). A tangle indeed.

On the proposal we are now considering, there is a natural way through this tangle: just recognize that talk about the essence of an individual is always going to be relativized to some kind—typically a kind made salient in the given context. Notice that when we consider some motley assortment of objects, and ask questions about *their fusion*, the salient way to categorize this object *just is* as a fusion of the given parts. So it would be natural for us to read a question about the 'essence' of this object as requesting information that will enhance our ability to answer mereological questions concerning this object. So of course we treat the information that it has those parts in particular as important. But when we ask about the essence of some individual that is described in ways that make salient some *other* category to which it belongs, we will provide a different answer. No surprise: on the present proposal, there just isn't any such thing as *the* essence of an individual.[15]

I hope that what I have presented up to this point has at least sparked your interest in the possibility of an 'epistemology first' account of essences of kinds. To develop such an account in sufficient detail, a few key issues need addressing. First and foremost, we need an account of the nature of and norms governing *inquiry*, one that will allow us to distinguish the main ways in which inquiry into a given subject matter can be conducted more or less skillfully. We need to draw on this account in order to illuminate the exact sense in which information can be 'inquiry-enhancing'. We need standards for selection that will allow us to see that, for each kind K that strikes us as having an 'essence', there is a reasonably precise and non-arbitrary way to pick out the information about Ks that is 'optimally' inquiry-enhancing. And finally, if our aim is a thoroughgoing Humean reductionism about essence, then these various accounts must take care *not* to posit the kinds of metaphysical structure whose rejection motivated Humeanism in the first place.

A tall order. Maybe it can't be carried out. (Likewise, maybe the unfinished business for Humeanism about laws of nature can't be completed.) But look, either way it will be *philosophically interesting* to find out. So let's start looking.

References

Armstrong, D. M. (1983). *What is a Law of Nature?* Cambridge: Cambridge University Press.

Beebee, H. (2000). 'The Non-Governing Conception of Laws of Nature', *Philosophy and Phenomenological Research*, 61(3), pp. 571–94.

---

[15] Compare again Lewis's deflationary treatment of *de re* modal ascriptions in his (1986), particularly the section titled 'Against Constancy'.

Cohen, J., and Callender, C. (2009). 'A Better Best System Account of Lawhood', *Philosophical Studies*, 145(1), pp. 1–34.

Curiel, E. (2021). 'Singularities and Black Holes', in Zalta, E. (ed.), *The Stanford Encyclopedia of Philosophy* (Fall 2021 ed.), https://plato.stanford.edu/archives/fall2021/entries/spacetime-singularities.

Dorst, C. (2018). 'Toward a Best Predictive System Account of Laws of Nature', *British Journal for the Philosophy of Science*, 70(3), pp. 877–900.

Fine, K. (1994). 'Essence and Modality', *Philosophical Perspectives* 8 (Logic and Language), pp. 1–16.

Hall, N. (2009). 'Humean Reductionism about Laws of Nature' https://philpapers.org/archive/HALHRA.pdf.

Hall, N. (forthcoming). 'Respectful Deflationism', in Andersen, H., and Mitchell, S. (eds.), *The Pragmatist Challenge*. Oxford: Oxford University Press.

Hicks, M. T. (2018). 'Dynamic Humeanism', *British Journal for the Philosophy of Science*, 69(4), pp. 983–1007.

Jaag, S., and Loew, C. (2020). 'Making Best Systems Best for Us', *Synthese*, 197, pp. 2525–50.

Lange, M. (2013). 'Grounding, Scientific Explanation, and Humean Laws', *Philosophical Studies*, 164(1), pp. 255–61.

Lewis, D. K. (1973). *Counterfactuals*. Oxford: Blackwell.

Lewis, D. K. (1983). 'New Work for a Theory of Universals', *Australasian Journal of Philosophy*, 61(4), pp. 343–77.

Lewis, D. K. (1986). *On the Plurality of Worlds*. Oxford: Blackwell.

Loewer, B. (1996). 'Humean Supervenience', *Philosophical Topics*, 24(1), pp. 101–27.

Loewer, B. (2012). 'Two Accounts of Laws and Time', *Philosophical Studies*, 160(1), 115–37.

Loewer, B. (2021). 'The Package Deal Account of Laws and Properties (PDA)', *Synthese*, 199, pp. 1065–89.

Maudlin, T. (2007). 'A Modest Proposal concerning Laws, Counterfactuals, and Explanations', in Maudlin, T., *The Metaphysics Within Physics*. Oxford: Oxford University Press, pp. 5–49.

Robertson, T., and Atkins, P. (2013). 'Essential vs. Accidental Properties', in Zalta, E. (ed.), *The Stanford Encyclopedia of Philosophy* (Winter 2013 ed.), http://plato.stanford.edu/archives/win2013/entries/essential-accidental/.

Schaffer, J. (2009). 'On What Grounds What', in Manley, D., Chalmers, D., and Wasserman, R. (eds.), *Metametaphysics: New Essays on the Foundations of Ontology*. Oxford: Oxford University Press, pp. 347–83.

Sider, T. (2011). *Writing the Book of the World*. Oxford: Oxford University Press.

Wikipedia contributors (2022). 'Black hole', in *Wikipedia, The Free Encyclopedia*, https://en.wikipedia.org/w/index.php?title=Black_hole&oldid=1082386839 (accessed Apr. 15, 2022).

Index

Access problem 6, 66–8, 76, 80, 84
Agents 57, 90–6, 103
  embedded 4, 43–4, 53, 63–4
  limited 1, 5, 10–1, 20–2, 26, 33, 46, 63,
    159 n.8, 195–6, 231, 275, 279
  situated, see embedded
Albert, David 7, 24–5, 43–4, 54–6, 57 n.15, 64,
  92, 96, 152 n.4, 172 n.11, 173, 186 n. 36
Anjum, Rani 74
Anti-Humeanism 1, 3, 5–6, 12–3, 59–63, 68, 72,
  74, 76–87, 98, 150, 176 n.17, 254
Armstrong, David 30, 76
Atkins, Philip 265 n.5
Axiom(s)
  of arithmetic 271–2, 274
  of best systems 1–2, 9, 17–9, 57, 62, 88, 96, 98,
    146, 148–9, 151–8, 169–70, 172,
    178–84, 188

Bayesianism 47, 58, 79
Beebee, Helen 73 n.6, 274–5, 279
Beings, see Agents
Best systems account (BSA), see laws of nature,
  Best systems account of
Better best systems account (BBSA), see laws of
  nature, Better best systems account of
Bhogal, Harjit 9, 95, 101, 110 n.10, 111 n.16,
  113 n.21, 187, 224 n.13, 225 n.15, 273
Bird, Alexander 74
Blanchard, Thomas 11, 199 n.10, 277
Bridge principles 67–8, 73–5, 80, 84, 152–5,
  158, 183–4
Busse, Ralf 76

Callender, Craig 4–5, 10, 20, 25–6, 43, 162 n.9,
  172 n.13, 173 n.15, 174, 186, 187 n.38
Canberra Plan 179
Carroll, John 159
Cartwright, Nancy 238, 246 n.18, 250, 256
Causal models 34–6
Causation 21, 33–7, 59, 250–1
Chalmers, David 154 n.6
Chance 2, 5–7, 43–65, 66–7, 78–81, 83, 86–104,
  112–5, 129–30, 186, 246

Classical mechanics 20, 50 n.7, see also
  Newtonian mechanics
Cohen, Jonathan 2–6, 43, 172 n.13, 173 n.15,
  174, 186, 187 n.38
Creatures, see agents
Combinatorial principle, see recombination
Conditionals
  material 75–79, 248
  counterfactual 7, 12, 19, 21, 75–9, 108,
    118–20, 186, 238–43, 247–8, 254, 277 n.11
  semifactual 12, 238–243
Counterfactuals, see conditionals, counterfactual
Credence(s) 5–7, 44–56, 77, 80–1, 86,
  88–97, 100–2
Creeping minimalism 4, 28, 30–3, 37–8

Dasgupta, Shamik 165
Deflationism 13, 267 n.6
  respectful 269–72, 274–5, 278–84
Demarest, Heather 7–8, 95 n.8, 150, 176 n.17
Descartes, Rene 22
Description theory of reference 176–7, 179–80
Dispositions, see properties, dispositional
Dorst, Chris 10–1, 25, 29, 159, 171 n.7, 217,
  218 n.4, 220

Eddon, Maya 147, 149, 150, 171 n.7
Effective method, see effective strategies
Effective strategies 12, 237–8, 240–56
Einstein 187, 217
Eliminativism 13
  modest 269
Emery, Nina 95, 132
Essence(s) 12–3, 99, 258–85
  dispositional 1, 68, 73–4, 77, 176
  modal, see essences(s), dispositional
  of individuals 261–7, 283–5
  of kinds 21, 260–7, 281, 283–5
Euthyphro problem 266–7
Explanation 8, 11, 19–21, 29–30, 99–101, 131,
  185, 259, 275–78, 280
  circularity 8, 131–4, 141, 224 n.13, 276
  Inference to the best, see inference to the best
    explanation

Explanation (*cont.*)
  inter–scientific 184
  metaphysical, *see* grounding
  scientific 8, 11, 99–101, 145, 154–5, 163–4
  by unification 134, 224–34, 277
Externalism, semantic 175–7, 189

Fernandes, Alison 6–7, 114 n.25
Fine, Kit 261, 266–7, 283–4
Foster, John 130 n.12, 134–6
Frege case 71
Friedman, Michael 226, 230
Functionalism, about modality 86–9, 100, 103–4
Fusion(s) 13, 284–5

Gibbard, Allan 27–9
Governing 130–1, 222
Grounding 74, 99–100, 146 n.1, 185–6, 239,
    242–3, 272, 275–6, *see also* explanation,
    metaphysical
Grue 73 n.7, 138, 145, 155–6, 169–71, 174,
    see also properties, unnatural

Hall, Ned 4, 12–3, 17–8, 21, 43, 45 n.3, 46, 86,
    97–8, 153 n.5, 159, 254–5, 273, 277
Hartle, James 61
Heil, John 74
Hicks, Mike 25, 29, 59 n.18, 60 n.20, 186 n.35,
    216–7, 227
Hoefer, Carl 90–1, 94, 97, 187
Hume, David 16–7, 27, 29, 33–6, 59, 128 n.1
Hume's dictum, *see* recombination
Humean mosaic 1, 4, 8, 9, 12, 17–21, 43–4, 46,
    47 n.4, 48–57, 61–3, 109, 128–9, 146–50,
    164, 168–9, 177–8, 189, 215, 251,
    253–5, 276
Humean Supervenience 1, 29–30 n.7, 46–50,
    109, 128–9, 146 n.1, 168–9, 275

Ideal advisor 4, 10, 23–30, 162 n.9, 195–7
Ideal observer 4, 9, 19–31, 35–6, *see also* ideal
    scientist
Ideal scientist 9, 147, 158–64, *see also* ideal
    observer
Ideology 6, 73–6
Indefinite extendability 43–4, 50–5, 63
Indifference reasoning 6, 90–1, 100–2, 136, 141
Induction, problem of 47 n. 5, 54–63, 80–1,
    134–7, 141, 164
Inference to the best explanation 66, 79, 230–2
Ismael, Jenann 5–8, 9, 25, 73, 93, 198, 210–1

Jaag, Siegfried 26, 29, 78 n.12, 152 n.3, 171 n.7,
    173 n.14, 176 n.18, 184 n.30, 216–7, 218 n.6

King, Jeff 123–4
Kitcher, Philip 224–5

Lange, Marc 12, 134 n.25, 249, 276
Laws of nature 1–4, 16–38, 42–64, 145–66,
    168–89, 237–8, 248–9, 253–5, 260–1,
    269, 273–9
  Best systems account (BSA) of 1–4,
      17–22, 24–6, 45, 50–5, 62, 88, 90,
      93, 98, 129, 146–52, 159–60, 164,
      169–72, 174–5, 194–5, 215, 237,
      251, 273–7
  Better best systems account (BBSA) of 3, 9,
      26, 43, 168–89, 221 n.11
  biological 9, 37, 166, 171–2, 180–5
  ceteris paribus 185
  chemical 9, 171–2, 181–2, 184–5
  conservation 178, 209 n.19
  Dretske–Tooley–Armstrong account
      of 30, 273
  Lotka–Volterra equations 178
  Mentaculus 92, 109–10, 115–18, 152 n.4,
      221–2, 232 n.26
  Mill–Ramsey–Lewis account of, *see* laws of
      nature, Best systems account
  Newtonian 152, 179, 187, 208, 228–32
  Package Deal Account of 3, 8, 109 n.5,
      138–41, 145–66, 176 n.17, 187
  Pauli exclusion principle 178
  physical 9, 59, 172, 181–2
  Pragmatic best systems account of 1–13, 29,
      172, 218–20
  Pragmatic Humean account of, *see* laws of
      nature, Pragmatic best systems account
  probabilistic 146 n.2, 169
  of Quantum mechanics 10, 18, 20–1,
      30, 58, 92
  of Quantum field theory 153
  relativistic 20, 37
  Second law of thermodynamics 115,
      178, 219–22
  special science 3, 7, 9–11, 26, 37, 118–24,
      166, 168–89
  usefulness of 3, 20–1, 24–6, 30, 87, 161–2,
      199–220, 230–1
Lazarovici, Dustin 56, 136–7
Lewis, David 1–4, 6, 8, 17, 19, 21, 26, 30, 43–6,
    48, 50, 53, 66, 72, 80–1, 87–8, 90, 96,
    98 n. 9, 128, 145–9, 151–3, 159–60,
    164–5, 168–76, 179–80, 182 n. 28,
    186 n. 35, 188–9, 206, 237, 253–4,
    267 n. 6, 270 n. 8, 273–7
Limited beings, *see* agents, limited
Limited creatures, *see* agents, limited

Loew, Christian 26, 29, 159, 171 n.7, 173 n.14, 216–17, 218 n.6
Loewer, Barry 3, 7, 8, 9, 26, 43–4, 54, 56, 63, 86–8, 90, 92–3, 95–6, 99, 102, 145–66, 170–1, 176 n.17, 180, 187–9, 275–7

Massimi, Michaela 208–9, 209 n.19
Maudlin, Tim 43 n. 1, 132, 149, 269, 275, 278
Meacham, Christopher 147, 149, 150, 171 n.7, 187
Mentaculus, see Laws of Nature, Mentaculus
Method of arbitrary functions 82
Mill, John Stuart 1
Miller, Elizabeth 7–8, 95 n.9, 110 n.10, 176 n.17
Minimalism about truth 4, 28, 30–2
Mismatch, problem 7, 114–5, 150
Moral Realism 16, 28, 30–2
Mumford, Steven 74

Natural kinds 13, 21, 260–7, 281, 283–5
Natural properties, see properties, natural
Necessary connections 5, 17, 34, 48, 50–2, 56, 58–9, 145–6, 149, 150, 164–5
Newtonian mechanics 151–2, 201, see also classical mechanics
Niiniluoto, Ilkka 173
Nominalism, see properties, nominalism about
Non–Humeanism, see Anti–Humeanism
Norms 6, 29, 32, 67, 80–2, 232–4, 279, 285

Ontological dependence, see grounding

Package deal account, PDA, see laws of nature, Package deal account of
Past Hypothesis, see laws of nature, Mentaculus
Peirce, Charles Sanders 11–2, 103, 250
Permutation argument 68–76
Powers, see properties, dispositional
Possibility, varieties of 111 n.16
Possible worlds 109, 113 n.23, 118, 123, 129, 151, 184 n.30, 254, 265
Pragmatism 1–13, 25–6, 29, 32, 103, 184 n. 30
Principal Principle 6, 44–8, 80–2, 84, 88–9, 92, 94, 96–9, 102–3
Predicate F Problem 2, 20–1, 25, 149, 151–5, 170
Price, Huw 210
Principle(s) of Indifference, see indifference reasoning
Projectivism
    Laws of nature 4–5, 26–32, 37–8, 96
    Meta–ethics 27, 30–1
Propensities 6, 66, 78–82, 89, 100 n.11, 102–3

Properties
    abundant 170–1, 174–5
    dispositional 6, 66–84, 110, 122, 131, 176, 263
    fundamental 8, 145, 148, 165, 169, 174, 176, see also properties, natural
    intrinsic 1, 3, 109 n.5, 148, 169
    natural 1–4, 8–9, 17, 21, 26, 43, 109 n.5, 138–9, 145–66, 169–72, 174, 273, see also properties, fundamental
    Nominalism about 175
    qualitative 74, 78
    quidditistic 169
    unnatural 145, 147, 149, 150, 155–8, 163, 166, 169–71, 174, see also grue

Quantum field theory 153, 162
Quantum mechanics 3, 10, 58, 81–2, 92, 197–8

Ramsey, Frank 1, 17, 28–9
Ratbag idealism 206–8, 253–6
Realism 5, 16
    About numbers 270, 272
    Moral 16, 28–32, 37
    Robust 269, 282
Recombination, of properties 1, 5, 8, 43–4, 49–50, 55–7, 60, 113 n.23, 175
Reference magnetism 176–7, 179, 189
Resiliency 81–2
Roberts, John 11–2, 218–9, 232 n.27
Robertson, Teresa 265 n.5

Schaffer, John 267
Schrenk, Markus 3, 9–10, 169 n.1, 273
Schwarz, Wolfgang 5–6, 88, 90, 92–4, 98
Semifactuals, see conditionals, semifactual; see also conditionals, counterfactual
Sider, Ted 268
Simplicity 1–4, 17, 19–26, 35 n. 8, 37, 43, 87–8, 90, 92–3, 98, 129, 146–51, 153–4, 157, 159–60, 169–72, 174, 179, 181–2, 184, 187, 254–5, 273, 276, 278–9, see also standards for lawhood
Skepticism, inductive 44, 49, 80, 134–41
Skyrms, Brian 81
Space(–)time 50 n.7, 146–9, 162, 174 n.16, 187, 198
Special science(s) 3, 7, 9, 10–1, 26, 37, 43, 166, 168–89
Standards for lawhood 17–22, 25–6, 37, 43, 87–8, 98, 169–72, 181–2, 195–7, 253–5

Statistical mechanics 58, 78, 81–2, 88, 92, 96, 98, 115–8
  Second law of, *see* Laws of nature, Second law of thermodynamics
Strawson, Galen 134–5
Subsystems 19–20, 25
Supervenience 1, 29–30 n.7, 46–50, 109, 146 n.1, 168–9, 181–4, 186, 275, *see also* Humean Supervenience
Symmetries 205

Theory of everything 11, 61, 150, 217
T'Hooft, Gerard 61
Truthmaker(s) 45, 53 n.12, 63, 78, 80, 110 n.11, 123–4, 164, 244 n.13

Undercutting, *see* undermining problem(s)
Undermining Problem(s) 7–8, 43–53, 97–101, 104, 108–9, 118–25
unHumean whatnots 6, 48, 66–8, 76–7, 82–3
Unviolatability 12, 199, 217, 248–9

Vagueness 13, 179, 188, 260, 283
Van Fraassen, Bas 150, 178
Vetter, Barbara 77 n.11, 176 n.17

Ward, Barry 29–30, 96, 101 n.14
Weslake, Brad 186 n.36
Williams, Neil 74